Explorations in College Algebra and Trigonometry
Using the TI-82/83/83 Plus/85/86

SECOND EDITION

Deborah Jolly Cochener

Bonnie MacLean Hodge

Austin Peay State University

Australia • Canada • Mexico • Singapore • Spain • United Kingdom • United States

Assistant Editor: *Rachael Sturgeon*
Marketing Team: *Leah Thompson, Samantha Cabaluna, Maria Salinas*
Editorial Assistant: *Lisa Jones*

Production Coordinator: *Dorothy Bell*
Cover Design: *Vernon T. Boes*
Print Buyer: *Micky Lawler*
Printing and Binding: *Globus Printing*

For more information about this or any other Brooks/Cole product, contact:
BROOKS/COLE
511 Forest Lodge Road
Pacific Grove, CA 93950 USA
www.brookscole.com
1-800-423-0563 (Thomson Learning Academic Resource Center)

Printed in the United States of America

10 9 8 7 6 5 4 3 2 1

ISBN 0-534-38196-0

This book is dedicated to
Dr. Aleeta Christian, our department chair,
and to Dr. Richard Hogan, our dean,
with thanks for their continued support
of our writing projects.

TABLE OF CONTENTS

PREFACE

This unique workbook/text provides the student the opportunity for guided exploration of topics in college algebra and trigonometry using Texas Instruments graphing calculators. Keystroking guidance and correlating concept charts are provided for the TI-82, TI-83, TI-83plus, TI-85, and TI-86 calculators. This enables the instructor to use the text in a classroom that requires any of the above listed calculators, as well as in the classroom where calculators are mixed. The text is intended for use as a supplemental text to a CORE classroom text, and is therefore arranged by topics. This enables the instructor to assign the appropriate *Explorations Unit(s)* that correlate(s) with the topic under discussion within the classroom. **The text is not meant to be worked in sequential order. Each unit has one or more prerequisite units.** *Only* **the prerequisite units are required for student success in working the assigned unit.** This allows the use of this ancillary text with virtually *any* core course textbook. Charts that correlate the *Exploration Units* with textbook sections are provided, upon request, from Brooks Cole.

Changes in the Second Edition

▸ The TI-83/83plus are the base calculators used in the text, with changes noted for the TI-82, TI-85, and TI-86 graphers.

▸ An index has been included at the end of the text.

▸ Writing exercises have been interspersed throughout the text and are clearly marked by the ✍ icon.

▸ This edition includes seven units which address trigonometric topics to include both polar and parametric graphing.

▸ Units have been reorganized to better reflect current pedagogy. For example, the TABLE feature is introduced in Unit 3 in conjunction with the use of the STOre feature, allowing the early integration of the use of tables.

▸ The units addressing complex numbers and business applications have been moved to the section entitled *Basic Calculator Operations*.

Features

*Each unit provides guided exploration of a topic. Units are not meant to be done in numerical order, but rather according to concept. A correlation chart at the beginning of each section correlates units to traditional text topics. Prerequisite units are listed at the beginning of each unit.

*The text requires NO instruction - but serves as a workbook for the student. Answers to questions/exercises within a unit appear at the end of the unit.

*Many units can be used in place of classroom instruction.

*A key correlation chart that shows the units in which keys are introduced is provided.

*The workbook may be used over a period of more than one semester/quarter as the student progresses through his/her mathematics sequence.

*As an institution changes textbooks, this supplement need not be changed.

*A Troubleshooting Appendix is provided. It contains common student errors as well as explanations of the error screens students most often encounter.

*Units are written in a manner that enables the student with little or no algebraic experience to read and explore independently.

*The units provide springboards for both classroom discussion and further investigation either individually or as a class.

Integration of technology into the mathematics classroom has grown significantly with the introduction of the graphing calculator. These calculators have brought relatively inexpensive technology into the students' hands. However, instructors are now faced with the problem of integrating the technology without sacrificing course content. These activities will enable the students to develop algorithms typically found in college algebra and trigonometry courses, improving both their understanding and retention of the material.

The workbook was written with the belief that students who are active contributors in the classroom increase their own understanding and their long term retention of material. Of primary importance, however, is the fact that the units have provided the means and the opportunity for students to create their own mathematics.

Organization

The workbook is divided into five sections. Each section contains both a concept correlation chart and a key introduction chart.

UNITS #1-6: These units introduce the student to the calculator and its capabilities in performing computational tasks.

UNITS #7-15: These units provide the link between the algebraic processes traditionally taught in college algebra courses and the graph associated with the algebraic equations and inequalities. The units can be worked before the text formally introduces graphing. However, instructors may wish to assign Units 16-18 prior to Unit 7.

UNITS #16-27: These units introduce the student to graphing as a formal process. It is in these units that adjusting view screens (WINDOWS) and interpreting graphs is emphasized. Emphasis in this section is on functions.

UNITS #28-34: These units explore trigonometric topics.

UNITS #35-39: These units explore the miscellaneous topics of matrices, combinatorics, probability, and basic statistics. The statistical units are *not* meant to provide a comprehensive coverage of the subject; they *are* intended to give the student the keystroking information necessary to introduce them to the basic capabilities of their calculator.

PREFACE TO THE STUDENT

This workbook was developed with YOU in mind. It is written in a style that is easily understood by students at varying levels of mathematical proficiency. The text provides the keystroking information necessary to use the calculator as a tool in your mathematics class. **The units are not meant to be worked in order.** Each unit lists one or more prerequisite units. Those are the *only* units required prior to completion of a unit assigned by your instructor. The following are suggestions for using the workbook:

*__READ__ and follow the instructions *slowly* and *carefully*. Pay close attention to detail.

*Do not skip the questions marked with a writing icon (✍). These questions help you bring your thoughts together and put the mathematics into language that is meaningful to YOU.

*Keep a log of those calculator techniques that have proven helpful.

*In this log, also write any questions and/or comments that occur to you as you work through the units. Form a study group with your classmates to discuss these entries. **WHEN YOU DISCUSS MATHEMATICS, YOU LEARN MATHEMATICS.**

*The calculator is a **TOOL** for learning and doing mathematics. If an answer does not seem reasonable, always double check your reasoning, keystrokes and screen.

*Answers appear at the end of each unit.

*The units are correlated to algebraic concepts - be sure you work the prerequisite units.

*Keep this workbook; it will serve as a <u>personalized</u> reference for future courses.

We hope that by learning to use the graphing calculator your confidence in your abilities to create and DO mathematics increases. Experiment frequently with the keys and menu options that are not covered in the workbook. Last, but not least, HAVE FUN!

ACKNOWLEDGEMENTS

We are grateful and appreciative for the input of time and expertise by those who reviewed this workbook:

Nancy G. Henry, Indiana University-Kokomo
Anthony Dodgen, Alcorn State University
Sheryl W. Sippel, Hillsborough Community College

We appreciate the guidance and support of Bob Pirtle, Stephanie Schmidt, and Rachael Sturgeon, our editors at Brooks/Cole publishing. Dr. Nell Rayburn at Austin Peay State University has provided invaluable help in the editing process, and we appreciate her support of our text. We would also like to thank our husbands, David and Blaine, and our children, Sherah, Blaire, and Trey, for their support and love. They have always believed in us and it is they who made the dream become reality. Our families' contributions to the editing, reviewing, wording, and the testing of material has been invaluable.

Cochener & Hodge: *Explorations in College Algebra and Trigonometry Using the TI-82/83plus/85/86, 2e*

Concept Correlation Chart for *College Algebra 7e* by Gustafson & Frisk

Units with * <u>introduce</u> concepts.	PRE-REQ. UNIT	CONCEPT	SECTION
#1 Getting Acquainted with Your Calculator		Basic arithmetic operations with signed numbers, absolute value, square roots and exponents. Shortcut keys are also introduced.	1.1
#2 Fractions and Integer Exponents	#1	The calculator is used to perform fraction arithmetic and the orders of operations. The effect of negative exponents is examined both algebraically and via the calculator.	1.1
#3 Evaluating Through tables and the STOre Feature *	#1,2	Evaluation of expressions and the checking of solutions is explored.	1.2
#4 Rational Exponents and Radicals *	#2	The relationship between rational exponents and radicals is discovered through the evaluation of expressions.	1.3
#5 TI-83/85/86 Complex Numbers *	#3	Complex number operations are addressed.	2.5
#6 TI-83 Business Applications	#1	Present value, future value and amortization schedules are examined.	Optional
#7 Graphical Solutions: Linear Equations	#3	Solving linear equations.	2.1
#8 Graphical Solutions: Absolute Value Equations	#7	Solving absolute value equations.	2.8
#9 Graphical Solutions: Quadratic and Higher Degree Equations	#7	This unit addresses solutions of quadratic and higher degree equations with real roots.	2.3
#10 Applications of Quadratic Equations	#9	Interpreting graphs of quadratic equations in real life situations.	2.4
#11 Graphical Solutions: Radical Equations	#9	Graphically solving equations containing radicals.	2.6
#12 Graphical Solutions: Linear Inequalities	#7	Solving Linear Inequalities	2.7
#13 Graphical Solutions: Absolute Value Inequalities	#12	This unit can be used to allow the student to "discover" the algorithms typically used for solving.	2.8
#14 Graphical Solutions: Quadratic Inequalities	#9	Solutions of quadratic inequalities are determined through graphs and tables.	2.7
#15 Graphical Solutions: Rational Inequalities	#14	Solutions of quadratic inequalities are determined through graphs and tables.	2.7
#16 How Does the Calculator Graph?	#3	This unit correlates calculator graphing with hand drawn graphs.	3.1
#17 Preparing to Graph: Calculator Viewing Windows	#16	This unit explores pre-set viewing windows on the calculator.	3.1

Activity	Prerequisite	Description	Section
#18 Where Did the Graph Go?	#17	Adjusting the viewing window to fit your graph.	3.1
#19 Functions	#17	An exploration of functions: notations, domain, range, inverses and evaluation.	4.1
#20 Discovering Parabolas*	#17	An exploration of the graphs of quadratic functions and the effects of constant values on graphs.	4.2
#21 Translating and Stretching Graphs	#17	Examines the effect of constants on families of graphs.	4.4
#22 Symmetry of Functions	#17	Examines the concept of symmetry about the origin and y-axis.	Optional
#23 Piecewise Functions	#18	Uses the TEST menu to graph functions with restricted domains.	4.3
#24 Rational Functions	#17	Links algebraic techniques and graphical displays in determining asymptotes.	4.5
#25 The Algebra of Functions	#19	Four arithmetic operations of functions and composition are graphically illustrated.	4.6
#26 Exponential and Logarithmic Functions	#19	The graphs and relationships between the exponential function and its inverse are examined.	5.1, 5.3
#27 Parabolas Revisited	#20	Examination of the graphs of parabolas that are not functions.	8.1
#28 Trigonometric Functions: Amplitude, Phase Shifts and Translations	#19	Examines the effects of constants on the graphs of trigonometric functions.	Optional
#29 Predict-A-Graph	#28	Provides practice in identifying different types of trigonometric graphs.	Optional
#30 Graphical Explorations: Trigonometric Identities	#28	Graphs of equations are examined to determine identities.	Optional
#31 Graphical Solutions: Trigonometric Equations	#30	Graphical solutions using the root/zero and/or intersect features.	Optional
#32 Polar Graphing	#28	Several polar graphs are examined, as well as the graphs of conics in polar form.	Optional
#33 Parametric Graphs of Conics	#32	Circles, ellipses, and hyperbola will be graphed using parametric equations.	p. 358
#34 Roots of Complex Numbers	#33	Use of the TABLE feature in finding the n nth roots of a complex number.	Optional
#35 Matrices	#3	Matrix operations and applications to complex numbers.	7.2
#36 Combinatorics and Probability	#18	A calculator/graphics approach to combinatorics and probability.	9.6
#37 Statistics: Plotting Paired Data	#17	Paired data and statistical plots are explored	Optional
#38 Frequency Distributions	#38	Frequency distributions: histograms and box-and-whisker plots	Optional
#39 Line of Best Fit	#39	A calculator approach to graphing linear regression equations and interpreting displayed data	optional

Cochener & Hodge: ***Explorations in College Algebra and Trigonometry Using the TI-82/83plus/85/86, 2e***

Concept Correlation Chart for *Algebra & Trigonometry* by Stewart, Redlin & Watson

Units with * <u>introduce</u> concepts.	PRE-REQ. UNIT	CONCEPT	SECTION
#1 Getting Acquainted with Your Calculator		Basic arithmetic operations with signed numbers, absolute value, square roots and exponents. Shortcut keys are also introduced.	1.1
#2 Fractions and Integer Exponents	#1	The calculator is used to perform fraction arithmetic and the orders of operations. The effect of negative exponents is examined both algebraically and via the calculator.	1.2
#3 Evaluating Through tables and the STOre Feature *	#1,2	Evaluation of expressions and the checking of solutions is explored.	1.2
#4 Rational Exponents and Radicals *	#2	The relationship between rational exponents and radicals is discovered through the evaluation of expressions.	1.3
#5 TI-83/85/86 Complex Numbers *	#3	Complex number operations are addressed.	3.4
#6 TI-83 Business Applications	#1	Present value, future value and amortization schedules are examined.	optional
#7 Graphical Solutions: Linear Equations	#3	Solving linear equations.	3.1
#8 Graphical Solutions: Absolute Value Equations	#7	Solving absolute value equations.	3.8
#9 Graphical Solutions: Quadratic and Higher Degree Equations	#7	This unit addresses solutions of quadratic and higher degree equations with real roots.	3.3
#10 Applications of Quadratic Equations	#9	Interpreting graphs of quadratic equations in real life situations.	3.3
#11 Graphical Solutions: Radical Equations	#9	Graphically solving equations containing radicals.	3.5
#12 Graphical Solutions: Linear Inequalities	#7	Solving Linear Inequalities	3.6
#13 Graphical Solutions: Absolute Value Inequalities	#12	This unit can be used to allow the student to "discover" the algorithms typically used for solving.	3.8
#14 Graphical Solutions: Quadratic Inequalities	#9	Solutions of quadratic inequalities are determined through graphs and tables.	3.7
#15 Graphical Solutions: Rational Inequalities	#14	Solutions of quadratic inequalities are determined through graphs and tables.	3.7
#16 How Does the Calculator Graph?	#3	This unit correlates calculator graphing with hand drawn graphs.	2.1
#17 Preparing to Graph: Calculator Viewing Windows	#16	This unit explores pre-set viewing windows on the calculator.	2.1

#18 Where Did the Graph Go?	#17	Adjusting the viewing window to fit your graph.	2.1
#19 Functions	#17	An exploration of functions: notations, domain, range, inverses and evaluation.	4.1
#20 Discovering Parabolas*	#17	An exploration of the graphs of quadratic functions and the effects of constant values on graphs.	4.5
#21 Translating and Stretching Graphs	#17	Examines the effect of constants on families of graphs.	4.5
#22 Symmetry of Functions	#17	Examines the concept of symmetry about the origin and y-axis.	4.5
#23 Piecewise Functions	#18	Uses the TEST menu to graph functions with restricted domains.	4.2
#24 Rational Functions	#17	Links algebraic techniques and graphical displays in determining asymptotes.	5.5
#25 The Algebra of Functions	#19	Four arithmetic operations of functions and composition are graphically illustrated.	4.7
#26 Exponential and Logarithmic Functions	#19	The graphs and relationships between the exponential function and its inverse are examined.	6.1,6.3
#27 Parabolas Revisited	#20	Examination of the graphs of parabolas that are not functions.	11.1
#28 Trigonometric Functions: Amplitude, Phase Shifts and Translations	#19	Examines the effects of constants on the graphs of trigonometric functions.	8.3
#29 Predict-A-Graph	#28	Provides practice in identifying different types of trigonometric graphs.	8.4
#30 Graphical Explorations: Trigonometric Identities	#28	Graphs of equations are examined to determine identities.	9.1
#31 Graphical Solutions: Trigonometric Equations	#30	Graphical solutions using the root/zero and/or intersect features.	9.5
#32 Polar Graphing	#28	Several polar graphs are examined, as well as the graphs of conics in polar form.	11.7
#33 Parametric Graphs of Conics	#32	Circles, ellipses, and hyperbola will be graphed using parametric equations.	11.8
#34 Roots of Complex Numbers	#33	Use of the TABLE feature in finding the n nth roots of a complex number.	9.6
#35 Matrices	#3	Matrix operations and applications to complex numbers.	10.4
#36 Combinatorics and Probability	#18	A calculator/graphics approach to combinatorics and probability.	13.2
#37 Statistics: Plotting Paired Data	#17	Paired data and statistical plots are explored	p.82
#38 Frequency Distributions	#38	Frequency distributions: histograms and box-and-whisker plots	Optional
#39 Line of Best Fit	#39	A calculator approach to graphing linear regression equations and interpreting displayed data	p. 120

Cochener & Hodge: *Explorations in College Algebra and Trigonometry Using the TI-82/83plus/85/86, 2e*

Concept Correlation Chart for *College Algebra* by Stewart, Redlin & Watson

Units with * <u>introduce</u> concepts.	PRE-REQ. UNIT	CONCEPT	SECTION
#1 Getting Acquainted with Your Calculator		Basic arithmetic operations with signed numbers, absolute value, square roots and exponents. Shortcut keys are also introduced.	1.1
#2 Fractions and Integer Exponents	#1	The calculator is used to perform fraction arithmetic and the orders of operations. The effect of negative exponents is examined both algebraically and via the calculator.	1.2
#3 Evaluating Through tables and the STOre Feature *	#1,2	Evaluation of expressions and the checking of solutions is explored.	1.6
#4 Rational Exponents and Radicals *	#2	The relationship between rational exponents and radicals is discovered through the evaluation of expressions.	1.3
#5 TI-83/85/86 Complex Numbers *	#3	Complex number operations are addressed.	3.4
#6 TI-83 Business Applications	#1	Present value, future value and amortization schedules are examined.	Optional
#7 Graphical Solutions: Linear Equations	#3	Solving linear equations.	3.1
#8 Graphical Solutions: Absolute Value Equations	#7	Solving absolute value equations.	3.8
#9 Graphical Solutions: Quadratic and Higher Degree Equations	#7	This unit addresses solutions of quadratic and higher degree equations with real roots.	3.3
#10 Applications of Quadratic Equations	#9	Interpreting graphs of quadratic equations in real life situations.	3.3
#11 Graphical Solutions: Radical Equations	#9	Graphically solving equations containing radicals.	3.5
#12 Graphical Solutions: Linear Inequalities	#7	Solving Linear Inequalities	3.6
#13 Graphical Solutions: Absolute Value Inequalities	#12	This unit can be used to allow the student to "discover" the algorithms typically used for solving.	3.8
#14 Graphical Solutions: Quadratic Inequalities	#9	Solutions of quadratic inequalities are determined through graphs and tables.	3.7
#15 Graphical Solutions: Rational Inequalities	#14	Solutions of quadratic inequalities are determined through graphs and tables.	3.7
#16 How Does the Calculator Graph?	#3	This unit correlates calculator graphing with hand drawn graphs.	2.2
#17 Preparing to Graph: Calculator Viewing Windows	#16	This unit explores pre-set viewing windows on the calculator.	2.3

#18 Where Did the Graph Go?	#17	Adjusting the viewing window to fit your graph.	2.3
#19 Functions	#17	An exploration of functions: notations, domain, range, inverses and evaluation.	4.1
#20 Discovering Parabolas*	#17	An exploration of the graphs of quadratic functions and the effects of constant values on graphs.	8.1
#21 Translating and Stretching Graphs	#17	Examines the effect of constants on families of graphs.	4.5
#22 Symmetry of Functions	#17	Examines the concept of symmetry about the origin and y-axis.	Optional
#23 Piecewise Functions	#18	Uses the TEST menu to graph functions with restricted domains.	4.2
#24 Rational Functions	#17	Links algebraic techniques and graphical displays in determining asymptotes.	5.5
#25 The Algebra of Functions	#19	Four arithmetic operations of functions and composition are graphically illustrated.	4.7
#26 Exponential and Logarithmic Functions	#19	The graphs and relationships between the exponential function and its inverse are examined.	6.1, 6.3
#27 Parabolas Revisited	#20	Examination of the graphs of parabolas that are not functions.	8.1
#28 Trigonometric Functions: Amplitude, Phase Shifts and Translations	#19	Examines the effects of constants on the graphs of trigonometric functions.	Optional
#29 Predict-A-Graph	#28	Provides practice in identifying different types of trigonometric graphs.	Optional
#30 Graphical Explorations: Trigonometric Identities	#28	Graphs of equations are examined to determine identities.	Optional
#31 Graphical Solutions: Trigonometric Equations	#30	Graphical solutions using the root/zero and/or intersect features.	Optional
#32 Polar Graphing	#28	Several polar graphs are examined, as well as the graphs of conics in polar form.	Optional
#33 Parametric Graphs of Conics	#32	Circles, ellipses, and hyperbola will be graphed using parametric equations.	Optional
#34 Roots of Complex Numbers	#33	Use of the TABLE feature in finding the n nth roots of a complex number.	Optional
#35 Matrices	#3	Matrix operations and applications to complex numbers.	7.4
#36 Combinatorics and Probability	#18	A calculator/graphics approach to combinatorics and probability.	10.2, 10.3
#37 Statistics: Plotting Paired Data	#17	Paired data and statistical plots are explored	p. 119
#38 Frequency Distributions	#38	Frequency distributions: histograms and box-and-whisker plots	Optional
#39 Line of Best Fit	#39	A calculator approach to graphing linear regression equations and interpreting displayed data	p. 119

Concept Correlation Chart for *Algebra and Trigonometry with Analytic Geometry 10e* by Swokowski and Cole

Units with * <u>introduce</u> concepts.	PRE-REQ. UNIT	CONCEPT	SECTION
#1 Getting Acquainted with Your Calculator		Basic arithmetic operations with signed numbers, absolute value, square roots and exponents. Shortcut keys are also introduced.	1.1
#2 Fractions and Integer Exponents	#1	The calculator is used to perform fraction arithmetic and the orders of operations. The effect of negative exponents is examined both algebraically and via the calculator.	1.2
#3 Evaluating Through tables and the STOre Feature *	#1,2	Evaluation of expressions and the checking of solutions is explored.	1.3
#4 Rational Exponents and Radicals *	#2	The relationship between rational exponents and radicals is discovered through the evaluation of expressions.	1.2
#5 TI-83/85/86 Complex Numbers *	#3	Complex number operations are addressed.	2.4
#6 TI-83 Business Applications	#1	Present value, future value and amortization schedules are examined.	Optional
#7 Graphical Solutions: Linear Equations	#3	Solving linear equations.	2.1
#8 Graphical Solutions: Absolute Value Equations	#7	Solving absolute value equations.	2.5
#9 Graphical Solutions: Quadratic and Higher Degree Equations	#7	This unit addresses solutions of quadratic and higher degree equations with real roots.	2.3
#10 Applications of Quadratic Equations	#9	Interpreting graphs of quadratic equations in real life situations.	2.3
#11 Graphical Solutions: Radical Equations	#9	Graphically solving equations containing radicals.	2.5
#12 Graphical Solutions: Linear Inequalities	#7	Solving Linear Inequalities	2.6
#13 Graphical Solutions: Absolute Value Inequalities	#12	This unit can be used to allow the student to "discover" the algorithms typically used for solving.	2.6
#14 Graphical Solutions: Quadratic Inequalities	#9	Solutions of quadratic inequalities are determined through graphs and tables.	2.7
#15 Graphical Solutions: Rational Inequalities	#14	Solutions of quadratic inequalities are determined through graphs and tables.	2.7
#16 How Does the Calculator Graph?	#3	This unit correlates calculator graphing with hand drawn graphs.	3.1
#17 Preparing to Graph: Calculator Viewing Windows	#16	This unit explores pre-set viewing windows on the calculator.	3.2

#18 Where Did the Graph Go?	#17	Adjusting the viewing window to fit your graph.	3.2
#19 Functions	#17	An exploration of functions: notations, domain, range, inverses and evaluation.	3.4
#20 Discovering Parabolas*	#17	An exploration of the graphs of quadratic functions and the effects of constant values on graphs.	3.6
#21 Translating and Stretching Graphs	#17	Examines the effect of constants on families of graphs.	3.5
#22 Symmetry of Functions	#17	Examines the concept of symmetry about the origin and y-axis.	3.5
#23 Piecewise Functions	#18	Uses the TEST menu to graph functions with restricted domains.	3.5
#24 Rational Functions	#17	Links algebraic techniques and graphical displays in determining asymptotes.	4.5
#25 The Algebra of Functions	#19	Four arithmetic operations of functions and composition are graphically illustrated.	3.7
#26 Exponential and Logarithmic Functions	#19	The graphs and relationships between the exponential function and its inverse are examined.	5.1
#27 Parabolas Revisited	#20	Examination of the graphs of parabolas that are not functions.	11.1
#28 Trigonometric Functions: Amplitude, Phase Shifts and Translations	#19	Examines the effects of constants on the graphs of trigonometric functions.	6.5
#29 Predict-A-Graph	#28	Provides practice in identifying different types of trigonometric graphs.	6.5
#30 Graphical Explorations: Trigonometric Identities	#28	Graphs of equations are examined to determine identities.	6.5
#31 Graphical Solutions: Trigonometric Equations	#30	Graphical solutions using the root/zero and/or intersect features.	7.2
#32 Polar Graphing	#28	Several polar graphs are examined, as well as the graphs of conics in polar form.	11.5-11.6
#33 Parametric Graphs of Conics	#32	Circles, ellipses, and hyperbola will be graphed using parametric equations.	11.4
#34 Roots of Complex Numbers	#33	Use of the TABLE feature in finding the n nth roots of a complex number.	8.6
#35 Matrices	#3	Matrix operations and applications to complex numbers.	9.6
#36 Combinatorics and Probability	#18	A calculator/graphics approach to combinatorics and probability.	10.6-10.8
#37 Statistics: Plotting Paired Data	#17	Paired data and statistical plots are explored	3.1
#38 Frequency Distributions	#38	Frequency distributions: histograms and box-and-whisker plots	Optional
#39 Line of Best Fit	#39	A calculator approach to graphing linear regression equations and interpreting displayed data	3.3

Cochener & Hodge: *Explorations in College Algebra and Trigonometry Using the TI-82/83plus/85/86, 2e*

Concept Correlation Chart for *Fundamentals of College Algebra 10e*
by Swokowski and Cole

Units with * <u>introduce</u> concepts.	PRE-REQ. UNIT	CONCEPT	SECTION
#1 Getting Acquainted with Your Calculator		Basic arithmetic operations with signed numbers, absolute value, square roots and exponents. Shortcut keys are also introduced.	1.1
#2 Fractions and Integer Exponents	#1	The calculator is used to perform fraction arithmetic and the orders of operations. The effect of negative exponents is examined both algebraically and via the calculator.	1.2
#3 Evaluating Through tables and the STOre Feature *	#1,2	Evaluation of expressions and the checking of solutions is explored.	1.3
#4 Rational Exponents and Radicals *	#2	The relationship between rational exponents and radicals is discovered through the evaluation of expressions.	1.2
#5 TI-83/85/86 Complex Numbers *	#3	Complex number operations are addressed.	2.4
#6 TI-83 Business Applications	#1	Present value, future value and amortization schedules are examined.	Optional
#7 Graphical Solutions: Linear Equations	#3	Solving linear equations.	2.1
#8 Graphical Solutions: Absolute Value Equations	#7	Solving absolute value equations.	2.5
#9 Graphical Solutions: Quadratic and Higher Degree Equations	#7	This unit addresses solutions of quadratic and higher degree equations with real roots.	2.3
#10 Applications of Quadratic Equations	#9	Interpreting graphs of quadratic equations in real life situations.	2.3
#11 Graphical Solutions: Radical Equations	#9	Graphically solving equations containing radicals.	2.5
#12 Graphical Solutions: Linear Inequalities	#7	Solving Linear Inequalities	2.6
#13 Graphical Solutions: Absolute Value Inequalities	#12	This unit can be used to allow the student to "discover" the algorithms typically used for solving.	2.6
#14 Graphical Solutions: Quadratic Inequalities	#9	Solutions of quadratic inequalities are determined through graphs and tables.	2.7
#15 Graphical Solutions: Rational Inequalities	#14	Solutions of quadratic inequalities are determined through graphs and tables.	2.7
#16 How Does the Calculator Graph?	#3	This unit correlates calculator graphing with hand drawn graphs.	3.1
#17 Preparing to Graph: Calculator Viewing Windows	#16	This unit explores pre-set viewing windows on the calculator.	3.2

#18 Where Did the Graph Go?	#17	Adjusting the viewing window to fit your graph.	3.2
#19 Functions	#17	An exploration of functions: notations, domain, range, inverses and evaluation.	3.4
#20 Discovering Parabolas*	#17	An exploration of the graphs of quadratic functions and the effects of constant values on graphs.	3.6
#21 Translating and Stretching Graphs	#17	Examines the effect of constants on families of graphs.	3.5
#22 Symmetry of Functions	#17	Examines the concept of symmetry about the origin and y-axis.	3.5
#23 Piecewise Functions	#18	Uses the TEST menu to graph functions with restricted domains.	3.5
#24 Rational Functions	#17	Links algebraic techniques and graphical displays in determining asymptotes.	4.5
#25 The Algebra of Functions	#19	Four arithmetic operations of functions and composition are graphically illustrated.	3.7
#26 Exponential and Logarithmic Functions	#19	The graphs and relationships between the exponential function and its inverse are examined.	5.1
#27 Parabolas Revisited	#20	Examination of the graphs of parabolas that are not functions.	8.1
#28 Trigonometric Functions: Amplitude, Phase Shifts and Translations	#19	Examines the effects of constants on the graphs of trigonometric functions.	Optional
#29 Predict-A-Graph	#28	Provides practice in identifying different types of trigonometric graphs.	Optional
#30 Graphical Explorations: Trigonometric Identities	#28	Graphs of equations are examined to determine identities.	Optional
#31 Graphical Solutions: Trigonometric Equations	#30	Graphical solutions using the root/zero and/or intersect features.	Optional
#32 Polar Graphing	#28	Several polar graphs are examined, as well as the graphs of conics in polar form.	Optional
#33 Parametric Graphs of Conics	#32	Circles, ellipses, and hyperbola will be graphed using parametric equations.	Optional
#34 Roots of Complex Numbers	#33	Use of the TABLE feature in finding the n nth roots of a complex number.	Optional
#35 Matrices	#3	Matrix operations and applications to complex numbers.	6.6
#36 Combinatorics and Probability	#18	A calculator/graphics approach to combinatorics and probability.	7.6-7.8
#37 Statistics: Plotting Paired Data	#17	Paired data and statistical plots are explored	3.1
#38 Frequency Distributions	#38	Frequency distributions: histograms and box-and-whisker plots	Optional
#39 Line of Best Fit	#39	A calculator approach to graphing linear regression equations and interpreting displayed data	3.3

CORRELATION CHART

UNIT * Instructors are encouraged to use Units marked by an * to introduce concepts.	PRE-REQ. UNIT	CORRELATING CONCEPT
#1: Getting Acquainted With Your Calculator, p.3		Basic arithmetic operations with signed numbers, absolute value, square roots and exponents. Shortcut keys are also introduced.
#2: Fractions and Integer Exponents, p.13	#1	The calculator is used to perform fraction arithmetic and the orders of operations. The effect of negative exponents is examined both algebraically and via the calculator.
#3: Evaluating Through Tables and the STOre Feature, p. 19 *	#1-2	Evaluation of expressions and the checking of solutions is explored.
#4: Rational Exponents and Radicals, p.29*	#2	The relationship between rational exponents and radicals is discovered through the evaluation of expressions.
#5: TI-83/83plus/85/86 Complex Numbers, p.33 *	#3	Complex number operations are addressed.
#6: TI-83/83plus Business Applications, p. 37	#1	Present value, future value and, amortization schedules are examined.

INTRODUCTION OF KEYS

Unit Title	TI-82 Keys	TI-83/83plus Keys	TI-85/86 Keys
#1: Getting Acquainted With Your Calculator, p.3	ON/OFF 2nd ▲ ▼ (to darken/lighten) MODE ENTER CLEAR (-) abs ( ) parentheses √ x^2 ∧ QUIT INS DEL π ENTRY ANS	ON/OFF 2nd ▲ ▼ (to darken/lighten) MODE ENTER CLEAR (-) CATALOG MATH ▸NUM 1: abs(( ) parentheses √ x^2 ∧ QUIT INS DEL π ENTRY ANS	ON/OFF 2nd ▲ ▼ (to darken/lighten) MODE ENTER CLEAR (-) CATALOG CUSTOM abs ( ) parentheses √ x^2 ∧ QUIT/EXIT INS DEL π ENTRY ANS
#2: Fractions and Integer Exponents, p. 13	MATH 1:▸Frac	MATH 1:▸Frac	CUSTOM▸Frac
#3: Evaluating Through Tables and the STOre Feature, p.19	STO▸ X,T,θ ALPHA : Y= TABLE	STO▸ X,T,θ,n ALPHA : Y= TABLE	STO▸ x-VAR ALPHA : y(x)= TABLE (TI-86) TI-85 (no table) use EVAL or evalF
#4: Rational Exponents and Radicals, p. 29	MATH 4:$\sqrt[3]{\ }$ 5:$\sqrt[x]{\ }$	MATH 4:$\sqrt[3]{\ }$(5:$\sqrt[x]{\ }$	CUSTOM $\sqrt[x]{\ }$
#5: TI-83/83plus/85/86 Complex Numbers, p. 33	DOES NOT APPLY	i MATH ▸ CPX MODE Real▸ a+bi	(a,b) = a+bi CPLX
#6: TI-83/83plus Business Applications, p. 37	NOT APPLICABLE	APPS 1:Finance… FINANCE▸VARS	N/A

UNIT 1
GETTING ACQUAINTED WITH YOUR CALCULATOR

*This unit is a prerequisite for all other units in the text. Answers appear at the end of the unit.

 If using the TI-82, go to the TI-82 guidelines which follow this unit (pg.9).

TI-85/86 IF USING THE TI-85/86, GO TO THE TI-85/86 GUIDELINES WHICH FOLLOW THIS UNIT (PG.11).

Touring the TI-83/83plus

Take a few minutes to study the TI-83/83plus graphing calculator. The keys are color-coded and positioned in a way that is user friendly. Notice there are dark blue, black, and gray keys, along with a single yellow key and a single green key.

Dark blue keys: On the right and across the top of the calculator are the dark blue keys. At the top right are four directional cursor keys. These may be used to move the cursor on the screen in the direction of the arrow printed on the key. The four arithmetic operation symbols are also in dark blue. Notice the key marked **ENTER**. This is used to activate entered commands, thus there is no key on the face of the calculator with the equal sign printed on it. Below the screen are five keys labeled [**Y =**], [**WINDOW**], [**ZOOM**], [**TRACE**], and [**GRAPH**]. These keys are positioned together below the screen because they are used for graphing functions. The TI-83plus has an additional blue key labeled **APPS**. It is used to access the finance menu and the CBL/CBR capabilities of the calculator.

Black keys: The majority of the keys on the calculator are black. Notice the [**X,T,θ,n**] key in the second row and second column. It will be used frequently in algebra to enter the variable **X**. The **ON** key is the black key located in the bottom left position.

Gray keys: The twelve gray keys that are clustered at the bottom center are used to enter digits, a decimal point, or a negative sign.

Yellow key: The yellow key is the **2nd** key located in the upper-left position. To access a symbol in yellow (printed above any of the keys) first press the yellow [**2nd**] key, and then the key BELOW the symbol (function) to be accessed. For example, to turn the calculator **OFF** notice that the word **OFF** is printed in yellow above the **ON** key. Therefore, press the keys [**2nd**] [**ON**] to turn the calculator off. These keystrokes are done *sequentially* - not simultaneously. Throughout this book the following symbolism will be used: symbols that appear on the key will be denoted in brackets, [ ], whereas symbols written above the key will be denoted in < >. Thus, the previous command for turning the calculator off would appear as [**2nd**] <**OFF**>. The symbols [ ] or < > will cue you <u>where</u> to look for a command - either printed on a key or above it.

Green key: Alphabet letters (printed in green above some of the keys), or any other symbol and/or word printed in green above a key are accessed by first pressing the green **ALPHA** key and then the key below the desired letter/symbol/word. The keystrokes are sequential.

Catalogue feature: Press **[2nd] <CATALOG>** to display an alphabetical list of available calculator operations. Use the **[▲]** and **[▼]** cursor keys to scroll through this list. Operations may be accessed by placing the pointer adjacent to the operation and pressing **[ENTER]**.

*Note: The TI-83/83plus has an **Automatic Power Down** (**APD**) feature which turns the calculator off when no keys have been pressed for several minutes. When this happens, press **[ON]** to access the last screen used.*

Let's Get Started!

Turn the calculator on by pressing **[ON]**. If the display is not clear, press **[2nd] [▲]** to darken the screen, or **[2nd] [▼]** to lighten the screen. Notice that when the **[2nd]** key is pressed, an arrow pointing up appears on the blinking cursor.

To ensure the calculator is in the desired mode, press **[MODE]**. All of the options on the far left should be highlighted. If not, use the **[▼]** to place the blinking cursor on the appropriate entry and press **[ENTER]**. Exit MODE by pressing **[CLEAR]**. This accesses the home screen which is where expressions are entered. Press **[CLEAR]** until the screen is cleared except for the blinking cursor in the top left corner.

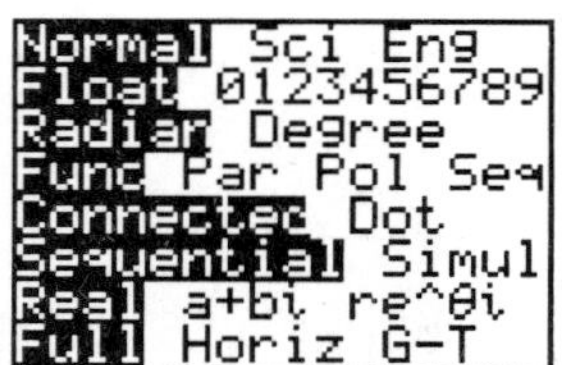

Integer Operations

When entering integers on the calculator, differentiation must be made between a subtraction sign and a negative sign. Notice that the subtraction sign appears on the right side of the calculator, with the other arithmetic operations. The negative sign appears to the left of the **[ENTER]** key and is labeled **(-)**.

Example: Simplify: -8 - 2

    Keystrokes:　　　　　　　　**Screen display:**

    [(-)] [8] [-] [2] [ENTER]

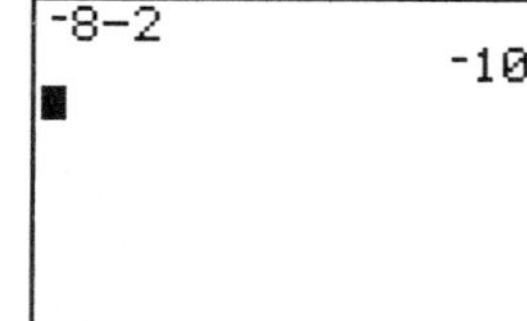

Observe the difference in <u>size</u> and <u>position</u> of the negative sign as compared to the subtraction sign.

TI-85/86	TI-85/86 USERS TURN TO "ABSOLUTE VALUE" IN THE GUIDELINES (PG.12).

Absolute Value

Absolute value is accessed by pressing **[MATH]** **[►](NUM)** **[1:abs( ]** or through the **[CATALOG]**. The absolute value operation is displayed as **abs(** . When absolute value is

accessed, a left parenthesis appears with the abbreviation *abs*. You must close the
parenthesis to ensure correct evaluation.

 Absolute value is located above the **[x⁻¹]** key. To access it, press **[2nd]**
<abs>. When finding the absolute value of a quantity, parentheses must be
used to enclose the quantity.

Example: Simplify: $|-3 - 2|$ *NOTE: This would be read as "the absolute value of the
quantity negative three minus two." Keep this in mind
as the expressions are entered.*

Keystrokes: **Screen display:**
[2nd] [MATH] [▶](NUM) [1:abs(] [(-)]
[3] [-] [2] [)] [ENTER]

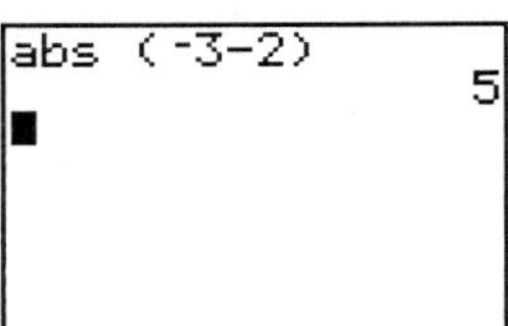

Square Roots

Square root is located above the **[x²]** key. To access it, press **[2nd]** $< \sqrt{\ } >$.

Example: Simplify: $\sqrt{9+16}$

Keystrokes: **Screen display:**

[2nd] $< \sqrt{\ } >$ [9] [+] [1] [6] [)] [ENTER]

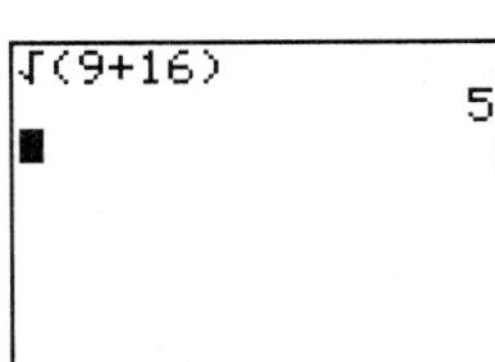

A number may be squared (raised to the second power) by either
pressing the **[x²]** key after entering the number, or by using the caret **[^]** key and entering
the desired exponent.

Example: Simplify: 4^2

Keystrokes: [4] [x²] [ENTER]   **OR**   **Keystrokes: [4] [^] [2] [ENTER]**

Screen display: **Screen display:**

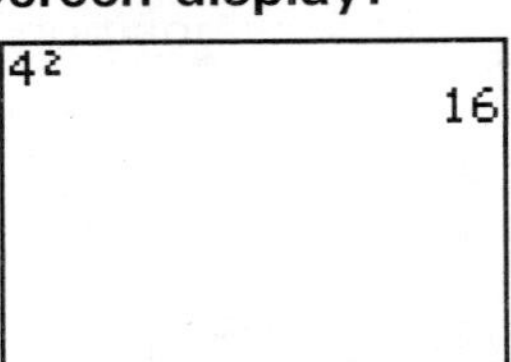

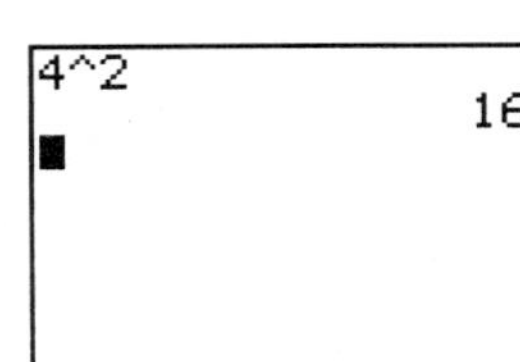

To raise to the third power (or higher), use the caret key.

Example: Simplify: $4(3)^5$
Keystrokes: **Screen Display:**

[4] [(] [3] [)] [^] [5]

Simplify the following expressions on the calculator.   Use the box provided below each problem to **RECORD THE CALCULATOR SCREEN <u>LINE BY LINE EXACTLY</u>** as it appears.

1. $|4^5 - (-6^2)|$

2. $|-4.7 - 5.28| - 18.3$

3. $-|-2^4 - 7|$

4. $\sqrt{5^3 - 10^2}$

5. $-\sqrt{169 - 25}$

6. $\sqrt{(-5)^2 - 4^2} + 16$

7. $(15 - 2)^3$

8. $|7 - 11| - \sqrt{64}$

✍ 9.   Explain the importance of using parentheses when entering expressions containing absolute value, roots, and exponents.

✍ 10. Explain why entering - 8, negative eight, as "minus 8" (using the subtraction key) is incorrect. Explain the result this incorrect keystroking produces.

Shortcut Keys

Below are descriptions of some keys that may be helpful in the efficient use of the calculator. You may want to reference this material in the future.

[QUIT]

To return to the home screen press **[2nd]** <QUIT>. This is helpful when stuck on a screen and pressing **[CLEAR]** does not return the home screen.

TI-85/86	THE **[EXIT]** KEY ON THE TI-85/86 WILL REMOVE MENUS FROM THE SCREEN.

[INS]

This key is helpful when data is entered incorrectly - particularly in expressions that are lengthy. When using the insert key, first place the cursor in the position in which the inserted digit/symbol should appear, press **[2nd]** <INS> (the character under the cursor will blink) and then the desired digit/symbol(s) to be inserted. The calculator will insert as many characters as desired as long as a cursor key is not pressed.

[DEL]

This key is helpful when an incorrect key is mistakenly pressed. Place the cursor over the character to be deleted and press **[DEL]**.

[π]

When evaluating formulas requiring the use of π , press **[2nd]** < π >. Although only nine decimal places are displayed, the calculator will evaluate the expression using an eleven decimal place approximation for π .

TI-85/86	The TI-85/86 displays an eleven decimal place approximation for π but uses a thirteen decimal place approximation for computation purposes.

[ENTRY]

Pressing **[2nd]** <ENTRY> accesses the ability of the calculator to recall the expression previously entered. Pressing **[2nd]** <ENTRY> repeatedly, performs "deep recall" by scrolling up the screen.

TI-85	The TI-85 does not have "deep recall." It will only recall the previous display line.

[2nd] [◄]

These keystrokes move the cursor to the beginning of the entry.

[2nd] [►]

These keystrokes move the cursor to the end of the entry.

[ANS]

The calculator will recall the answer from a previous computation. Access this function by pressing **[2nd]** <ANS>. ANS is located above the gray key used to enter a negative sign. It will also be activated if you press an operation key (+, -, X, ÷) before entering a number.

<u>Solutions:</u>

1. 1060   2. - 8.32   3. - 23   4. 5   5. - 12   6. 19   7. 2197

8. - 4   9. Absolute value and root symbols are grouping symbols, therefore, parentheses are required to designate the grouped quantity. An exponent only applies to a single number/character unless grouping is used.

10. "Minus 8" indicates the operation of subtracting 8 from another quantity, whereas "negative 8" specifies the opposite of the number 8. Therefore, when a minus sign is used the calculator will retrieve the previous answer (ANS) for the 8 to be subtracted from.

TI-82 GUIDELINES UNIT 1

Take a few minutes to study the TI-82 graphing calculator. The keys are color-coded and positioned in a way that is user friendly. Notice there are dark blue, black, and gray keys, along with a single light blue key.

Dark blue keys: On the right side of the calculator are the dark blue keys. At the top are four directional cursor keys. These may be used to move the cursor on the screen in the direction of the arrow printed on the key. The four operation symbols (addition, subtraction, multiplication, and division) are also in dark blue. Notice the key marked **ENTER**. This will be used to activate commands that have been entered; thus there is no key on the face of the calculator with the equal sign printed on it.

Gray keys: The twelve gray keys that are clustered at the bottom center are used to enter digits, a decimal point, or a negative sign. Notice the gray key beneath the light blue key in the upper left position, labeled **ALPHA**. We will come back to this key shortly; because its position is different from the other gray keys, it serves a different function.

Black keys: The majority of the keys on the calculator are black. Below the screen are five black keys labeled [Y=], [**WINDOW**], [**ZOOM**], [**TRACE**], and [**GRAPH**]. These keys are positioned together below the screen because they are used for graphing functions. Notice the [X,T,θ] key in the second row and second column. It will be used frequently in algebra to enter the variable **X**. The **ON** key is the black key located in the bottom left position.

Light blue key: The only light blue key is the **2nd** key located in the upper-left position.

 Above most of the keys are words and/or symbols printed in either light blue or white. To access a symbol in light blue (printed above any of the keys) first press the light blue [**2nd**] key, and then the key BELOW the symbol (function) to be accessed. For example, to turn the calculator **OFF** notice that the word **OFF** is printed in light blue above the **ON** key. Therefore, press the keys [**2nd**] [**ON**] to turn the calculator off. These keystrokes are done *sequentially* - not simultaneously. Throughout this book the following symbolism will be used: symbols that appear on the key will be denoted in brackets, [ ], whereas symbols written above the key will be denoted in < >. Thus, the previous command for turning the calculator off would appear as [**2nd**] <**OFF**>. The symbols [ ] or < > will cue you <u>where</u> to look for a command - either printed on a key or above it.

Alphabet letters and other symbols printed in white above some of the keys are accessed by first pressing the gray **ALPHA** key and then the key *below* the desired letter or symbol. Again, the keystrokes are sequential.

*Note: The TI-82 has an **Automatic Power Down (APD)** feature which turns the calculator off when no keys have been pressed for several minutes. When this happens, press [ON] to access the last screen used.*

Turn the calculator on by pressing **[ON]**. If the display is not clear, press **[2nd] [▲]** to darken the screen, or **[2nd] [▼]** to lighten the screen. Notice that when the **[2nd]** key is pressed, an arrow pointing up appears on the blinking cursor.

To ensure the calculator is in the desired mode, press **[MODE]**. All of the options on the far left should be highlighted. If not, use the **[▼]** to place the blinking cursor on the appropriate entry and press **[ENTER]**. Exit MODE by pressing **[CLEAR]**. This accesses the home screen which is where expressions are entered. Press **[CLEAR]** until the screen is cleared except for the blinking cursor in the top left corner.

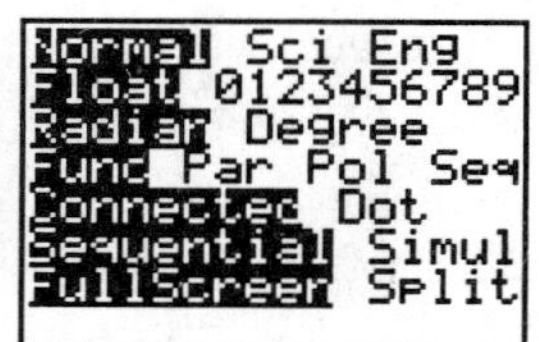

☞ Return to the core unit section entitled "Integer Operations" (pg.4) and complete the unit.

Touring the TI-85/86

Take a few minutes to study the TI-85/86 graphing calculator. The keys are color-coded and positioned in a way that is user friendly.

Gray keys: The twelve gray keys that are clustered at the bottom center are used to enter digits, a decimal point, and a negative sign. At the top right are four directional cursor keys. These may be used to move the cursor on the screen in the direction of the arrow.

Black keys: The majority of the keys on the calculator are black. The four operation symbols (division, multiplication, subtraction and addition) are located in the column on the far right. Notice the key marked **ENTER** at the bottom right position. This will be used to activate commands that have been entered. The equal sign on the face of the calculator is NOT used for computation. The key marked **x-VAR** in the second row, second column will be used frequently to enter the variable **x**. The **ON** key is in the bottom left position.

Below the screen are five black keys labeled **F1**, **F2**, **F3**, **F4** and **F5**. These are menu keys and will be addressed as they are needed.

Yellow-orange key: This key is located in the top-left position and is labeled **2nd**.

Blue key: This key is labeled **ALPHA** and is located at the top left of the key pad.

Above most of the keys are words and/or symbols printed in either yellow-orange or blue. To access a symbol in yellow-orange (printed above any of the keys), first press the yellow-orange [**2nd**] key, and then the key BELOW the symbol (function) you wish to access. For example, to turn the calculator **OFF** notice that the word **OFF** is printed in yellow-orange above the **ON** key. Therefore, press the **2nd** key and the **ON** key to turn the calculator off. These keystrokes are done *sequentially* - not simultaneously. Throughout this book the following symbolism will be used: symbols that appear on the key will be denoted in brackets, [ ], whereas symbols written above the key will be denoted in < >. A function that is accessed from a menu will be written in parentheses, ( ). Thus, the previous command for turning the calculator off would appear as [**2nd**] <**OFF**>. The symbols [ ], < >, or ( ) will cue you <u>where</u> to look for a command - either printed on a key, above it, or as a menu option. When a menu key is indicated, the current function of the key will follow in parentheses to correspond to the display at the bottom of the screen. For example, press [**GRAPH**] to display the graph menu. The notation [**F2**](**RANGE**) would denote access of the range submenu (if using the TI-86, [**F2**] (**WIND**) is displayed and denotes access to the WINDOW submenu). Users should be aware that the menu denote as RANGE on the TI-85 and WIND on the TI-86 corresponds to the WINDOW menu referred to for the TI-82/83. To remove the menu at the bottom of the screen, press [**EXIT**].

To access a symbol printed in blue above some of the keys, first press the blue **ALPHA** key and then the key *below* the desired letter or symbol. Again, the keystrokes are sequential. *NOTE: The **A**utomatic **P**ower **D**own (APD) feature turns the calculator off when no keys have been pressed for several minutes. When this happens, press [ON] to access the last screen used.*

Turn the calculator on by pressing **[ON]**. If the display is not clear, press **[2nd]** and hold
down **[▲]** to darken the screen or **[2nd]** **[▼]** to lighten. Notice that when the **[2nd]** key is
pressed an arrow pointing up appears on the blinking cursor.

To ensure the calculator is in the desired mode, press **[2nd]** <MODE>.
All of the options on the far left should be highlighted. If not, use the
[▼] to place the blinking cursor on the appropriate entry and press
[ENTER]. Exit **MODE** by pressing **[EXIT]**, **[CLEAR]**, or **[2nd]** <QUIT>.
This accesses the home screen where expressions are entered.

Integer Operations

When entering integers on the calculator, differentiation must be made between a
subtraction sign and a negative sign. Notice the subtraction sign is in the right column
whereas the negative sign appears as a gray key next to the **[ENTER]** key and is labeled **(-)**.

 STUDY THE EXAMPLE IN THE SECTION "INTEGER OPERATIONS" IN THE CORE UNIT (PG.4) AND
CONTINUE UNTIL THE NEXT TI-85/86 PROMPT.

Absolute Value

The TI-85/86 does not have absolute value on its keypad. However, a key can be created
using the **[CUSTOM]** key. Up to fifteen frequently used functions can be customized (and
accessed with only two key strokes) provided the function desired is listed under
CATALOG. To customize a function, press **[2nd]** <CATALOG> (if using the TI-86, press
[F1] (CATLG) next) followed by **[F3](CUSTM)**. Place the arrow next to **abs** in the list, then
press **[F1]**. The function **abs** is now listed under **PAGE↓**. The same procedure can be used
to add other functions to the **CUSTOM** menu as necessary. Each time select an open slot
in the menu by choosing the menu key below the open slot. Pressing **[MORE]** accesses
additional slots for customizing. When finished, press **[EXIT]** until the menus at the bottom
of the screen clear, and the home screen is displayed.

Pressing **[CUSTOM]** reveals the customized menu; to access a customized function, simply
press the **F** key below the desired function. Pressing **[EXIT]** removes this menu.

Example: Simplify │ -3 - 2 │
  (Ensure you are at the home screen - press **[EXIT]** if necessary.)

Keystrokes: **Screen display:**
[CUSTOM] **[F1](abs)** **[(]** **[(-)]** **[3]** **[-]**
[2] **[)]** **[ENTER]**
  Note: Absolute value can also be accessed by pressing **[2nd]**
  <MATH> **[F1](NUM)**   **[F5](abs)**.

☞   RETURN TO THE "ABSOLUTE VALUE" SECTION OF THE CORE UNIT (PG.4) AND CONTINUE UNTIL
THE NEXT TI-85/86 PROMPT. KEYSTROKES FOR YOUR CALCULATOR WILL DIFFER FROM THOSE
GIVEN FOR THE TI-82/83/83PLUS; MAKE APPROPRIATE CHANGES IN THE MARGIN.

UNIT 2
FRACTIONS AND INTEGER EXPONENTS

*Unit 1 is a prerequisite for this unit.

TI-85/86   TI-85/86 USERS TURN TO "FRACTIONS AND THE TI-85/86" IN THE GUILDELINES (PG. 17).

Fractions and the TI-82/83/83plus

The calculator can be used to perform arithmetic operations with fractions. Pressing the **[MATH]** key reveals a math menu. Take a minute to use the **[▾]** cursor to scroll down the menu. Notice that there are 10 options available. Use of the first option, **1:▶ Frac**, will now be illustrated.

TI-82 Screen

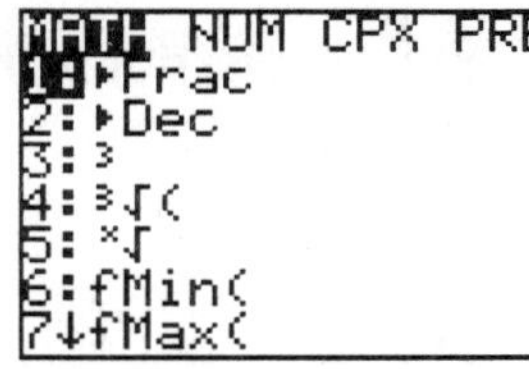

TI-83/83plus Screen

Enter the decimal number 0.42 by pressing **[.] [4] [2] [MATH]**. Under the MATH menu **1:▶Frac** is highlighted. Because it is highlighted, press **[ENTER]** to select option number **1**. Press **[ENTER]** again to activate the "convert to a fraction" command. Notice that the calculator displays the fraction reduced to lowest terms.

Recall that a mixed number is actually the sum of an integer and a fraction. Therefore, to enter mixed numbers on the TI-82/83/83plus simply indicate the addition of the integer and the fraction. To enter $-3\frac{1}{2}$, press **[(-)] [3] [+] [(-)] [1] [÷] [2] [MATH] [ENTER]** (to

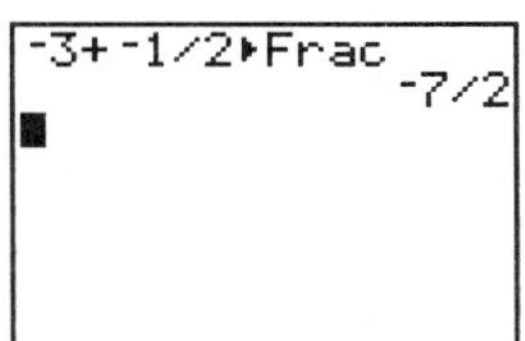

select option number **1**) **[ENTER]**. Your calculator display should correspond to the one at the right. Because the entire fraction is negative, both the integer part and the rational part must be entered as negative numbers.

Example 1: Simplify $\frac{2}{3} + \frac{1}{5}$.

    Keystrokes: **Screen display:**

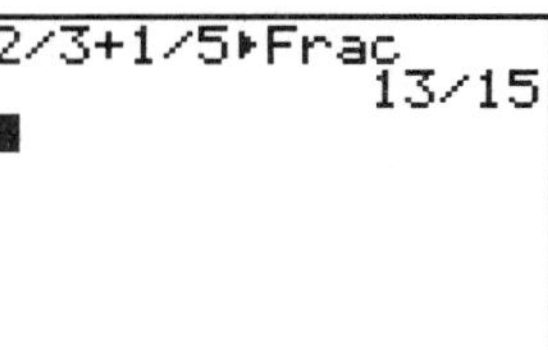

    [2] [÷] [3] [+] [1] [÷] [5] [MATH] [ENTER] [ENTER]

NOTE: If a denominator is more than four digits, the decimal equivalent of the fraction will be returned. Enter the fraction $\frac{17}{10,000}$ and activate the convert to Frac option. The calculator will only display the decimal form of the number.   ◆

Now consider the following example which involves the division of fractions. Observe the use of parentheses in the problem.

Example 2: Simplify $\dfrac{2}{3} \div \dfrac{1}{5}$.

Keystrokes:
[(] [2] [÷] [3] [)] [÷] [(] [1] [÷] [5] [)]
[MATH] [ENTER] [ENTER]

Screen display:

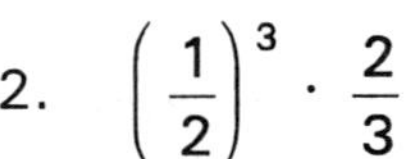

Parentheses were necessary in this division problem to ensure that the order of operations was followed. Addition and subtraction problems involving fractions do not require the use of parentheses because the calculator will follow the order of operations and perform the division (fractions) operations first and then conclude with addition/subtraction. However, with division or multiplication, the operations are performed strictly from left to right following the order of operations. Therefore, to multiply or divide fractions parentheses are necessary to override the order of operations. ◆

EXERCISE SET

Simplify the following expressions on the calculator; all answers should be expressed as fractions. **RECORD THE CALCULATOR SCREEN <u>LINE BY LINE EXACTLY</u>** as it appears.

1. $\left| -\dfrac{3}{5} \right| \cdot \sqrt{\dfrac{1}{4}}$

2. $\left(\dfrac{1}{2} \right)^3 \cdot \dfrac{2}{3}$

3. $\sqrt{\dfrac{4}{25}} \div 2$

4. $\left(\dfrac{3}{5} \right)^2 \cdot \left(\dfrac{2}{3} \right)^3$

5. $\dfrac{3}{5} \div \dfrac{1}{2} - \dfrac{5}{8}$

6. $\dfrac{5 - |1 - 3| + 7}{6 - 9 \div 3}$

7. $\dfrac{3 - |-15| \div 5}{-3\left(5 - |-2|\right) + 15}$

8. $\dfrac{6 + 12 \div 2 \cdot 3 - 5}{3^2 - 5^2 \cdot 3 + 6 \cdot 11}$

Note: Be sure to translate any error messages to the appropriate mathematical notation.

9. Enter the expression $\dfrac{16 + 8}{4}$ on your calculator in each of the following ways:

16 + 8 / 4 and (16 + 8)/4. Which is the correct format and why?

Integer Exponents

Enter each of these groups of expressions on your calculator, expressing the result as a fraction. Can you generalize the result when a base is raised to a negative power?

a. $5^{-2} =$ _______ $5^{-3} =$ _______ $5^{-4} =$ _______

b. $\left(\dfrac{2}{3}\right)^{-2} =$ _____ $\left(\dfrac{2}{3}\right)^{-3} =$ _____ $\left(\dfrac{2}{3}\right)^{-4} =$ _____

Whenever a negative exponent is applied to a base, regardless of the type of number, the effect of the negative exponent is to take the reciprocal of the base.

EXERCISE SET

Each of the following problems should be worked in two ways:

a)   arithmetically, using the Laws of Exponents - **SHOW** each step of your computation. Be sure this *hand computation* agrees with the *calculator computation*, and

b)   with the calculator, copying the screen display to justify your work - before computing the result, use the ▶FRAC option from the MATH menu so that <u>all</u> answers that are fractions will be displayed as such.

10. $2^{-3} \cdot 2^3$

Arithmetically: *Calculator display:*

11. $2^{-3} + 2^3$

    Arithmetically:

Calculator display:

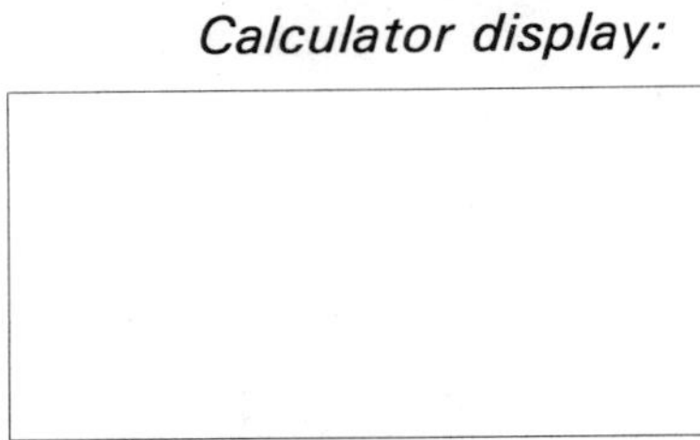

12. $\dfrac{2^{-3} \cdot 2^3}{2^3}$

    Arithmetically:

Calculator display:

13. $\dfrac{2^{-3} + 2^3}{2^3}$

    Arithmetically:

Calculator display:

Solutions:

1. 3/10   **2.** 1/12   **3.** 1/5   **4.** 8/75   **5.** 23/40   **6.** -10/3

7. 0   **8.** undefined   9. The numerator must be grouped in parentheses, (16 + 8), in order for the calculator to divide the denominator, 4, into the entire numerator.

10. 1   **11.** 65/8   **12.** 1/8   **13.** 65/4

Fractions and the TI-85/86

Since the ►**Frac** option is used frequently, it will be added to the CUSTOM menu. To do this, press **[2nd]** **<CATALOG>** **[F3](CUSTM)**. Cursor down the list to place the arrow next to ►**Frac**, found near the bottom of the list. Press **[F2]** or another open **F** key to create the custom key. HINT: Since ►**Frac** was found at the bottom of the list, pressing **[▲] [2nd]** **<M2>(PAGE↑)** until the arrow is positioned next to the desired selection would be quicker than scrolling through the entire list.

Note: To delete a customized entry, press **[2nd] [CUSTOM] [F4](BLANK)** and then the **F** key that contains the command to be deleted.

Press **[EXIT]** until the blinking cursor is displayed at the HOME screen.

Enter the decimal number 0.42 by pressing **[.] [4] [2] [CUSTOM] [F2](►Frac) [ENTER]**. Notice that the calculator displays the fraction reduced to lowest terms.

Recall that a mixed number is actually the sum of an integer and a fraction. Therefore, to enter a mixed number, simply enter the expression as an indicated sum. To enter $-3\frac{1}{2}$ and express it as a fraction, press **[(-)] [3] [+]** **[(-)] [1] [÷] [2] [CUSTOM] [F2](►Frac)**. Because the entire fraction is negative, both the integer part and the rational part must be entered as negative numbers. Your display should correspond to the one at the right.

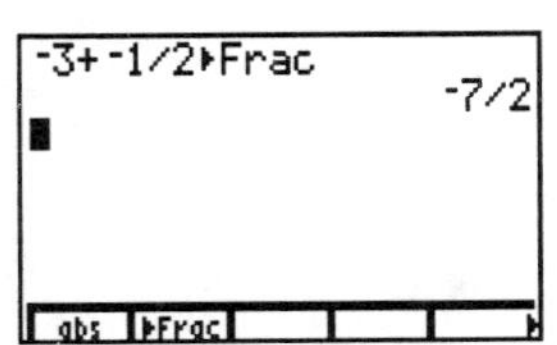

☞   Return to Example 1 in the section "Fractions and the TI-82/83/83plus" of the core unit (pg.13) and complete the unit.

UNIT 3
EVALUATING THROUGH TABLES AND THE STOre FEATURE

*Units 1- 2 are prerequisites for this unit. Answers appear at the end of this unit.

Evaluating an expression means to find the value of the expression for given values of the variables. To analytically evaluate expressions, the substitution property is applied when constant values are substituted for specified variables. The expression is then simplified following the order of operations. The STOre feature and the TABLE of the calculator can be used to accomplish the same result.

Using the STOre Key

TI-85/86   IF USING THE TI-85/86, GO TO THE GUIDELINES (PG.27) WHICH FOLLOW THIS UNIT.

To store a value, simply enter the value, press **[STO▸]** and then the desired variable. For example, to store X = 5, press **[5]** **[STO▸]** **[X,T,θ,n]** **[ENTER]**. The value 5 is now stored under the variable X. This value will remain stored in X until another value replaces it. To check and see what value is actually stored under a specific variable, enter the variable at the prompt (blinking cursor) and then press **[ENTER]**. The value is then displayed on the screen.

> NOTE: Because **X** is used as a variable so often in algebra, it has its own key on the calculator. The calculator is in function MODE; therefore when **[X,T,θ,n]** is pressed the screen displays the variable **X**.
>
> **TI-82**   This key appears as **[X,T,θ]** on the TI-82.

Example 1:   Evaluate $3X^2 + 6X + 2$ when X = -11.

Solution: Store X = -11, by pressing **[(-)]** **[1]** **[1]** **[STO▸]** **[X,T,θ,n]**. The colon ":" key (located above the gray decimal key) is used to separate commands that are to be entered on the same line, so now press **[ALPHA]** <:> before entering the expression to be evaluated.

> **TI-82**   TI-82 users should press **[2nd]**< : >.

Press **[3]** **[X,T,θ, n]** **[x²]** **[+]** **[6]** **[X,T,θ,n]** **[+]** **[2]** to enter the expression. At this point, the calculator has been instructed to store -11 in X and then evaluate $3X^2 + 6X + 2$ for this value of X. The calculator will not perform the instructions until **[ENTER]** is pressed. The polynomial evaluates to 299. Your screen should look like the one at the right.

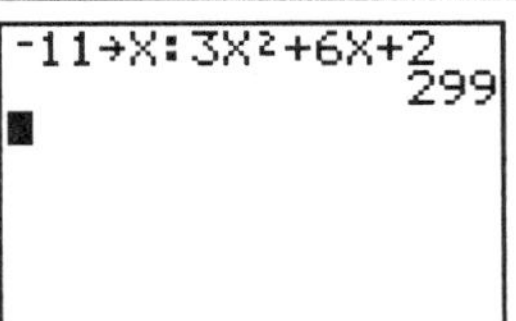

To store a value under a variable other than **X**, the **[ALPHA]** key must be used to access the letters of the alphabet. Pressing the **[ALPHA]** key, followed by another key, allows access to the upper case letters and symbols written above the key pads in the same color as the alpha key.

Example 2:   **Evaluate** $a^2 - 3a + 5$ when $a = -7$.

Solution: See screen at right.

NOTE: Be sure to use the [(-)] key for the negative sign on a number and use the subtraction key, [-] , to indicate subtraction.

◆

EXERCISE SET

✍1.   Look at the screen displayed at the right. What have you told the calculator to do?

   a. -2→X means _______________________

   b. 3→Y means _______________________

   c. $X^2 - 2X + Y^2$ means ____

   d. 17 means ____

   e. The purpose of the colon (:) is to_____________________________________

Exercises 2-6 should be completed as follows:
(a) evaluate each expression analytically by substituting the values for the variables into the given expression and following the order of operations, and
(b) evaluate with the calculator, using the **STO▸** feature. Record the screen display as a means of showing your work. You must copy the screen display **exactly** as it appears, line by line.

2.   Evaluate $2X^2 + 3X + 1$ when $X = -3$.

   Algebraically: *Calculator display:*

3.   Evaluate $(X - 2)^2 + 3X$ when $X = \frac{1}{2}$.

   Algebraically: *Calculator display:*

4.   Evaluate $4(X - 3) + 5(Y - 3) - 4$ when $X = -1$ and $Y = -3$.

Algebraically: *Calculator display:*

5.   Evaluate $-X^2 + 3XY^2 - 6Y^3$ when $X = 3$ and $Y = 4$.

TI-85/86   IF USING THE TI-85/86, GO TO THE GUIDELINES (PG.28) WHICH FOLLOW THIS UNIT.

Algebraically: *Calculator display:*

6.   Evaluate $\dfrac{2X^3Y - 2X^2Y + X}{5X - Y}$ when $X = 2$ and $Y = \frac{1}{2}$.

Algebraically: *Calculator display:*

7.   Troubleshooting: Each of the problems below has been entered **<u>incorrectly</u>** on the calculator. Make the necessary corrections so that the calculator display accurately represents the problem given. Be sure to verify with your calculator!

   a. Evaluate $2X - 5$ when $X = -3$.

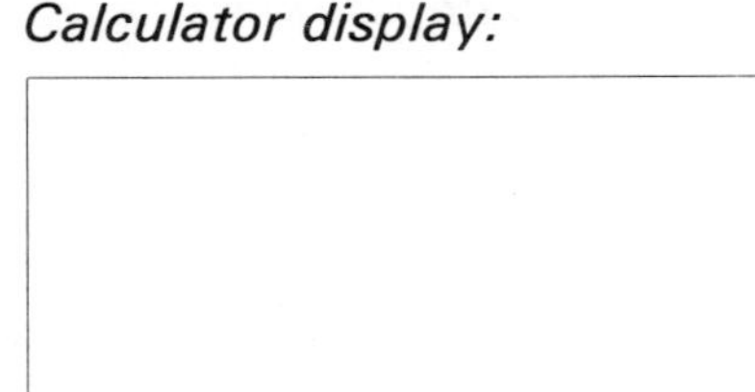

   b. Evaluate $\dfrac{\sqrt{B^2 - 4AC}}{2A}$ when $B = 4$, $A = 5$, $C = -1$.

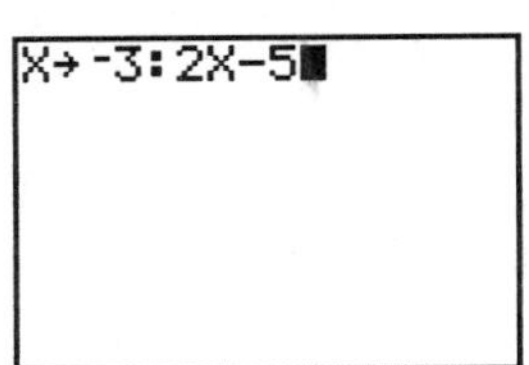

✎ 8. Discuss the use of parentheses when entering expressions containing radicals or fractions.

Checking Solutions

The graphing calculator can be used to check solutions to equations and to confirm that expressions are equivalent.

Example 3:   Is 6 a solution to the equation $4(2X - 1) - 8 = 42 - X$?

Solution: If 6 is a solution to the equation then the left side, $4(2X - 1) - 8$, will have the same value as the right side, $42 - X$, when 6 is substituted for X. Store 6 in X and then find the value of each side of the equation. Your screen should look like the one at the right.

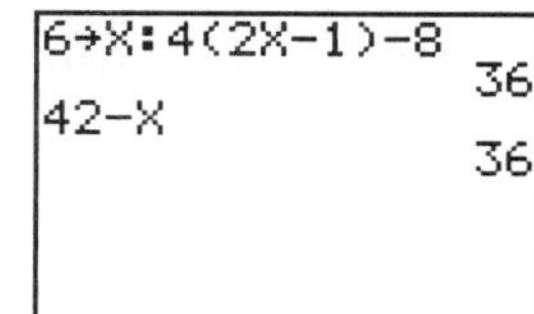

<table>
<tr><td>TI-85</td><td>Because the TI-85 does not have a TABLE feature, TI-85 users should use the STO▸ key to duplicate TABLE results in the following examples. You may want to explore the WWW.TI.COM/CALC/DOCS/ARCH.HTM web site for TABLE programs for your calculator.</td></tr>
</table>

A proposed solution can also be verified through the TABLE feature of the TI-82, TI-83/83plus, or TI-86 graphing calculators.

Example 4:   Consider the equation $3(5x + 1) = 8x + 24$. Is 3 a solution to this equation?

Solution: To confirm that 3 is a solution via the TABLE, we will *store* the expressions on each side of the equation at the **y-edit** screen. To do this, press [Y=] and CLEAR all entries.

<table>
<tr><td>TI-86</td><td>Press [GRAPH] [F1](y(x)=).</td></tr>
</table>

Enter the left side of the equation at the **Y1=** prompt. Pressing [ENTER] will move the cursor to the **Y2=** prompt, which is where the right side of the equation is to be entered. To display the appropriate TABLE value, press [2nd] [TblSet].

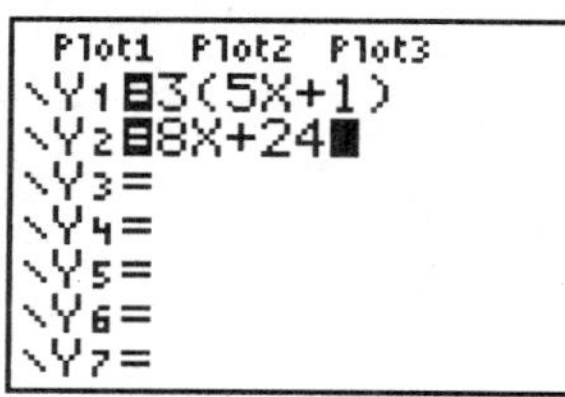

<table>
<tr><td>TI-86</td><td>TI-86 users press [TABLE] [F2](TBLST).</td></tr>
</table>

The cursor is now at the **TblStart=** prompt (TblMin= on the TI-82). Enter 3, since that is the value to be verified as a solution. Press [ENTER] to move the cursor to the **ΔTbl=** prompt. Enter a 1 (one) to increment the X- value of the TABLE by one unit. (Since we will only be concerned with one entry line in the TABLE, the table increment is not critical. It is suggested that the increment be left at 1.) Access the TABLE by pressing [2nd] [TABLE].

<table>
<tr><td>TI-86</td><td>TI-86 users press [F1](TABLE).</td></tr>
</table>

Your screen should look like the one pictured to the right. Notice the first entry under the X column is 3. Recall the expression $3(5x + 1)$ was entered at the $Y1 =$ prompt and $8x + 24$ was entered at the $Y2 =$ prompt. The values in the Y1 and Y2 column are equal (i.e. both 48) verifying that when $x = 3$, each of the expressions entered at the $Y1 =$ and $Y2 =$ prompts evaluate to 48.

X	Y1	Y2
3	48	48
4	63	56
5	78	64
6	93	72
7	108	80
8	123	88
9	138	96

X=3

Use the [▲] and [▼] cursors to scroll through the TABLE. Notice that the values in the Y1 and Y2 columns are equal ONLY when $x = 3$.

◆

The TABLE has provided numerical evidence that 3 is a solution to the equation $3(5x + 1) = 8x + 24$. In the space below, use the Substitution Property to evaluate both sides of the equation analytically (by hand) to confirm that $x = 3$ is a solution.

The **STO►** feature can also be used to confirm that when 3 is substituted into the expression on each side of the equation, the expressions both evaluate to 48.

STEPS FOR CONFIRMING A SOLUTION VIA THE TABLE

1. Press **[Y =]** and enter the left side of the equation at **Y1 =** and the right side at **Y2 =**.

    | TI-86 | TI-86 USERS PRESS **[TABLE] [F2](TBLST)**. |

2. Press **[2nd] [TblSet]** to set the TblStart (TblMin on the TI-82) to the value believed to be the solution of the equation. The value for **ΔTbl** may be any value. It is suggested that the number one be used.

3. Access the TABLE by pressing [2nd] <TABLE> (TI-86 users press **[F1](TABLE)**.) The solution is confirmed if and only if the Y1 and Y2 columns are identical for the designated value of X.

EXERCISE SET

Directions: Verify that each number below is a solution to the given equation. Copy the line(s) of the TABLE that verify the solution(s).

9. Verify that 3/2 is a solution to the equation $X + 5 = 3X + 2$

X	Y1	Y2

10. Verify that 2 is a solution to the equation $\dfrac{3}{2}(X + 4) = \dfrac{20 - X}{2}$

X	Y1	Y2

11. Verify that 4 **and** -3 are both solutions to the equation $X^2 - X - 12 = 0$.

X	Y1	Y2

12. What two values of X can be verified as solutions to an equation from the displayed table?

    X = _______ and X = ________

✍ What assures you that these numbers are solutions?

✍13. The solutions to the equation $3X^2 = 5X + 12$ are -4/3 and 3. When a student attempts to confirm the solution of -4/3, he instructs the calculator to display the TblStart as -4/3 and to be incremented by 1. Note that the TblStart of -4/3 is converted to a decimal approximation for display in the TABLE. The top row of the TABLE confirms -4/3 as a solution. How can the solution X = 3 be confirmed?

Example 5:   Simplify the following expression algebraically: $3X(X - 2) + 5X^2$

Solution: The simplified expression is $8X^2 - 6X$. To determine if $3X(X - 2) + 5X^2$ equals $8X^2 - 6X$, enter the original expression at the **Y1 =** prompt and the simplified expression at the **Y2 =** prompt. Set the TABLE to have a start value of O and increment the TABLE by 1. This start value and the increment are arbitrary. Any values could be used in either position. Your screen should match the one displayed.

X	Y1	Y2
0	0	0
1	2	2
2	20	20
3	54	54
4	104	104
5	170	170
6	252	252

X=0

WARNING: This procedure confirms that two expressions are equivalent. It does not determine that the expression is completely simplified.

◆

EXERCISE SET CONTINUED

14. Explain **how**, in Example 5, the use of the TABLE confirms that the given expressions are equal.

15. The expression $(3X^2 - 5X + 2) - (8X^2 - 3X - 1)$ simplifies to $-5X^2 - 2X + 3$. Confirm that these two expressions are equivalent by using the TABLE feature of the graphing calculator.
Describe the entries on the TABLE screen.

16. The expression $(4X - 5)(3X + 7)$ expands to $12X^2 + 13X - 35$. Confirm that these two expressions are equivalent by using the TABLE feature of the graphing calculator.
Describe the entries on the TABLE screen.

17.   In attempting to perform the division on the expression $\dfrac{4x^2 - 8x - 16}{8x^2}$, a student produced $\dfrac{1}{2} - \dfrac{1}{x} - \dfrac{4}{2x^2}$.

a. When he tried to use the calculator to confirm his work, he selected O as his TblStart for X and the calculator display an error message in **both** the Y1 and Y2 columns. What does this mean in terms of the two expressions?

25

b. Can he conclude that his simplification is correct?

18. A student expanded the quantity $(2X + 5)^2$ to $4X^2 + 25$. Confirm or deny with the TABLE feature that this is the correct expansion.

<u>Solutions</u>:

1. a. store -2 for the variable x   **b.** store 3 for the variable y   **c.** evaluate the expression for the stored value of the variables   **d.** the value of the expression is 17 when x = -2 and y = 3

e. separate commands

2. 10   **3.** 15/4   **4.** -50   **5.** -249   **6.** 12/19

7. a. Your screen display should indicate that -3 is stored in X: -3→X   **b.** After the values are stored in the given variables, the display for the expression should look like the following:
$\sqrt{(B^2 - 4AC)} / (2A)$

8. answers may vary

9.

X	Y1	Y2
1.5	6.5	6.5

10.

X	Y1	Y2
2	9	9

11.

X	Y1	Y2
-3	0	0
4	0	0

12. X = -1, X = 3; The Y columns are equal for these X values.

13 Change the table minimum (start) to 3.

14. answers may vary

15. Entry by entry, all values of Y1 equal all values of Y2.

16. Entry by entry, all values of Y1 equal all values of Y2.

17a. The expression is undefined when X = 0.

17b. It is correct, but not completely simplified.

18. The expansion is <u>not</u> <u>correct</u> because the TABLE values for Y1 and Y2 are not equal.

The **[x-VAR]** key is used to display the variable **x**. Note that the variable **x** is displayed as a lowercase letter on the TI-85/86, instead of an uppercase **X** as the **[X,T,θ,_n_]** key on the TI-82/83/83plus. When using this text, each **X** printed as a variable for the TI-82/83/83plus should appear as **x** on the TI-85/86.

To store a value, simply enter the value, press **[STO▸]** and then the desired variable. For example, to store X = 5, press **[5] [STO▸] [x-VAR] [ENTER]**. The value 5 is now stored under the variable **x**. This value will remain stored in **x** until another value replaces it. To determine the value that is actually stored under a specific variable, enter the variable at the prompt (blinking cursor) and then press **[ENTER]**. The value is then displayed on the screen.

NOTE: Pressing the **STO▸** key automatically initiates the ALPHA key, allowing access to the remaining letters of the alphabet. Press **STO▸** and observe that the blinking cursor has an **A** in it for **ALPHA**. Letters will automatically be recorded in uppercase format. To disengage the ALPHA feature, press the **[ALPHA]** key.

Example 1:   Evaluate $3X^2 + 6X + 2$ when X = -11.

Solution: Store X = -11, by pressing **[(-)] [1] [1] [STO▸] [x-VAR]**. The colon ":" key (located above the gray decimal key) is used to separate commands that are to be entered on the same line. Press **[2nd] <:>** before entering the expression to be evaluated. At this point you should note that the blinking cursor has an **A** in it for ALPHA. Disengage the ALPHA feature by pressing **[ALPHA]** now. Press **[3] [x-VAR] [x²] [+] [6] [x-VAR] [+] [2]** to enter the expression that is to be evaluated. At this point, the calculator has been instructed to store -11 in **x** and to evaluate $3x^2 + 6x + 2$ at this value of **x**. The calculator will not perform the instructions until the **[ENTER]** key is pressed. The polynomial evaluates to 299. Your screen should look like the one above.

```
-11→x:3 x²+6 x+2
                        299
■
```

The TI-85/86 has both uppercase and lowercase alphabet letters that can be used as variables. To access a lowercase letter when using the **STO▸** feature, press **[2nd] <ALPHA>** after pressing the **[STO▸]** key. However, realize that an expression containing a lower case variable could be entered as an expression using an upper case variable as long as you are consistent. Use lower case letters with care. As noted at the end of this guideline, some lower case letters are reserved for system variables.

When not using the **STO▸** feature, uppercase letters are accessed by **[ALPHA]** and lowercase letters are accessed by **[2nd] <ALPHA>**.

Example 2:   Evaluate $a^2 - 3a + 5$ when a = -7.

Solution: See screen at right.

```
-7→A:A²-3A+5
                        75
■
```

NOTE: Be sure to use the gray [(-)] key for the negative sign on a number and use the black [-] key to indicate subtraction.

The keystrokes would need to be [(-)] [7] [STO▸] [A] [2nd] <:> [A]*(observe that the cursor is still blinking with an uppercase A inside of it)* [ALPHA] *(the ALPHA feature is now disengaged)* [x²] [-] [3] [ALPHA] [A] [+] [5] [ENTER].

◆

☞ RETURN TO THE EXERCISE SET IN THE CORE UNIT (PG.20) AND WORK THE EXERCISES.

The TI-85/86 has the capability of recognizing combinations of letters as independent variables. For example, the expression **4AC** means "four times A times C". However, the TI-85/86 would recognize **AC** as representing only <u>one</u> unknown, not a product of two unknowns. Thus the expression "four times A times C" must be displayed on the TI-85/86 as **4A*C** or **4A C**. Variables can be separated with a space [⌴] to denote the multiplication of two different quantities. By being able to combine letters to form new variables, the TI-85/86 has a limitless list of variables available for use. Uppercase letters are reserved for user variable names. Note that the TI-85/86 would recognize **4a*c** as an entirely different product since the variables are lowercase. Thus Exercise #5 on page 21 should be entered with either a multiplication sign or a **space** between the two variables in the middle term; i.e. $3X * Y^2$ or $3X\ Y^2$, but not $3XY^2$.

Lowercase letters should be used with care as some are reserved for system variables. Examples of these are:

a	coefficient of regression
b	coefficient of regression
c	speed of light
e	e natural log base
g	force of gravity
h	Planck's constant
k	Boltzman's constant
n	number of items in a sample
u	atomic mass unit

The variables r,t,x,y, and θ are updated by graph coordinates based on the graphing mode.

☞ RETURN TO THE CORE UNIT EXERCISE 5 (PG.21) AND COMPLETE THE UNIT.

UNIT 4
RATIONAL EXPONENTS AND RADICALS

*Unit 2 is a prerequisite for this unit. Answers appear at the end of the unit.

TI-83/83plus users should press [MODE] and verify that **Real** is highlighted in the left column. The TI-83/83plus have complex number capabilities as is indicated by the "a + bi" selection adjacent to **Real**. Complex number operations will be discussed in the unit entitled "TI-83/83plus/85/86 Complex Numbers." For now, the calculator should be set to compute only with real numbers.

Radicals

The inverse operation of raising a number to a power is extracting a root. For example, $4^3 = 64$ and $\sqrt[3]{64} = 4$.

The only type of radical that has been addressed thus far in this text is $\sqrt{}$ or principal square root. We will now examine $\sqrt[3]{}$, $\sqrt[4]{}$, $\sqrt[6]{}$, etc., and the relationship of these radicals to rational exponents.

TI-85/86

PRESS [2ND] <MATH> TO DISPLAY THE FIRST FIVE SUB-MENUS OF THE TI-85/86'S MATH FEATURE. THE "▶" AFTER **MISC** INDICATES THERE ARE MORE SUB-MENUS. PRESS THE [MORE] KEY TO DISPLAY THE REMAINING MENU AND THEN PRESS [MORE] ONE MORE TIME TO RETURN TO THE INITIAL SET OF SUB-MENUS. (SUGGESTION: AFTER COMPLETING THIS UNIT, EXAMINE THE CONTENTS OF EACH OF THESE SIX SUB-MENUS WHICH ARE ACCESSED BY PRESSING THE APPROPRIATE **F** KEY.) THE RADICAL SYMBOL, $\sqrt[x]{}$, THAT WILL BE USED IN THIS UNIT IS FOUND UNDER THE **MISC** SUBMENU. PRESS [F5](MISC) FOLLOWED BY [MORE] TO DISPLAY THE $\sqrt[x]{}$. AT THIS POINT, IT WOULD BE A GOOD IDEA TO RETURN TO THE CUSTOM MENU AND CUSTOMIZE THE $\sqrt[x]{}$ USING THE CATALOG.

THERE IS NO $\sqrt[3]{}$ AVAILABLE ON THE TI-85/86; CONTINUE THE UNIT ON THE NEXT PAGE.

Begin by looking at the MATH menu. Press [MATH] and the appropriate screen is displayed at the right. To scroll through the entire menu, press the down arrow key. This unit will use the ▶FRAC option, as well as the $\sqrt[3]{}$ and $\sqrt[x]{}$ options.

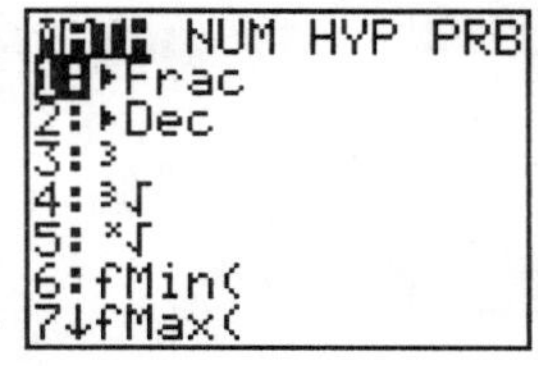

TI-82 Screen   TI-83/83plus Screen

To find the cube root of -27, $\sqrt[3]{-27}$, press [MATH] [4] (to select the $\sqrt[3]{(}$ option) [(-)] [2] [7] [)] [ENTER]. The result displayed should be -3. Recall, the inverse of extracting a root is raising to a power. Thus -3 is the correct result for $\sqrt[3]{-27}$ because $(-3)^3 = -27$.

To compute seven times the cube root of 5, $7\sqrt[3]{5}$, press [7] [MATH] [4:$\sqrt[3]{(}$] [5]. The result displayed is an approximate answer rounded to nine decimal places: 11.96983163.

To compute roots other than square roots ($\sqrt{}$ is found on the face of the calculator) or cube roots, the $\sqrt[x]{}$ selection on the MATH menu must be used. Designate a value for x by entering the index first and then the $\sqrt[x]{}$ symbol.

To compute the sixth root of 64, $\sqrt[6]{64}$, press [6] [MATH] [5: $\sqrt[x]{}$] [6] [4] [ENTER]. Your screen should match the one at the right.

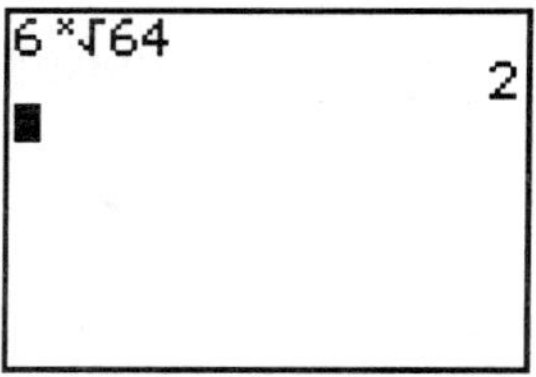

Because the display is 6 $\sqrt[x]{64}$, care must be taken when entering expressions like 4 $\sqrt[6]{64}$. In a problem such as 7 $\sqrt[3]{5}$, it is unnecessary to "tell" the calculator to multiply 7 times $\sqrt[3]{5}$ because 7 $\sqrt[3]{5}$ indicates <u>implied</u> multiplication. However in 4 $\sqrt[6]{64}$ implied multiplication cannot be used because of the calculator notation. Had the expression 4 6 $\sqrt[x]{64}$ been entered, the calculator would compute $\sqrt[46]{64}$. A multiplication symbol, " * ", must be entered after the digit 4 for clarification. Your screen display for 4 $\sqrt[6]{64}$ should match the one at the right.

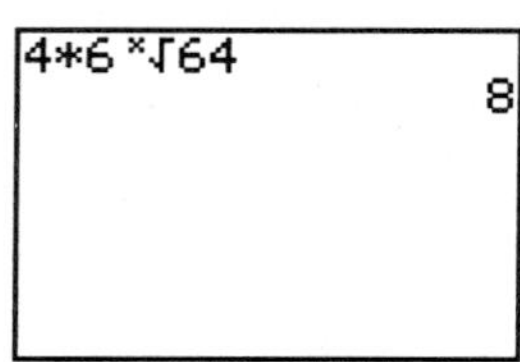

NOTE: TI-83 users must be particularly careful when using the $\sqrt[x]{}$. Up to this point, a parenthesis has been automatically entered after the square root or cube root symbol. The $\sqrt[x]{}$ symbol does not include this parenthesis. Any fraction or expression containing more than one term must be enclosed in parentheses when placed under this radical symbol.

Use the calculator to find the value of $\sqrt[4]{\dfrac{16}{625}}$. The correct root is 2/5 (or 0.4). If you got 2/625 (or 0.0032) then you failed to group the fraction with parentheses. Parentheses are necessary for the calculator to extract the fourth root of the <u>quantity</u> 16/625.

EXERCISE SET

Directions: Evaluate each radical expression. Record the screen display, being sure that the indicated root is displayed as an integer or fraction - rather than a decimal.

1. $\sqrt[3]{-125}$ 2. $\sqrt[4]{4096}$ 3. $\sqrt[6]{5^6}$

4. $\sqrt{\dfrac{4}{25}}$ 5. $\sqrt[5]{\dfrac{1}{32}}$ 6. $\sqrt{-625}$

Rational Exponents

Positive exponents increase the size of a whole number and negative exponents decrease the size. Negative exponents have the effect of taking the reciprocal of a number. We now want to consider the effect of rational (fractional) exponents.

Rational exponents are entered into the calculator in the same manner as positive and negative integer exponents. Be aware of the order of operations as you enter the expression. An expression has been entered on the calculator and is displayed on the screen at the right. Two operations are indicated, raising a base to a power and division. Because of the order of operation rules, the powers will be performed

before the division. This calculator display is not an illustration of $25^{\frac{1}{2}}$ but is an illustration

of $\dfrac{25^1}{2}$. For the calculator to evaluate the expression $25^{\frac{1}{2}}$, parentheses must be inserted to override the order of operations.

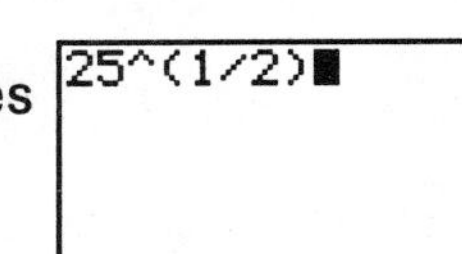

EXERCISE SET

The following problems have been placed in groups of four to more easily compare the effect of differing exponents on the same base. **Be sure all rational exponents are enclosed in parentheses.** Enter final results on the appropriate blank.

7. $25^2 =$ _____ $25^{1/2} =$ _____ $25^{-1/2} =$ _____ $25^{-2} =$ _____

8. $8^3 =$ _____ $8^{1/3} =$ _____ $8^{-1/3} =$ _____ $8^{-3} =$ _____

9. $\left(\dfrac{16}{81}\right)^2 =$ ___ $\left(\dfrac{16}{81}\right)^{\frac{1}{2}} =$ ___ $\left(\dfrac{16}{81}\right)^{-\frac{1}{2}} =$ ___ $\left(\dfrac{16}{81}\right)^{-2} =$ ___

The following exercises allow you to determine the relationship between $\sqrt[n]{a^m}$ and $a^{m/n}$. The problems are grouped in such a way that patterns can be discovered. Record the result on the blank.

10. $\sqrt[3]{4^6} = $ ______ $4^{\frac{6}{3}} = $ ______ $4^2 = $ ______

11. $\sqrt[3]{-8} = $ ______ $(-8)^{\frac{1}{3}} = $ ______ $\sqrt[3]{(-2)^3} = $ ______

12. $\sqrt{16} = $ ______ $16^{\frac{1}{2}} = $ ______ $\left(\dfrac{1}{16}\right)^{-\frac{1}{2}} = $ ______

13. $27^{\frac{2}{3}} = $ ______ $\left(\sqrt[3]{27}\right)^2 = $ ______ $\sqrt[3]{27^2} = $ ______

14. $\sqrt{-36} = $ ______ $(-36)^{\frac{1}{2}} = $ ______

✍15. In #14 the TI-82/83/83plus calculator displayed a DOMAIN ERROR. Explain what is wrong with the problem.

✍16. Explain the relationship between $\sqrt[n]{a^m}$ and $a^{m/n}$.

Solutions: **1.** -5   **2.** 8   **3.** 5   **4.** 2/5   **5.** ½

6. Did you get an error message? The DOMAIN error message is displayed because the TI-82 only computes values of real numbers and TI-83/83plus users were instructed to set the calculator to compute only with real numbers. In the real number system, $\sqrt{-625}$ does not exist. There is no real number that when raised to the second power will equal -625.

The TI-85/86 will display (0,25) instead of an ERROR message as the TI-82/83 does. This is because the TI-85/86 can perform complex number operations. The $\sqrt{-625}$ simplifies to the complex number $0 + 25i$. The calculator notation for $0 + 25i$ is (0,25). The use of the TI-85/86 in the simplification of complex number expressions is discussed in detail in the unit entitled "TI-83/85/86 Complex Numbers." Until the instructor and/or textbook addresses complex numbers, results displayed in the form (a, b) should be recorded as *no real number*.

7. 625, 5, 1/5, 1/625   **8.** 512, 2, 1/2, 1/512   **9.** 256/6561, 4/9, 9/4, 6561/256
10. 16, 16, 16   **11.** -2, -2, -2   **12.** 4, 4, 4   **13.** 9, 9, 9
14. not a real number, not a real number

TI-85/86	TI-85/86 USERS WILL HAVE (0,6) FOR BOTH RESULTS IN 14. REMEMBER, THIS IS THE CALCULATOR'S NOTATION FOR COMPLEX NUMBERS WHICH WILL BE ADDRESSED IN A LATER UNIT.

15. The square root function is not defined for negative radicands.

16. In the fractional exponent $\dfrac{m}{n}$, the n designates the index of the radical and the m is the exponent applied to the radical.

UNIT 5
TI-83/83plus/85/86 COMPLEX NUMBERS

*Unit 3 is a prerequisite for this unit. Answers appear at the end of the unit.
NOTE: The TI-82 does not address complex numbers.
Complex numbers are numbers that can be expressed in the form a + bi where **a** represents the real part and **b** represents the imaginary part.

TI-83/83plus The TI-83/83plus has the value "i" on the calculator face, located above the decimal point. This allows the easy entry and arithmetic manipulation of complex number expressions.

1. Complex numbers can be represented by variables, that is to say they can be stored as variables. Let X = 4 - 2i and evaluate $3X^2 + 2X - 5$. Using the **STO►** key, your display should look like the one at the right. The interpretation of the display is that replacing X by the complex number 4 - 2i in the expression $3X^2 + 2X - 5$, yields a value of 39 - 52i.

```
4-2i→X:3X²+2X-5
            39-52i
■
```

2. Access the complex number menu by pressing **[MATH] [►] [►]** to highlight **CPX**. This menu indicates that for any complex number, a + bi, the calculator will identify the conjugate, determine the real part, determine the imaginary part, and determine the absolute value. The **►FRAC** option can be used with any of the operations.

```
conj(1/2-3i)
          .5+3i
real(1/2-3i)
             .5
imag(1/2-3i)
             -3
```

```
abs(1/2-3i)
      3.041381265
```

3. Arithmetic operations with complex numbers can be performed easily with this calculator. For example, to perform the division $\dfrac{2 + 3i}{3 + 2i}$ by hand, the numerator and denominator of the fraction must be multiplied by the conjugate of the denominator in order to simplify the expression. Be sure to use parentheses around both the numerator and denominator when entering the fraction in the calculator and use the **►FRAC** option to express the quotient in fraction form. The displayed result should correspond to the one pictured.

```
(2+3i)/(3+2i)►Fr
ac
       12/13+5/13i
```

TI-85/86 COMPLEX NUMBERS ARE DISPLAYED AS **(REAL, IMAGINARY)** ON THE TI-85/86. THUS 2 - 3i WOULD BE DISPLAYED AS (2,-3) SINCE 2 IS THE REAL PART AND -3 IS THE IMAGINARY PART. **NOTE:** (REAL, IMAGINARY) IS RECTANGULAR FORMAT, NOT POLAR IN THE UNIT ENTITLED "EVALUATING THROUGH TABLE AND THE STOre FEATURE", PG.28, THE USE OF LOWERCASE LETTERS AS SYSTEM RESERVED VARIABLES IS DISCUSSED. THE LOWERCASE i CAN BE DESIGNATED AS A USER DEFINED VARIABLE IN THE TI-85/86 BY USING THE CONSTANT EDITOR. USE OF THE CONSTANT EDITOR TO DEFINE i AS (0,1), REPRESENTING 0 + i, PREVENTS YOU FROM STORING AN ALTERNATE VALUE IN i AT A LATER DATE.

PRESS [2ND] <CONS> [F2](EDIT) AND ENTER A LOWERCASE **i** AFTER **Name** = BY PRESSING [2ND] <ALPHA> <I>. CURSOR DOWN TO **Value** = AND ENTER (0,1) TO DEFINE THE VALUE AS THE COMPLEX NUMBER $0 + i$. COMPLEX NUMBERS MAY NOW BE ENTERED ON THE HOME SCREEN IN A + Bi FORMAT INSTEAD OF (A,B) FORMAT IF DESIRED. HOWEVER, ANSWERS WILL BE RETURNED BY THE CALCULATOR IN THE (A,B) FORMAT.

1. COMPLEX NUMBERS CAN BE REPRESENTED BY VARIABLES, THAT IS TO SAY THEY CAN BE STORED AS VARIABLES. LET $x = 4 - 2i$ AND EVALUATE $3x^2 + 2x - 5$. USING THE **STO▶** KEY, THE DISPLAY SHOULD LOOK LIKE THE ONE AT THE RIGHT. INTERPRETING THE DISPLAY, WE SEE THAT WHEN $x = 4 - 2i$, $3x^2 + 2x - 5$ HAS A VALUE OF $39 - 52i$.

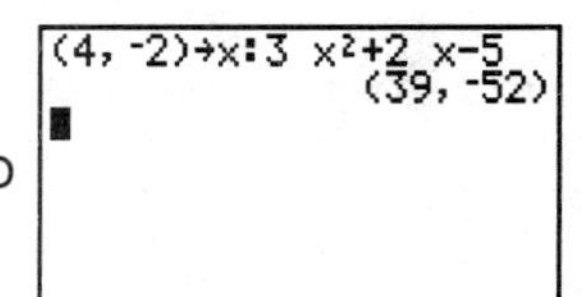

2. PRESS [2ND] <CPLX> TO ACCESS THE COMPLEX NUMBER MENU. FOR ANY COMPLEX NUMBER, A + Bi, THE CALCULATOR WILL IDENTIFY THE CONJUGATE (F1), DETERMINE THE REAL PART (F2), DETERMINE THE IMAGINARY PART (F3), AND DETERMINE THE ABSOLUTE VALUE (F4). THE REMAINING OPTIONS ARE USED WITH POLAR FORMAT. THE DISPLAYS AT THE RIGHT INDICATE THE APPLICATIONS OF EACH OF THESE OPTIONS WHEN APPLIED TO ½ - 3i. THE **▶FRAC** OPTION CAN BE USED WITH ANY OF THESE OPERATIONS.

3. ARITHMETIC OPERATIONS WITH COMPLEX NUMBERS CAN BE PERFORMED EASILY WITH THE TI-85/86. FOR EXAMPLE, TO PERFORM THE DIVISION $\dfrac{2 + 3i}{3 + 2i}$ BY HAND, YOU WOULD NEED TO MULTIPLY THE NUMERATOR AND DENOMINATOR BY THE CONJUGATE OF THE DENOMINATOR IN ORDER TO SIMPLIFY THE EXPRESSION. THE CALCULATOR ALLOWS YOU TO ENTER THE PROBLEM AS DISPLAYED ON THE SCREEN AT THE RIGHT. BE SURE TO USE THE **▶FRAC** OPTION TO EXPRESS THE QUOTIENT IN FRACTION FORM. THE DISPLAYED RESULT IS $\dfrac{12}{13} + \dfrac{5}{13}i$ IN STANDARD FORM.

EXERCISE SET

Directions: Use the complex number format and menu on the calculator to perform the complex number operations below. Copy your screen display and record all answers in standard a + bi form.

1. $(3 + 2i) + 4(-2 + 5i)$

        1.________

2. $(3 + 2i)(-2 + 5i)$

        2.________

3. $\dfrac{5 + i}{2 - 3i}$ 3._________

4. $(2 - 3i)^3$ 4._________

5. i^{23} 5._________

6. i^{114} 6._________

7.   Use the **STO►** feature to show that $\pm 5i\sqrt{2}$ is a solution to $x^2 + 50 = 0$.

8. $\left| \dfrac{4}{5} + \dfrac{2}{3}i \right|$ 8._________

Can the result be converted to a fraction?_____________

✐ Compute this same problem by hand and explain why.

TI-83/83plus users should enter $\sqrt{-4}$ on their calculator. If the **MODE** is still set to highlight **Real** then the calculator will display an error message: nonreal answer. At this point change the **MODE** setting to highlight **a + bi** so the calculator will compute the value of $\sqrt{-4}$. The result should now be 2i. When computing even roots of negative numbers, the calculator's MODE must be set for complex numbers.

9. Simplify the following on the calculator:

   a. $4 + \sqrt{-4}$ 9a.__________

   b. $\dfrac{2}{5} + \sqrt{-\dfrac{4}{9}}$ (Be sure the answer is expressed in fractional form.)   9b.__________

10. Evaluate $\sqrt{-2}$ on the calculator. Express the result as a fraction, or explain why the calculator will not convert it to a fraction.

Solutions: 1. $-5 + 22i$ **2.** $-16 + 11i$ **3.** $\dfrac{7}{13} + \dfrac{7}{13}i$ **4.** $-46 - 9i$

5. TI-83: 3E -13 -i which is equivalent to 0 - i
   TI-85: (3E -13, -1) which is equivalent to (0,-1) which is 0 - i in standard form

6. TI-83: -1 + 1.4E -12 which is equivalent to -1 + 0i or -1
   TI-85: (-1, 1.4E -12) which is equivalent to (-1,0) which is -1 + 0i in standard form or -1.

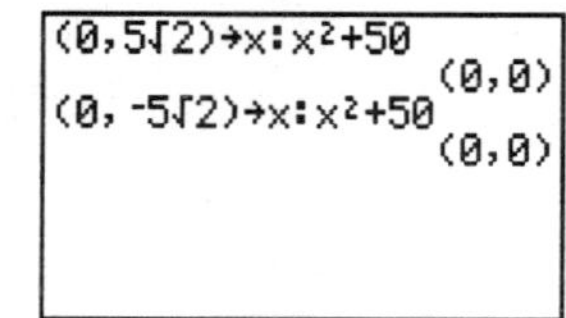
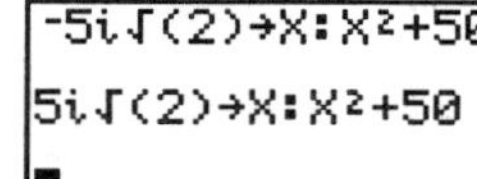

7.

   TI-85/86 Screen   TI-83/83plus Screen

8. 1.04136662345: This number cannot be converted to a fraction because it is an irrational number.

9a. $4 + 2i$ **9b.** $\dfrac{2}{5} + \dfrac{2}{3}i$ **10.** The number cannot be converted to a fraction because it is an irrational number.

UNIT 6
TI-83/83plus BUSINESS APPLICATIONS

*Unit 1 is a prerequisite for this unit. Answers appear at the end of the unit.

This unit examines some of the financial functions of the TI-83/83plus. These functions will be illustrated through the use of the following problem.

Justin purchased a new car for $22,248.72 (tax, title, and license included). In order to save on the amount of money paid out to interest, his father loaned him the money at an annual rate of 3.75%. Justin is to repay the money in monthly installments.

Example1:   If Justin can afford only $400 per month for the payment, how many months will he pay on the car?

Solution: a. Begin finance problems by fixing the decimal to two places. Press [MODE]; cursor down and over to highlight the number **2** on the float line, then press [ENTER]. The decimal is fixed to two places because computation will involve dollars and cents.

b. Press [APPS] [1:Finance] to display the menu at the right. (On the TI-83 press [2nd] <FINANCE>.) To select [1:TVM Solver...] press either [ENTER] (since the number 1 is highlighted) or press [1]. The notation indicated on this screen is:

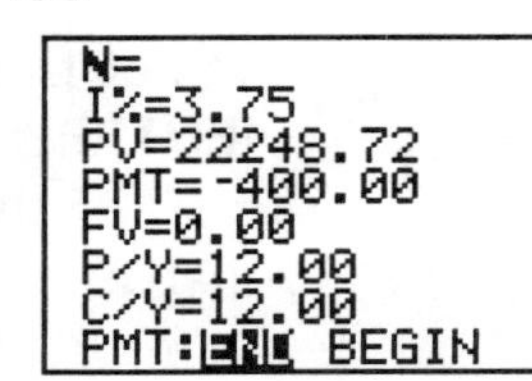

     PV - present value of the money
     PMT - payment
     FV - future value of the money
     P/Y - number of payments per year
     C/Y - number of compounding periods per year
     PMT:END - payments occur at the end of each period

Cursor down to each entry on the **TVM Solver...** screen and enter the information from the problem above. (Note: You must have at least a 0 entered at the **N**= location or the calculator will not allow you to cursor down to make the subsequent entries.) Outgoing cash (in this case the payment) should be entered as a negative value and incoming cash should be entered as a positive value. When all information has been entered, place the cursor at the **N**= variable prompt. The cursor must be placed beside the designated unknown when using the **SOLVE** feature to determine a missing variable quantity.

Note: When a value is entered for P/Y, C/Y is automatically entered at the same value. If P/Y ≠ C/Y, enter C/Y's unique value.

Your screen should look like the one pictured. With the cursor at **N**=, press [ALPHA] <SOLVE> to initiate the solver. The calculator will then compute the number of payments, N.

Answer: Justin will make 61 full payments of $400 each with the last payment (at the end of the sixty-second month) being less than $400.   ◆

NOTE: Return to the home screen by pressing **[2nd]** **<QUIT>**.

Example 2:   Determine Justin's outstanding loan balance at the end of 15 months and the amount of money paid to principal and interest thus far.

Solution: a. To determine the balance for an amortization schedule, values must be stored in **PV, I%,** and **PMT.** This has already been done in the first example, as is indicated by the last screen displayed. The balance is computed by pressing **[APPS] [1:Finance]** , cursoring down to select (press) option 9, **bal(,** and pressing **[ENTER].** (On the TI-83, press **[2nd]** **<FINANCE>** **[9:bal( ]** .) The notation bal(T) computes the outstanding balance after T months. Thus, to compute the outstanding balance after 15 months, enter 15 after the **bal(** displayed on the home screen, close the expression with a right parenthesis and press **[ENTER].** Your screen should correspond to the one at the right which indicates the outstanding balance is $17181.71 after 15 months.

```
bal(15)
            17181.71
```

b. To determine the sum of the principal or the sum of the interest paid during this 15 month time period, use the functions $\sum$ **Prn** and $\sum$ **Int** listed in the **FINANCE** menu. Each of these functions will be entered on the home screen and must be followed by the information "starting payment, ending payment" enclosed in parentheses. Press **[APPS] [1:Finance]** .(On the TI-83, press **[2nd]** **<1:Finance>**.) Then cursor down to highlight **0:** $\sum$ **Prn(** and press **[ENTER]** to select, followed by **[1] [,] [1] [5] [)] [ENTER].** Remember, outgoing cash is represented by a negative number. Thus, -5067.01 indicates that $5067.01 has been paid out in principal. Now compute the sum of the interest by pressing **[APPS] [1:Finance]** and cursoring down to highlight option **A:** $\sum$ **Int(** , press **[ENTER]** to select this option and then press **[1] [,] [1] [5] [)]** **[ENTER].** (On the TI-83, press **[2nd]** **<FINANCE>** and proceed as above.) The sum of the interest payments over the first 15 months is $932.99. The display at the right indicates the computation of both principal and interest sums.

```
ΣPrn(1,15)
            -5067.01
ΣInt(1,15)
             -932.99
```

Answer: Justin has an outstanding balance of $17181.71. At this time, he has paid $5067.01 towards the principal and $932.99 in interest.   ◆

Example 3:   Use the TABLE feature of the calculator to set up an amortization schedule.

```
N=61.20
I%=3.75
PV=22248.72
PMT=-400.00
FV=0.00
P/Y=12.00
C/Y=12.00
PMT:END BEGIN
```

Solution:

a. In order to determine the amortization table and the corresponding graph, the information at right must be entered on the **TMV Solver...** screen. Another approach would be to enter each individual value at the home screen using the **STO►** feature and the finance variables located on the **FINANCE/VARS** menu, found by pressing **[APPS] [1:Finance] [►]** to highlight **VARS**. (On the TI-83, press **[2nd]** <**Finance**> and then cursor over to highlight **VARS**..) If the **STO►** feature is used then your screen will correspond to the one pictured. When the **STO►** feature is used, values are automatically copied to the **TMV Solver...** screen.

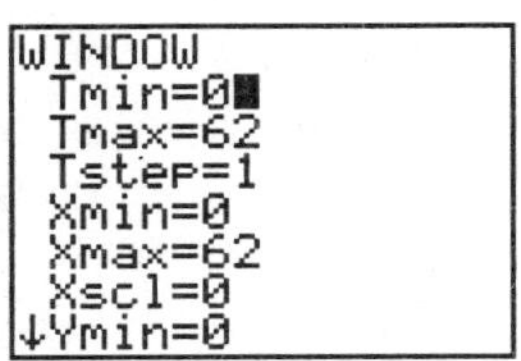

b. The TABLE and graph that represent the relationship between the elapsed time and the outstanding balance will be determined by a parametric equation. Change **MODE** from function to parametric by pressing **[MODE]**, cursoring down to **Func** and over to highlight **Par** and pressing **[ENTER]**. The variable **T** will be the independent variable of time. Press **[Y=]** and enter **T** after $X1_T =$ by pressing the **[X,T,θ,n]** key. Next define $Y1_T =$ to be the balance of outstanding payments at any given time **T**; enter **bal(T)** after $Y1_T =$ by pressing **[APPS] [1:Finance] [9: bal( ] [X,T,θ,n] [ )]**. (On the TI-83, press **[2nd]** <**FINANCE**>.) The authors are aware that some readers may not have been exposed to parametric graphing at this point. However, since **T** is always used to define $X_T =$ and **bal(T)** is always used to define $Y_T =$ in creating these amortization tables, there should not be any difficulties.

Start the TABLE at 15 with increments of one (for each monthly payment) by pressing **[2nd]** <**TBLSET**> and entering the appropriate values. Now press **[2nd]** <**TABLE**> to display the amortization schedule at the right. Observe that a **T** value of 15 indicates that $Y1_T$ = bal(T) = 17182. The balance outstanding after 15 months is \$17182, which is \$17181.71 rounded to the nearest dollar. Scroll through the table to month 61 and determine the final payment: ________

Before displaying the graph that relates this information, the **WINDOW** values must be established. Press **[WINDOW]** and cursor down to enter each of the values displayed at the right. An explanation follows as to **why** each value was selected:

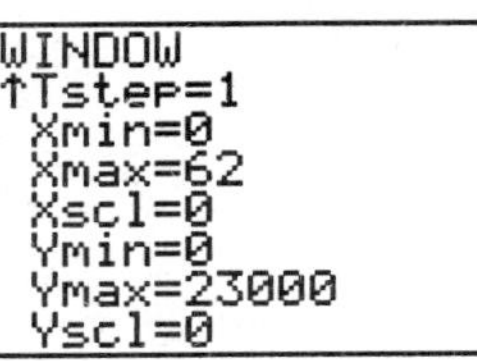

Tmin = 0, Tmax = 62: the time spans over 62 payment months
Tstep = 1: payments are made monthly
Xmin = 0, Xmax = 62: the same as Tmin and Tmax because $X1_T =$ was
                    defined to be equal to T
Xscl = 0: deletes tic marks from the horizontal axis
Ymin = 0: the loan balance will be \$0 at the end of the 62 months
Ymax = 23000: \$22,248.72 was the amount borrowed
Yscl = 0: deletes tic marks from the vertical axis

Press [**TRACE**] and TRACE along the graph to X = 20 T = 20. When 20 months have elapsed, the outstanding balance is $15439.32. Compare this figure to the information displayed in the TABLE when T = 20. The graph provides for better accuracy because figures are rounded to the nearest hundredth whereas in the TABLE only five significant digits are displayed.

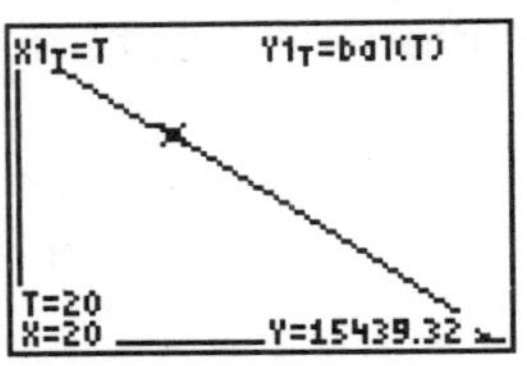

◆

EXERCISE SET

Directions: Use the **FINANCE** menu on the calculator to work the following problems. Copy the screen display for each problem.

After deducting the amount of the downpayment, the MacLean's financed $112,000 to purchase a new home. They obtained a simple interest amortized loan from their bank at 8.5% for 30 years.

1. What is the amount of their monthly payment?______________

2. a. What is their outstanding balance at the end of the first year? _________

   b. What is the sum of the amount paid to the principal at this point? _________

   c. What is the sum of the amount paid to interest at this point? _________

   d. What is the sum of the amount paid to the bank at this point? _________

3. a. What is their outstanding balance at the beginning of the last year? _________

   b. What is the sum of the amount paid to the principal during the last 12 months?
   ______

   c. What is the sum of the amount paid to interest during the last 12 months?
   ____

4.   Compute the total amount paid to interest over the 30 year period. ________

5.   Construct an amortization table for this loan and record the values displayed in the indicated section of the table.

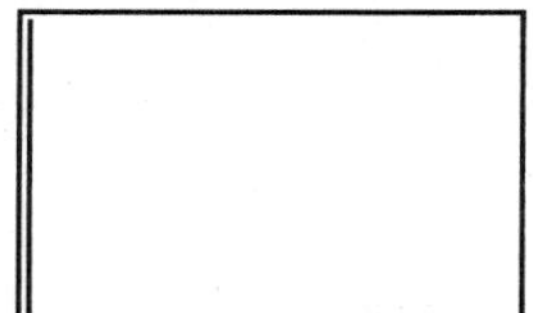

6.   Enter the WINDOW values necessary for displaying the graph of the amortization table and sketch the graph display.

Tmin = _____   Tmax = _____   Tstep = _____
Xmin = _____   Xmax = _____   Xscl = _____
Ymin = _____   Ymax = _____   Yscl = _____

7.   TRACE on the graph to determine how many payments will have been made before the balance drops below $100,000. ________

✍8.   Explain how to establish Tmin/Tmax, Tstep, Xmin/Xmax, and Ymin/Ymax.

<u>**Solutions:**</u>   **Example 3.** $78.40   **1.** $861.18

2a bal: $111153.36, **2b.** Prn: $846.64, **2c.** Int: $9487.52, **2d.** total pmt: $10334.16

3a.bal: $9878.47, **3b.** Prn: 9873.25, **3c.** Int: 460.91   **4.**198030.02

5.

T	X₁ᴛ	Y₁ᴛ
30.00	30.00	109741
31.00	31.00	109657
32.00	32.00	109573
33.00	33.00	109488
34.00	34.00	109402
35.00	35.00	109316
36.00	36.00	109229

T=30

6. 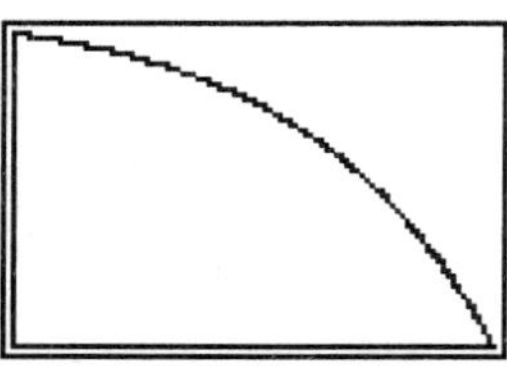

7. 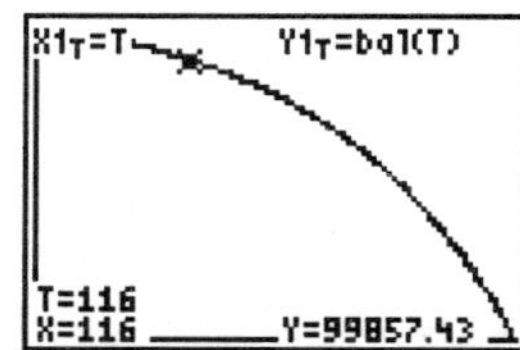

116 payments or 9 yrs. 8 mos.

8. answers may vary

CORRELATION CHART

Note: These units can be done with no formal introduction to graphing. If the course text introduces graphing early, Units 16 and 17 may be worked prior to Unit 7.

UNIT * Instructors are encouraged to use Units marked by an * to <u>introduce</u> concepts.	PRE-REQ. UNIT(S)	CORRELATING CONCEPT
#7 Graphical Solutions: Linear Equations, p.45	#3	Solving linear equations.
#8 Graphical Solutions: Absolute Value Equations, p.53	#7	Solving absolute value equations.
#9 Graphical Solutions: Quadratic and Higher Degree Equations, p.57	#7	This unit addresses solutions of quadratic and higher degree equations with real roots.
#10 Applications of Quadratic Equations, p.67	#9	Interpreting graphs of quadratic equations in real life situations
#11 Graphical Solutions: Radical Equations, p.73	#9	Graphically solving equations containing radicals
#12 Graphical Solutions: Linear Inequalities, p.79	#7	Solving linear inequalities
#13 Graphical Solutions: Absolute Value Inequalities, p.87	#12	This unit can be used to allow the student to "discover" the algorithms typically used for solving.
#14 Graphical Solutions: Quadratic Inequalities, p.93	#9	Solutions of quadratic inequalities are determined through graphs and tables.
#15 Graphical Solutions: Rational Inequalities, p.99	#14	Solutions of quadratic inequalities are determined through graphs and tables.

Unit Title	TI-82 Keys	TI-83/83plus Keys	TI-85/86 Keys
#7: Graphical Solutions: Linear Equations, p.45	Y = WINDOW TRACE GRAPH CALC 5:intersect QUIT ZOOM 6:ZStandard	Y = WINDOW TRACE GRAPH CALC 5:intersect QUIT ZOOM 6:ZStandard	GRAPH y(x) = RANGE TRACE GRAPH MATH/ISECT EXIT MORE GRAPH/ZOOM ZSTD
#8: Graphical Solutions: Absolute Value Eq., p.53	No New Keys	No New Keys	No New Keys
#9: Graphical Solutions: Quadratic and Higher Degree Equations, p.57	ZOOM 1:ZBox 2 :Zoom In CALC 2:root	ZOOM 1:ZBox 2 :Zoom In CALC 2:zero	GRAPH/ZOOM BOX ZIN GRAPH/MATH ROOT
#10: Applications of Quadratic Equations, p.67	CALC 4:maximum 1:value (EVAL X) ZOOM 8:ZInteger	CALC 4:maximum 1:value (EVAL X) ZOOM 8:ZInteger	GRAPH/MATH FMAX GRAPH/MORE/ EVAL ZOOM: ZINT
#11: Graphical Solutions: Radical Equations, p.73	No new keys	No new keys	No new keys
#12: Graphical Solutions: Linear Inequalities, p.79	Y-vars 1:Function.. 1:Y1 TEST menu	VARS Y-VARS 1:Function.. 1:Y1 TEST menu Graph Style Icon	y1: USE ALPHA KEY TO ENTER TEST menu Graph Style Icon
#13: Graphical Solutions: Absolute Val. Ineq., p.87	No new keys	No new keys	No new keys
#14: Graphical Solutions: Quadratic Inequalities, p.93	WINDOW ▶FORMAT AxesOn/AxesOff	FORMAT AxesOn/AxesOff	GRAPH FORMAT AxesOff
#15: Graphical Solutions: Rational Inequalities, p. 99	MODE Connected/Dot	MODE Connected/Dot	GRAPH FORMAT DrawDot

UNIT 7
GRAPHICAL SOLUTIONS: LINEAR EQUATIONS

*Unit 3 is a prerequisite for this unit. However, depending on the organization of your textbook you may choose to work units 16-18 prior to this unit. Answers appear at the end of the unit.

This unit explores the use of the INTERSECT feature for solving equations whereas Unit 9 explores the use of the ROOT/ZERO feature. Recall, an equation is a statement that two algebraic expressions are equal. Solving an equation means finding a replacement value for X that will produce the same value for both expressions. This unit will consider <u>first</u> degree <u>conditional</u> equations in which there will be one correct solution, <u>identities</u> for which there are infinitely many solutions and <u>contradictions</u> for which there are no solutions.

Solve 5X - 1 = -3, algebraically: Check the solution by substitution:

TI-85/86	TI-85/86 USERS GO TO THE GUIDELINES WHICH FOLLOW THIS UNIT (PG.51).

To graphically solve this same equation the graphical representation of the algebraic expression on each side of the equation will be examined. Press [**Y =**] and [**CLEAR**] to delete any expressions. The left side of the equation, 5X - 1, will be designated as Y1. Enter 5X - 1 after "Y1 = " (remember to use the [X,T,θ,n] key). The right side of the equation, -3, will be designated as Y2. Enter -3 after "Y2 = " (be sure to use the gray (-) sign). We want to determine graphically <u>where</u> Y1 = Y2.

Press [**WINDOW**] and cursor down to each line to enter the values displayed on the adjacent screen. The resolution (Xres) at the bottom of the screen should be equal to 1. If not, cursor down to change it.

```
WINDOW
Xmin=-9.4
Xmax=9.4
Xscl=1
Ymin=-6.2
Ymax=6.2
Yscl=1
Xres=1
```

(**TI-82**) When [**WINDOW**] is pressed on the TI-82, the blinking cursor appears on WINDOW. Cursor down to enter the values displayed.

```
WINDOW FORMAT
Xmin=-9.4
Xmax=9.4
Xscl=1
Ymin=-6.2
Ymax=6.2
Yscl=1
```

These values designate the size of the viewing window when graphing Y1 and Y2. These WINDOW values were chosen to give " friendly" numbers when the TRACE feature is activated later in this unit. Press [**GRAPH**]. The graphical solution to the equation is the intersection of Y1 and Y2. At the intersection point, Y1 is equal to Y2. Circle the intersection point.

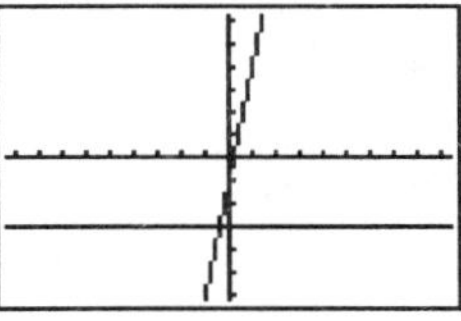

To interpret this graph, we want the **<u>X value</u>** that is the solution to the equation. The X value can be found by pressing **[TRACE]** and tracing along the graph to the intersection point. The left and right arrow keys will move the cursor along the path of the graph. The equation in the upper left corner indicates which graph is being traced, either Y1 or Y2.

At the intersection point, X = -0.4, the equation 5X - 1 = -3 evaluates to a true arithmetic sentence. (Does this agree with the solution computed analytically?)

To ensure the cursor is on the <u>exact</u> point of intersection, record the values indicated at the bottom of the screen: ____________________. Press the up (or down) arrow key once to move to the other graph. The equation (or number for TI-82 users) at the top of the screen will change, even though the cursor will not appear to move. Compare the numbers that are now displayed at the bottom of the screen to those recorded above. If they are <u>both</u> the same, then the cursor is on the exact point of intersection. If not, adjust the TRACE cursor and test again.

Solve 3 - 2X = 2X + 7 graphically. Press **[Y=]** (TI-85/86 users press **[GRAPH]** to redisplay the menu) and clear the expressions after Y1 = and Y2 =. Enter 3 - 2X after Y1 = and 2X + 7 after Y2 =. Press **[TRACE]** and trace along the line to the point of intersection. The solution is X =-1. Be sure that the cursor is on the exact point of intersection.

Solve 3X + 4 = 2 - X graphically. Press **[Y=]** and clear the expressions after Y1 = and Y2 =. Enter 3X + 4 after Y1 = and 2 - X after Y2 =. The TRACE feature is unable to reach the <u>exact</u> point of intersection. The calculator's INTERSECT option will be used to find the point of intersection.

The INTERSECT option will be used exclusively from this point on. It is not dependent on the window values entered. The only requirement is that the point of intersection be visible on the screen. Check your WINDOW values to ensure they are the same as those displayed on the screen at the right. Press **[WINDOW]** and use the down arrow key to move down the screen and change the values, if necessary.

```
WINDOW
 Xmin=-10
 Xmax=10
 Xscl=1
 Ymin=-10
 Ymax=10
 Yscl=1
 Xres=1
```

TI-85/86   TI-85/86 USERS SHOULD RETURN TO THE GUIDELINES WHICH FOLLOW THIS UNIT (PG.51).

To access the INTERSECT option, press **[2nd] <CALC>**. (CALC is located above TRACE.) Press **[5:intersect]** to select INTERSECT. Move the cursor along the first curve to the approximate point of intersection and press **[ENTER]**. At the *second curve* prompt press **[ENTER]** again because the cursor will still be close to the point of intersection. At the *guess* prompt, press **[ENTER]**. This is instructing the calculator that our *guess* is the approximate point of intersection that was designated at the *first curve* prompt. The solution for X is -0.5. Your screen should look like the one at the right.

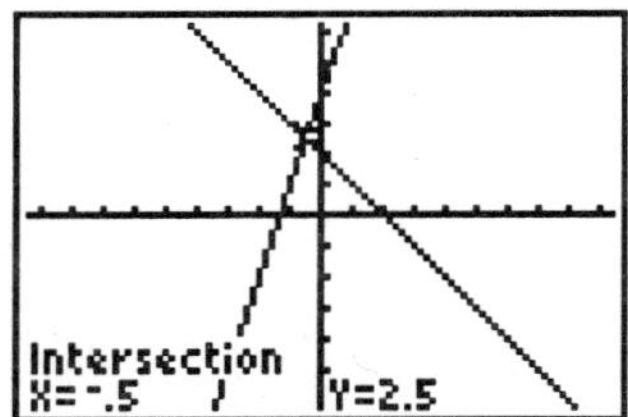

Solve 3X + 4 = 2 - X algebraically.

Is the result equivalent to the result displayed on the calculator screen? Return to the Home Screen, **[2nd] <QUIT>**. The value of **X** computed by the INTERSECT feature is stored in X. To retrieve this value, enter X (press **[X,T,θ,_n_]**) and then convert its value to a fraction, if necessary, by pressing **[MATH] [1:▶Frac] [ENTER]**.

Now check the solution. Refer back to the unit entitled "Evaluating Through Tables and the STOre Feature," if you do not remember how to do this. Compare the results of your check work to the information displayed on the INTERSECT screen.

EXERCISE SET

Directions: Using the INTERSECT option, solve these equations graphically. Follow the same procedure as outlined. Sketch the INTERSECT screen that yields the solution and use ▶Frac to convert all decimal answers to fractions.

1. $5 = 2 - 7X$

     X = _______
     Converted to a fraction, X = _____

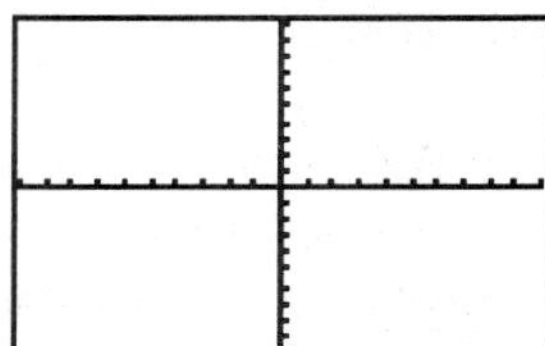

2. $(-4/3)X = -2$

     X = _______
     Converted to a fraction, X = _____

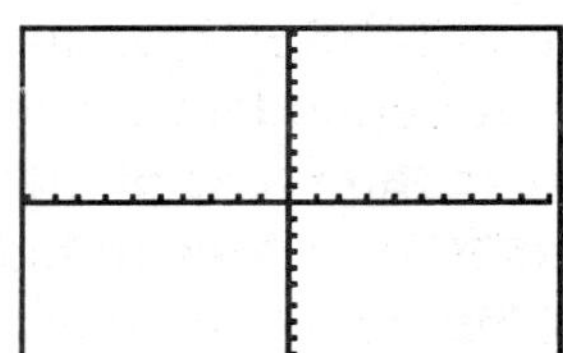

     *(In your textbook, this problem would look like this: $-\dfrac{4}{3}x = -2$)

3. $\dfrac{-4X}{3} + 6 = -1$

     X = _______
     Converted to a fraction, X = _____

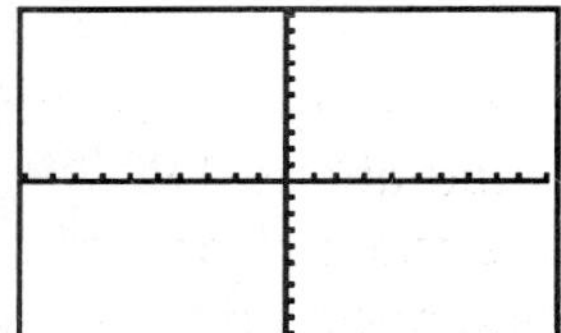

4. $\dfrac{2X - 1.2}{0.6} = \dfrac{4X + 3}{-1.2}$

     X = _______
     Converted to a fraction, X = _____

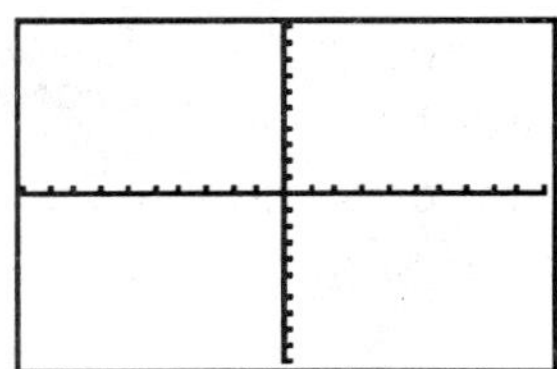

Did you get -3/40 (i.e. -.075)? If you did not, check the way the equation was entered. Parentheses will need to be inserted in the appropriate places.

✐5. When the equation 7x + 8 = -9 - 3x is solved graphically, the bottom of the screen displays the message "INTERSECTION   x = -1.7   y = -3.9". Explain the meaning of the X and Y values in the context of the equation that was being solved.

NOTE: The equations in this unit were specifically written to conform to the WINDOW values displayed on the screen at the right. This set of window values is designated as the standard viewing window. In order to maintain a consistent point of reference, all graphical solutions to equations will begin by displaying the graph in the standard viewing window. This can be easily set by pressing [ZOOM] [6:ZStandard]. (TI-85/86 users press [GRAPH] [F3](ZOOM)   [F4](ZSTD).)

```
WINDOW
 Xmin=-10
 Xmax=10
 Xscl=1
 Ymin=-10
 Ymax=10
 Yscl=1
 Xres=1
```

Recall that the axes seen displayed on the graph screen are simply two number lines placed perpendicularly at their origins. The horizontal number line is the X-axis and the vertical number line is the Y-axis. They are oriented so that right is the positive and up is also the positive direction. The term Xmax on the WINDOW screen refers to the maximum distance from the origin (0) to the right side of the viewing window, whereas Xmin refers to the distance from the origin to the left side of the window. Ymax is the distance from the origin to the top of the viewing window, whereas Ymin is the distance from the origin to the bottom of the screen.
It is best that the point(s) of intersection be visible in the viewing window if using the INTERSECT option. When not visible, one or both of the axes can be stretched.

Example 1: Solve the equation 3(X - 4) = 6 + X graphically, using the INTERSECT feature.

Solution: After entering the left and right sides of the equation at the appropriate prompts and pressing **[GRAPH]** , the displayed graph appears. The viewing WINDOW is not sufficiently large to allow the display of the point of intersection.
Which part of which axis must be stretched? by how much?

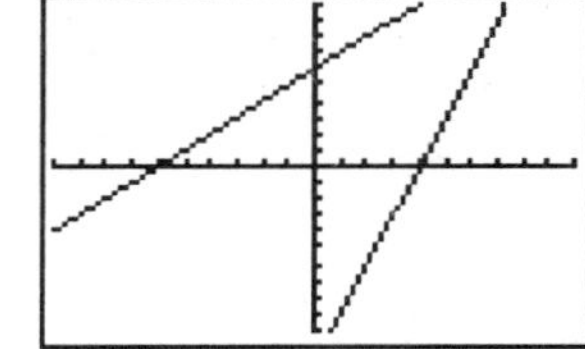

The Ymax must be made larger, and it is suggested that the student simply make an educated guess. After accessing the WINDOW menu, change Ymax to 20. The point of intersection should be clearly visible as illustrated by the second screen. Use of the INTERSECT option displays the correct solution of 9.

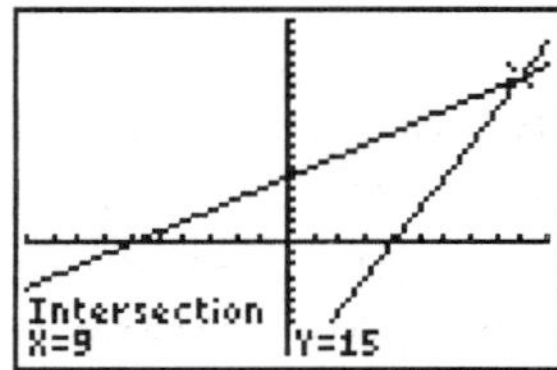

◆

<u>EXERCISE SET CONT'D</u>

Directions: Solve each equation using the INTERSECT feature of the calculator. Sketch the graph displayed, including the axes. Record window values in the spaces provided. It is suggested that the standard viewing window, given above, be used when first graphing, and then modifications can be made as appropriate.

6. $3(X + 2) - 9 = 15$

X = _______

WINDOW values:

XMin: _______   XMax: _______   YMin: _______   YMax: _______

7. $3(X - 4) + 7 = 2(X + 4)$

X = _______

WINDOW values:

XMin: _______   XMax: _______   YMin: _______   YMax: _______

8. $\dfrac{X}{2} + \dfrac{X}{2} = 20$

X = _______

WINDOW values:

XMin: _______   XMax: _______   YMin: _______   YMax: _______

NOTE: When using the graph screen to solve equations/inequalities, you should be aware that the displayed coordinate values approximate the actual mathematical coordinates. The accuracy of the displayed values is determined by the height and width of the pixel space being displayed. The space height/width formulas are discussed in detail in the Unit entitled "Preparing to Graph: Viewing Windows".

Special Cases

Example 2: Solve $4(X - 1) = 4X - 4$ graphically.

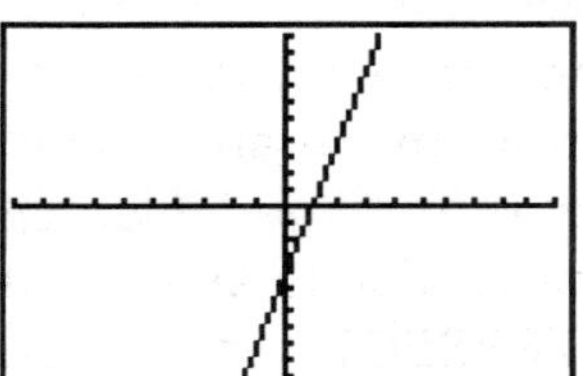

Solution: Press **[Y=]** and enter 4(X - 1) after Y1= and 4X - 4 after Y2=. Press **[TRACE]**. ONLY ONE graph is displayed!! Trace along this line and observe the equation in the upper left corner of the screen. Which graph are you tracing on? Now use the up (or down) arrow key (press only one) and move the TRACE cursor to the other graph. Again check the equation displayed in the upper left corner or the number in the upper right corner, depending on the calculator. Which graph are you on now? Both graphs are the same! When both graphs are the same, then both sides of the equation must be equivalent expressions. Equivalent expressions produce identical values for all replacement values of the variable. Look at the table of values at the right that displays both the left and right side of the equation. Clearly, both sides of the equation are equivalent for each value of x. Simplifying each side of the equation algebraically justifies that the equation is an identity. Identities are true for all values of X that are acceptable replacement values for the variable in the equation. Thus the solution to this equation is the set of all real numbers.

◆

Example 3: Solve 2X - 5 = 2(X + 1) graphically.

Solution: Press **[Y=]** enter 2X - 5 after Y1= and 2(X + 1) after Y2=. Press **[TRACE]**. Observe that the two lines appear to be parallel. Parallel lines never intersect and hence there is no solution. To indicate that this equation has no solution, the symbol for the empty set is written. Upon simplifying both sides of the original equation it is apparent algebraically there is no solution. This equation is called a contradiction.

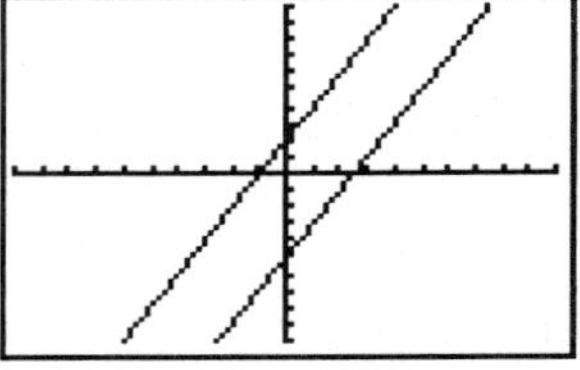

◆

EXERCISE SET CONT'D

9. Graphical representations of the three types of equations are displayed below. In the blank provided, identify each type of equation.

   a. 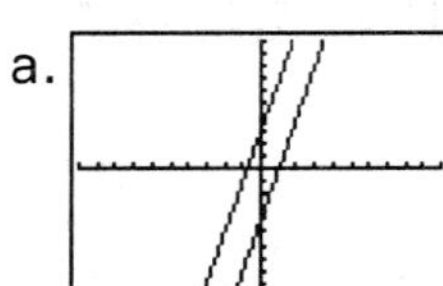   b. 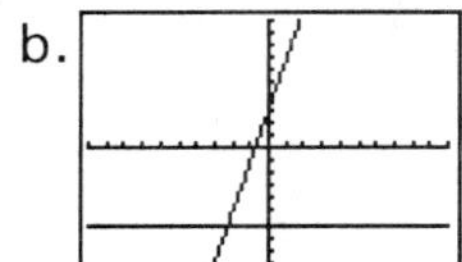   c. 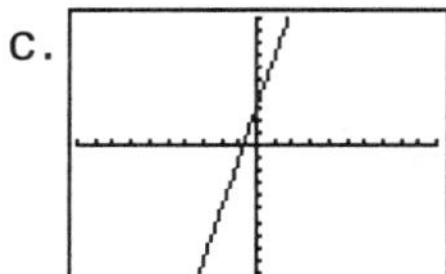

   _____________   _____________   _____________

10. The table below represents one of the three types of equations. Determine the type of equation *and* the solution to the equation.

   type:_______________ solution:____________

X	Y1	Y2
-2	-5	11
-1	-1	11
0	3	11
1	7	11
2	11	11
3	15	11
4	19	11

X= -2

Solutions to Exercise Sets: 1. -.4285714, $-\dfrac{3}{7}$ 2. 1.5, $\dfrac{3}{2}$ 3. 5.25, $\dfrac{21}{4}$ 4. -.075, $-\dfrac{3}{40}$

5. The X-value is the solution to the equation. The Y-value is the quantity each side of the equation will evaluate to when the variable is replaced with the solution value.

6. 6; Xmin = -10, Xmax = 10, Ymin = -10, Ymax = at least 16

7. 13; Xmin = -10, Xmax = at least 14, Ymin = -10, Ymax = at least 35

8. 20; Xmin = -10, Xmax = at least 21, Ymin = -10, Ymax = at least 21

9a. contradiction **9b.** conditional **9c.** identity

10. Type: conditional; Solution: 2

To graphically solve this same equation (5x - 1 = -3), the graphical representation of the algebraic expression on each side of the equation will be examined.

To graph, press **[GRAPH]**. This will bring up a menu at the bottom of the screen. These options correspond to the five buttons at the top of the TI-82/83/83plus, with RANGE corresponding to WINDOW if you are using a TI-85, specified as WIND on the TI-86.

To graphically solve 5x - 1 = -3, press **[F1](y(x)=)**. Enter 5x - 1 (the left side of the equation) at the y1 = prompt that is displayed (enter the variable x by pressing the **[x-VAR]** key). Pressing **[ENTER]** after typing in the expression will automatically display y2 = . Enter -3 at this prompt. Be sure to enter a "negative three" rather than a "minus three".

Press **[2nd] <M2>(RANGE)**, **WIND** on the TI-86, and cursor down to enter the following values for a "friendly" viewing window. TI-86 users be sure that xRes is set equal to 1.

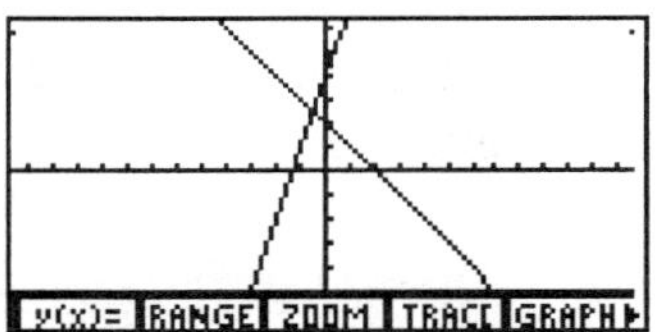

Press **[F5](GRAPH)**.

☞   RETURN TO PG.45 IN THE CORE UNIT AND CONTINUE READING BELOW THE BOX WHICH IS DESIGNATED FOR THE TI-82.

After entering the expression 3x + 4 at the y1 = prompt and 2 - x at the y2 = prompt, press **[2nd] <M5>(GRAPH)**. Your display should match the one pictured. The solution of the equation 3x + 4 = 2 - x is the point on the graph where the two lines intersect. We will use the **ISECT** option (intersect) of the calculator to find the x-value of this point.

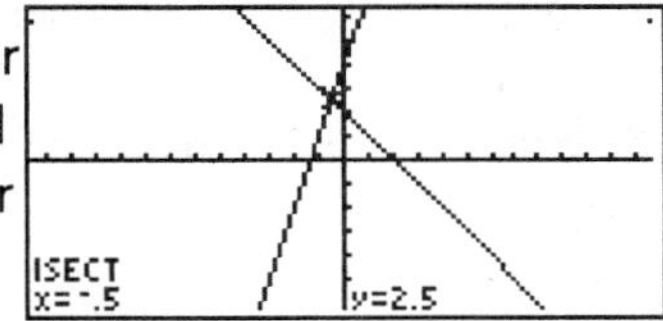

Press **[MORE]** to see more options on the menu. The left option is **MATH**. Press **[F1](MATH)** to select the **MATH** menu, then **[MORE]** and select **(ISECT)** for intersection. (TI-86 users should return page 46 in the core unit and follow the prompts.) With the cursor near the desired point of intersection, press **[ENTER]**. The cursor actually moves from one graph to the other with this first press of **ENTER**. If necessary, again move the cursor close to the point of intersection and press **[ENTER]** for the second time. This press of **ENTER** activates the **ISECT** computation. Your screen should correspond to the one at the right.

     Note: If there is more than one point of intersection, the **ISECT** function must be completed for each point of intersection.

(Pressing **[EXIT]** or **[CLEAR]** removes the menus displayed at the bottom of the graph.)

☞   RETURN TO PG.47 AND ALGEBRAICALLY SOLVE THE EQUATION AT THE TOP OF THE PAGE.

UNIT 8
GRAPHICAL SOLUTIONS: ABSOLUTE VALUE EQUATIONS

*Unit 7 is a prerequisite for this unit. Answers appear at the end of this unit.

In the prerequisite unit, linear equations were solved graphically using the INTERSECT feature of the calculator. This same approach will be used to solve equations containing absolute value. Remember, absolute value is located by pressing **[MATH] [▶]** for the **NUM** menu or by pressing **[2nd] <CATALOG>**. (TI85/86 users should have this feature customized).

Remember, absolute value is located above the x^{-1} key and is accessed by **[2ⁿᵈ] <x⁻¹>**.

Consider the equation $|X + 3| = 6$. To solve graphically, enter the left side of the equation as **abs(X + 3)** at the Y1 = prompt and the right side of the equation at Y2 = . (Remember, $|X + 3|$ is read as "the absolute value of the quantity X plus three".) Press **[ZOOM] [6:ZStandard]** (TI-85/86 users press **[GRAPH] [F3](ZOOM) [F4](ZSTD)**) to automatically enter the values on the WINDOW screen

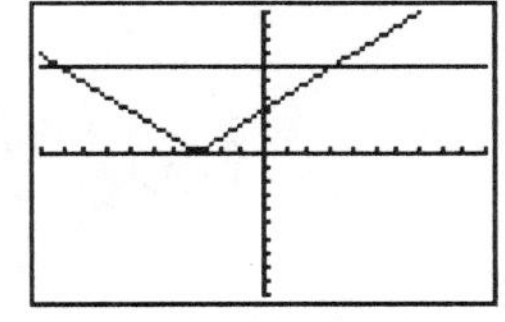

at the right . The keystrokes display the graph; confirm these WINDOW values by pressing **[WINDOW]**.

Press **[GRAPH]** to display the graphical representation of the absolute value equation. See the display at the right.

    a.   Circle the two points of intersection that are displayed.

    b.   Use the calculator's INTERSECT option to find the X values at the points of intersection. You will need to use the INTERSECT option <u>twice</u> since there are two points of intersection. Copy the screen display to show where you found <u>one</u> of the two solutions, but record both of the solutions here:
    X = _____ or X = _____

NOTE CONCERNING THE USE OF THE INTERSECT OPTION: In the future there may be other equations that have more than one solution. There will be an intersection point for each of these solutions that is a real number. The INTERSECT option will have to be completed for each point of intersection. To justify your work you will only be required to sketch one of the INTERSECT screens that you used and merely record the solutions derived from the other screens.

Confirm the solutions of 3 and -9 in the following ways:
    a. analytically through substitution   b. via the TABLE

Directions: Graphically solve each of the equations below. Sketch the screen. Circle the points of intersection. Use the INTERSECT option to find the intersections. REMEMBER: Because there are two points of intersection, the process will have to be done twice. Record both of the solutions on the blanks provided.

1. $|2X - 1| = 5$

    X = _____ or X = _____

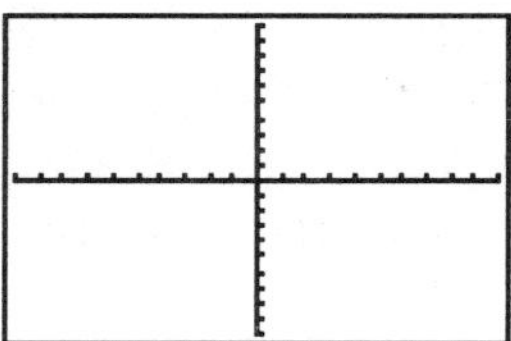

2. $\left|\dfrac{1}{2}X + 1\right| = 3$

    X = _____ or X = _____

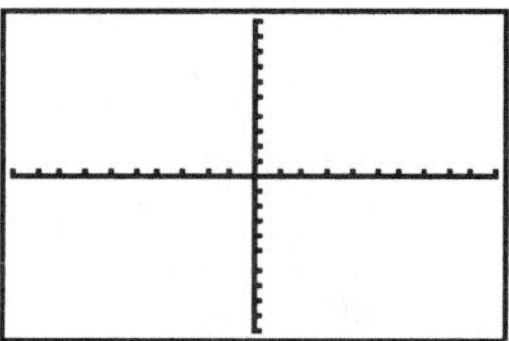

3. $\left|\dfrac{4 - X}{2}\right| = \dfrac{8}{5}$

    X = _____ or X = _____

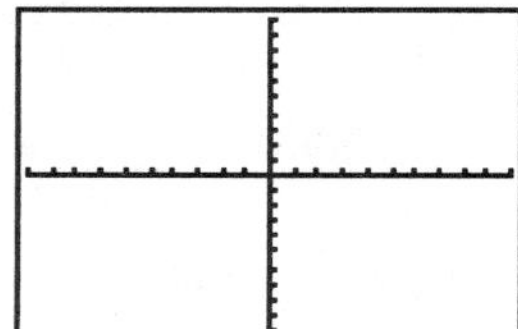

4. $|X + 1| = |2X - 3|$

    X = _____ or X = _____

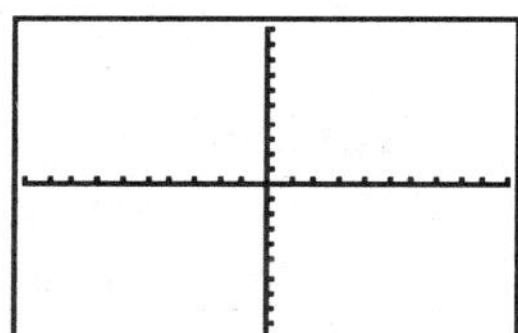

5. $|2 - X| = |3X + 4|$

    X = _____ or X = _____

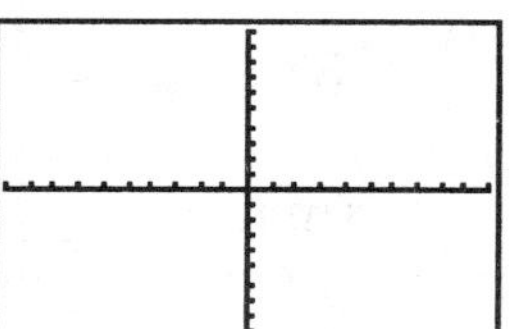

6. $|4X + 5| = -2.$

    Do the graphs intersect?_____

    What is the solution? ______

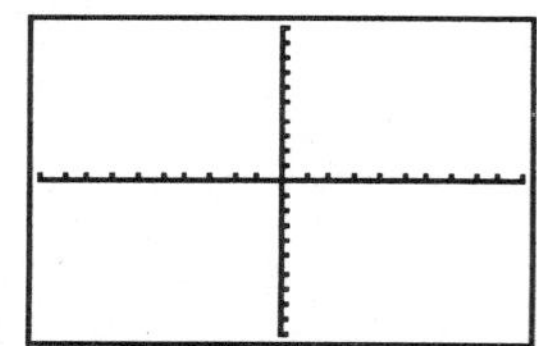

✍ 7. You should be able to determine the solution to $|4X + 5| = -2$ by merely <u>looking</u> at the problem. What clue lets you know that there is no solution?

Directions: Graphically solve each of the equations below. Sketch the screen (axes will need to be adjusted and WINDOW values recorded in the spaces provided). Circle the points of intersection. Use the INTERSECT option to find the intersections. REMEMBER: Because there are two points of intersection, the process will have to be done twice. Record both of the solutions on the blanks provided.

8. $|2(X - 5) - 9| = 5$

Xmin = _____ , Xmax = ______ , Ymin = _____ , Ymax = ____

X = _____ or X = _____

9. $|2X + 7| = 11$

Xmin = _____ , Xmax = ______ , Ymin = _____ , Ymax = ____

X = _____ or X = _____

10. $|4X - 3| = |2x + 5|$

Xmin = _____ , Xmax = ______ , Ymin = _____ , Ymax = ____

X = _____ or X = _____

11. $|3X - 1| = |7 + 4X|$

Xmin = _____ , Xmax = ______ , Ymin = _____ , Ymax = ____

X = _____ or X = _____

12. $|X - 3| = -|X + 20|$

Xmin = _______ , Xmax = _______, Ymin = _______, Ymax = _______

Do the graphs intersect?_______

What is the solution?_______

Hint: Try zooming out by pressing **[ZOOM] [3:Zoom Out] [ENTER]**. TI-85/86 users press **[GRAPH] [F3] (ZOOM) [F3] (ZOUT) [ENTER]**.

Solutions: **1b.** X = 3 or X = -9

Exercise Set: 1. X = -2 or X = 3, **2.** X = -8 or X = 4, **3.** X = 0.8 or X = 7.2, **4.** X = 2/3 or X = 4

5. X = -3 or X = -0.5 **6.** No, null set **7.** The solution is the empty set because the two graphs do not

intersect.   **8.** X = 7 or X = 12, Xmax should be larger than 12

9. X = -9 or X = 2, Ymax at least 12 **10.** X = -1/3 or X = 4, Ymax at least 14

11. X = -8 or X = -6/7, Ymax at least 26   **12..** Xmin = -30, no, ϕ,

UNIT 9
GRAPHICAL SOLUTIONS: QUADRATIC
AND HIGHER DEGREE EQUATIONS

*Unit 7 is a prerequisite for this unit. Answers appear at the end of this unit.

FACTORABLE EQUATIONS

Polynomial equations can be solved by the INTERSECT method that was used in previous units. Both sides of the equation can be graphed and the solution(s) determined from the point(s) of intersection. There is, however, another graphic approach to solving equations that will be considered in this unit. This method is the ROOT or ZERO method and can be used to find the REAL roots/zeroes of all the equations you have learned to solve graphically thus far and for any equation encountered in the future. Calculator techniques introduced early in the unit focus on quadratic equations. Later exercises expand the techniques to higher degree polynomial equations. Thus, the unit can easily be divided into two parts if your text addresses quadratic equations in a section separate from higher order equations.

INTERSECT METHOD

First, a review of the INTERSECT method: This method is best applied to problems in which the equation is not set equal to zero. Remember, it is critical when using this method that all point(s) of intersection are visible. These points of intersection are the solutions/roots/zeroes of the equation.

Example 1: Graphically solve $2(X^2 - 3) - 2X = 3(X + 2)$.

Solution: To solve $2(X^2 - 3) - 2X = 3(X + 2)$ you will need to press **[Y=]** and enter $2(X^2 -3) - 2X$ after Y1= and $3(X + 2)$ after Y2=. Display the graphs in the standard viewing window. One point of intersection is obvious, but the other point is out of the viewing window. (See the screen at the right.) Since the graph appears to have another intersection point "above" the viewing window, press **[WINDOW]** and adjust the size of the viewing window.   Change the Ymax to 20 (an educated guess) instead of 10. Press **[GRAPH]** and your screen should look like the one displayed at the right.

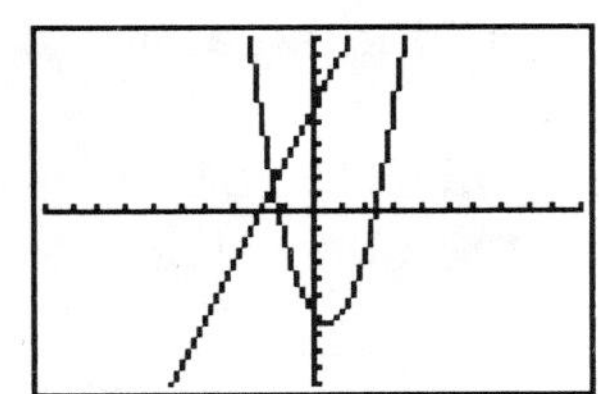

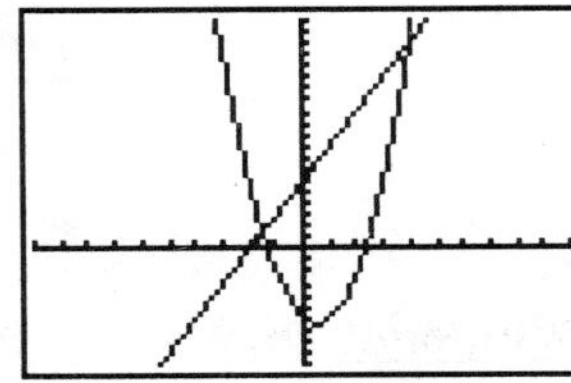

Because the INTERSECT option works independently of the viewing window, the WINDOW may be adjusted as necessary for the given equation. As long as all points of intersection are visible on the display, the calculator will compute the value for each point of intersection. Two points of intersection are displayed. The INTERSECT option discussed in the next paragraph will need to be performed for each point of intersection.

TI-85/86 IF MORE THAN ONE LINE OF MENUS ARE DISPLAYED, PRESS **[EXIT]** TO ENSURE ONLY ONE LINE OF MENU OPTIONS IS DISPLAYED. PRESS **[MORE] [MATH] [MORE]** AND THE APPROPRIATE F KEY TO ACCESS THE **ISECT** OPTION.

To access the INTERSECT option, press **[2nd] <CALC>**. (**CALC** is located above **TRACE**.)
Press **[5:intersect]** to select INTERSECT. Move the cursor along the first curve to the
approximate location of one point of intersection and press **[ENTER]**.
At the "second curve" prompt, press **[ENTER]** again because the
cursor will still be close to the point of intersection. At the "guess"
prompt, press **[ENTER]**. This is instructing the calculator that our
"guess" is the approximate point of intersection that was
designated at the "first curve" prompt. Your screen should
correspond to the one adjacent when you compute the far right
point of intersection. Repeat the process for the remaining point of
intersection to determine that the two solutions are 4 and -1.5.

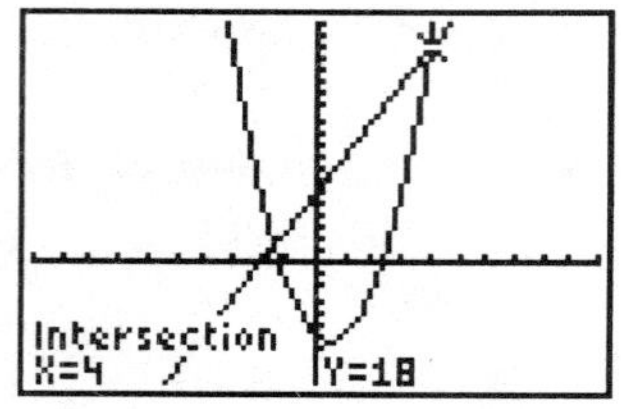

Example 2: Use the INTERSECT option to solve
$2(X + 2)(X - 2) + 4 = (X + 4)(X - 1) - 2X$ graphically.

Solution: After the expressions have been entered at the Y1 =
and Y2 = prompts, view them in the standard viewing window.
Your screen should look like the screen at the right.

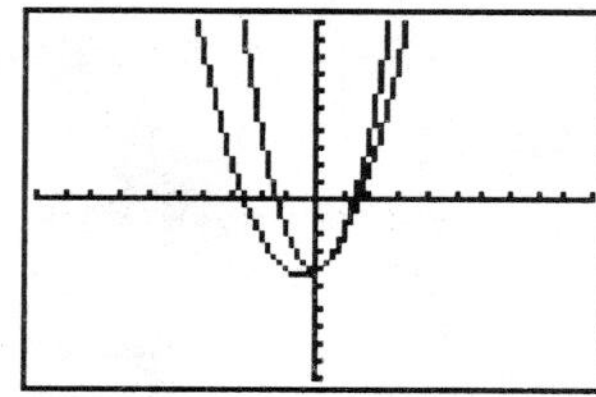

To help us determine the points of intersection, we will use the
ZOOM IN option to get a closer look at the section where the two
graphs appear to intersect. Press **[ZOOM] [2:Zoom In]**.

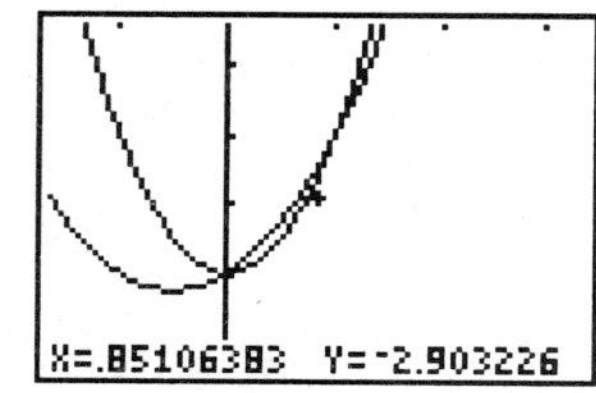

TI-85/86	TI-85/86 USERS PRESS **[F3](ZOOM) [F2](ZIN)**.

Use the down arrow key to cursor down to the third tic mark on the
vertical axis (where Y = -3) and then the right arrow key to locate
the cursor over the graph. (NOTE: The cursor is moved to a point in the neighborhood of
the intersection of the two graphs so the calculator will ZOOM IN on that specific region.)
Now press **[ENTER]**. Your graph display should look very much like the one above. <u>Two</u>
points of intersection are now visible.

Use the INTERSECT option to compute both points of intersection and thus determine the
two solution values for X (i.e. roots). The roots are 0 and 1.

ROOT/ZERO METHOD

(TI-82)	This option is called "root" and is the second option under the **CALC** menu (i.e. the same relative position as the "zero" option on the TI-83/83plus).

There is an alternate method for solving equations instead of the INTERSECT option. This
method uses the ROOT (TI-82/85/86) or ZERO (TI-83/83plus) option. If an equation is set
equal to zero, when the non-zero side of the equation is graphed, the real roots or zeroes
are the X-values at the point(s) where the graph crosses the horizontal axis (X-axis). This
method is <u>perfect</u> for equations of any type that are already set equal to zero.

If an equation is not set equal to zero, then you should try the INTERSECT method first.
WHY? Because often when you begin to manipulate the terms of an equation with paper
and pencil you make careless errors. By merely entering the existing equation into the
calculator you run less risk of error. If the points of intersection are not clearly visible then
you can set the equation equal to zero and use the ROOT option.

Example 3: Solve the previous equation by the ROOT/ZERO method.

Solution: Begin by setting $2(x + 2)(x - 2) + 4 = (x + 4)(x - 1) - 2x$ equal to zero. We will perform the **least** amount of algebra possible to accomplish this:

$$2(x+2)(x-2)+4 \;=\; (x+4)(x-1)-2x$$

$$2(x+2)(x-2)+4 \;\underline{+\; \textbf{2x}} \;=\; (x+4)(x-1)-2x \;\underline{+\; \textbf{2x}}$$
a. add 2x to both sides

$$2(x+2)(x-2)+4 \;+\; 2x \;=\; (x+4)(x-1)$$

$$2(x+2)(x-2)+4 \;+\; 2x \;\underline{-\,(x+4)(x\text{-}1)} \;=\; (x+4)(x-1) \;\underline{-\,(x+4)(x\text{-}1)}$$
b. subtract $(x+4)(x-1)$ from both sides

$$2(x+2)(x-2)+4 \;\underline{+\; \textit{\textbf{2x}} \,-\, \textit{\textbf{(x+4)(x-1)}}} = 0$$
c. compare the underlined part to the original equation

It is this algebraic process of setting the expression equal to zero that encourages us to use the INTERSECT option if at all possible when the equation is not already set equal to 0!

OPTIONAL NOTE: If $2(X+2)(X-2)$ is entered at Y1= and $(X+4)(X-1) - 2x$ is entered at Y2=, then Y1-Y2 represents the non-zero side of the above equation. Entering Y1-Y2 after Y3= accomplishes the same result as the above algebraic manipulation. The advantage is that all possibility of error by hand computation is eliminated. If this approach is used, be sure to "turn off" the graphs of Y1 and Y2 by placing the cursor over the equal sign and pressing [ENTER]. (TI-85/86 users press [F5](SELCT) from the y(x)= menu.) This ensures that only the graph of Y3 is displayed.

Enter the NON-zero side of the equation after Y1= and display in the standard viewing window. The graph is displayed at the right.

The curve appears to "dip" slightly below the horizontal axis. To get a better view of this section of the curve, use the calculator's **ZBox** option to box in this section. This will alter the WINDOW values.

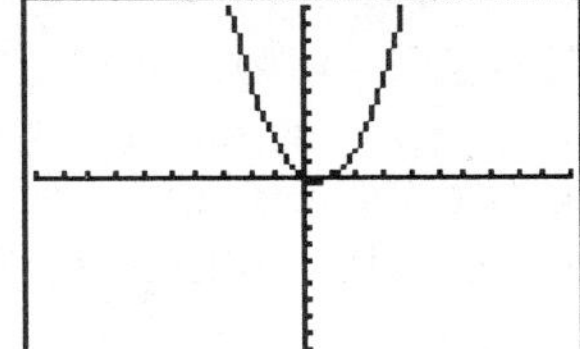

ZBOX IS DESIGNATED AS "BOX" ON THE ZOOM MENU. PRESS [ZOOM] [F1](BOX) TO ACCESS THE BOX OPTION AND THEN FOLLOW THE DIRECTIONS IN THE FOLLOWING PARAGRAPH.

Press **[ZOOM] [1:ZBox]**. To box in the area around the root(s), use the arrow keys to move the blinking cursor to the upper left hand corner of the area to be boxed in. Press **[ENTER]**. Use the right arrow key to establish the width of the box followed by the down arrow to establish the height of the box. Your screen should be similar to the one displayed at the right.

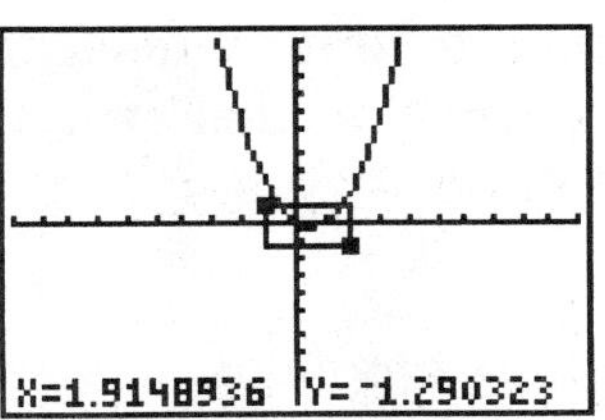

The calculator will now enlarge the boxed in area. Press [ENTER] to activate. Your screen should be similar to the one at the right.

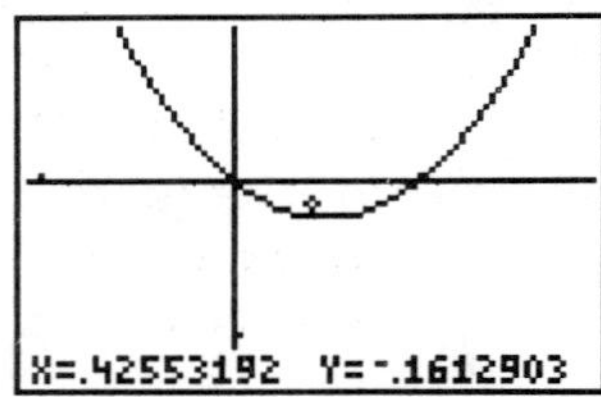

NOTE: Rather than the **ZBox** option, the WINDOW values could have been adjusted using the methods described in Example 1, or the ZOOM IN option could have been used. You may want to explore both the ZBOX and the ZOOM IN options of the ZOOM menu.

The solutions of this equation will be found by accessing the ROOT/ZERO option on the calculator.

> BEFORE ACCESSING THE ROOT OPTION, BE SURE THAT ONLY ONE LINE OF MENU OPTIONS IS DISPLAYED. IF MORE THAN ONE LINE IS DISPLAYED, PRESS [EXIT]. NOW, PRESS [MORE] [F1](MATH) AND PRESS THE APPROPRIATE F KEY TO SELECT (ROOT). THE CURSOR SHOULD BE PLACED NEAR THE DESIRED ROOT. PRESS [ENTER] TO SEE THE DESIRED ROOT WHICH IS THE X-VALUE DISPLAYED AT THE BOTTOM OF THE SCREEN. READ THE TROUBLE SHOOTING NOTE AT THE TOP OF THE NEXT PAGE AND THEN BEGIN THE EXERCISE SET.

To access the ROOT/ZERO option, press [2nd] <CALC> [2:zero] (TI-86 users access the root feature as discussed for the TI-85 above and then follow the boxed steps below).

> TI-82 users are reminded that this option is listed as [2:root] and should note that the prompts for lower and upper bounds will appear as "left bound" and "right bound."

STEPS FOR SOLVING AN EQUATION USING THE ROOT/ZERO FEATURE

a. **Set lower bound:** The screen display asks for a lower bound. A lower bound is an X value smaller than the expected root; move the cursor to the left of the left-hand root and press [ENTER]. Because the roots are determined on the horizontal axis, a lower bound is always determined by moving the cursor to the <u>left of the root</u>. Notice at the top of the screen a ▸ marker has been placed to designate the location of the lower bound. TI-83/83plus/86 users have the option of not moving the cursor to bound the root, but rather to enter an appropriate value for X at the "lower bound" prompt. Care should be taken that the value entered for a lower bound is *smaller* than the expected solution. Press [ENTER] after entering the value.

b. **Set upper bound:** Similarly the upper bound is always determined by moving the cursor to the <u>right of the root</u>. At the upper bound prompt, move the cursor to an X value larger than the expected root and press [ENTER]. Again, a ◂ marker is at the top of the screen to designate the location of the bound. Again, TI-83/83plus/86 users have the option of not moving the cursor to bound the root, but rather to enter an appropriate value for X at the "upper bound" prompt. Care should be taken that the value entered for an upper bound is *larger* than the expected solution. Press [ENTER] after entering the value.

NOTE: If the bound markers do not point toward each other, ▸ ◂, then you will get an "ERROR:bounds" message. If this happens, start the ROOT/ZERO calculation over. Care must be taken when setting bounds that you bound *only* the specific root for which you are searching.

c. **Locate first root:** Move the cursor to the approximate location where the graph crosses the X-axis for your guess. When you press [ENTER] the calculator will search for the root, within the area marked by ▸ and ◂. The root is X = 0. Make sure you read the "Trouble Shooting Note" at the top of the next page. It addresses round-off error.

d. **Locate subsequent roots:** Repeat the entire process outlined above to determine the right-hand root. This root is X = 1.

TROUBLE SHOOTING NOTE: There will be times when the calculator will be very close to ZERO but will not display ZERO exactly. For example X = -7.65 E -15 is the calculator's version of the scientific notation -7.65×10^{-15} which is equivalent to X = -.00000000000000765. For all practical purposes this value is ZERO. To verify that the X value is actually zero, scroll through the TABLE to X = 0 (or use EVAL) and note that the Y value is -4, the same as was displayed on the INTERSECT screen. Therefore, when using the graph screen to solve equations/inequalities with the ROOT/ZERO feature, you should be aware that the display coordinate values sometimes approximate the actual mathematical coordinates.

EXERCISE SET

Directions: Solve each of the following quadratics, using the ROOT/ZERO option. So that everyone's graph will look alike, display the graph in the standard viewing window. Sketch your graph display, circle the two roots and record their values in the blanks provided. Beneath each problem, factor the quadratic that you graphed.

1. $X^2 + 8X - 9 = 0$

     The roots are X = _______ and X = _______.

     Factorization:_________________________________

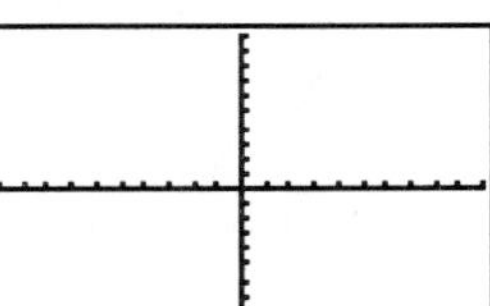

2. $(X-2)^2 + 3X - 10 = 0$

     The roots are X = _______ and X = _______.

     Factorization:_________________________________

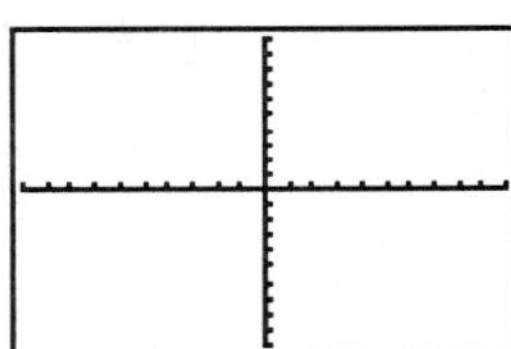

3. $X^2 + 6X + 9 = 0$

     The root(s) is/are X = _______ .

     Factorization:_________________________________

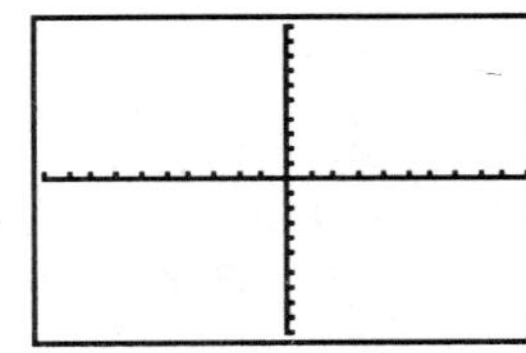

     (Note: the point at which the graph touches, but does not cross the X-axis, produces two identical roots - often called a double root.)

4.   Compare the factorization of the polynomial to the real roots that were determined in each of the problems above. How do they compare?

5.   If you know that the roots to an equation are 4 and -2, you should be able to write an <u>equation</u>, in factored form that has these roots. Write an <u>equation</u>, in factored form:

     ____

6.   The equations that have been solved thus far in this unit have all been second degree equations. Based on the exercises, how many roots should you expect to have with a second degree quadratic equation?

✏ 7. **ONE** of the equations does not conform to the pattern. Which one is it and why is the number of roots different from the rest of the problems?

Directions: The next set of equations contain polynomials that are not second degree. However, these polynomials can be factored.
a. First use the ROOT/ZERO method to solve the equation.
b. Copy your screen display.
c. Circle the real roots.
d. Record the value of the roots in fractional form.
e. Factor the polynomial.

8. $X^3 - 7X^2 + 10X = 0$

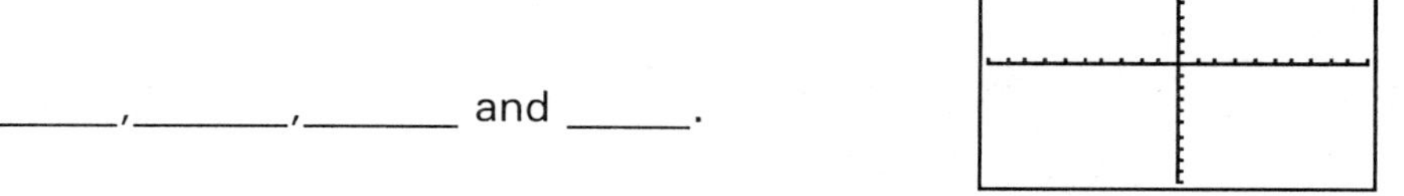

The roots are X = _______, _______ and _______.

Factorization:_______________________

9. $X^4 - 5X^2 + 4 = 0$

The roots are X = _______,_______,_______ and _______.

Factorization:_______________________

10. $3X^4 + 2X^3 - 5X^2 = 0$
Suggestion: Change the WINDOW values (Xmin = -5, Xmax = 5) OR use the **ZBox** option.

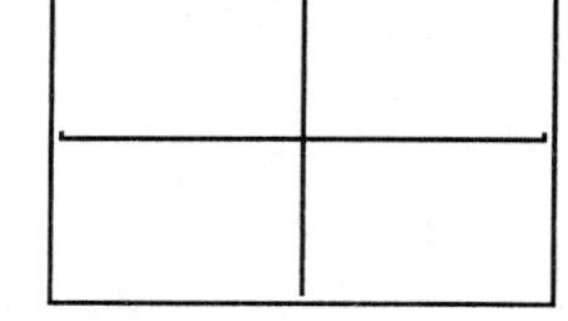

The roots are X = _______,_______ and _______.

Note: There is a double root in this problem. Which root is the double root?_______

Factorization:_______________________

✏11. What conclusions can be drawn about the number of real solutions and the degree of the polynomial equations that have been solved thus far?

Previously, factorable polynomial equations were solved. Several observations were made: 1) for each factor there was a root 2) the number of factors corresponded to the degree of the polynomial 3) the number of real roots was equal to or less than the degree of the polynomial. CONCLUSION: A polynomial equation of degree n will have <u>at most n real</u> <u>roots</u>. We will now examine polynomial equations that do not factor over the rational numbers and hence may not have any real roots, or at best roots that are irrational.

Example 4: Solve the equation $3x^2 - 18x + 25 = 0$.

Solution: This trinomial does not factor, so it would be solved using either the quadratic formula or the method of "completing the square".

<table>
<tr><td>

a. <u>QUADRATIC FORMULA</u>:

$$3x^2 - 18x + 25 = 0$$

$$a = 3,\ b = -18,\ c = 25$$

$$\frac{-b \pm \sqrt{b^2 - 4ac}}{2a} =$$

$$\frac{-(-18) \pm \sqrt{(-18)^2 - 4(3)(25)}}{2(3)} =$$

$$\frac{18 \pm \sqrt{324 - 300}}{6} =$$

$$\frac{18 \pm \sqrt{24}}{6} =$$

$$\frac{18 \pm 2\sqrt{6}}{6} = \frac{9 \pm \sqrt{6}}{3} = 3 \pm \frac{\sqrt{6}}{3}$$

</td><td>

b. <u>COMPLETING THE SQUARE</u>:

$$3x^2 - 18x + 25 = 0$$

$$\frac{1}{3}(3x^2 - 18x + 25) = \frac{1}{3}(0)$$

$$x^2 - 6x + \frac{25}{3} = 0$$

$$x^2 - 6x = -\frac{25}{3}$$

$$x^2 - 6x + 9 = -\frac{25}{3} + 9$$

$$(x - 3)^2 = \frac{2}{3}$$

$$x - 3 = \pm \sqrt{\frac{2}{3}}$$

$$x = 3 \pm \sqrt{\frac{2}{3}}$$

$$x = 3 \pm \frac{\sqrt{6}}{3}$$

</td></tr>
</table>

Enter each of the two roots found above into the calculator to determine a decimal approximation. Record the two roots **exactly** as they are displayed on the screen. DO NOT round.

X = ___________________ and X = _________________

Enter $3X^2 - 18X + 25$ after Y1 = and graph in the standard viewing window. Use the ROOT/ZERO option twice to compute both roots of the equation. The screens displayed indicate both roots. These roots should be comparable to the approximate values that were determined above.

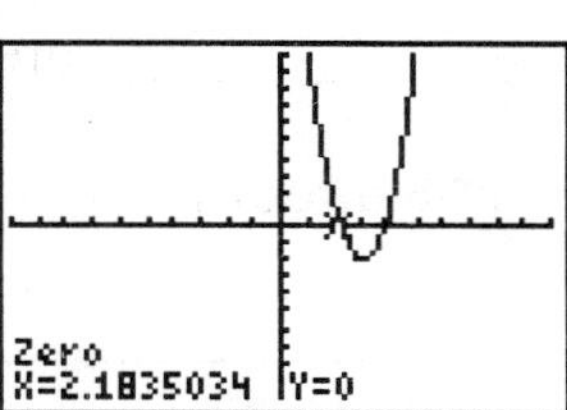

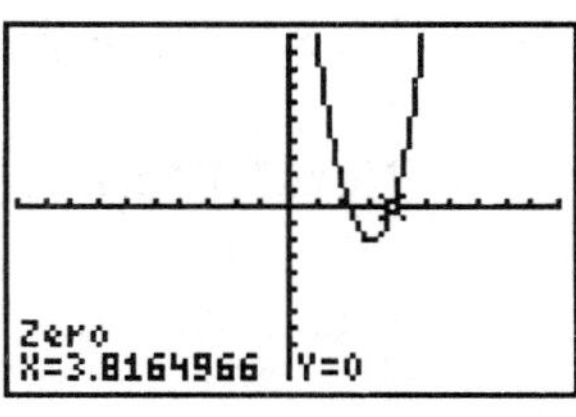

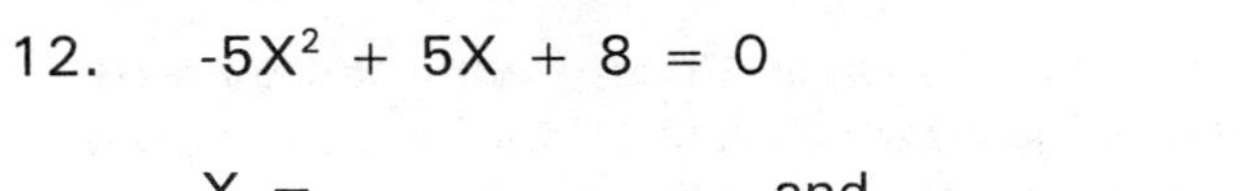

EXERCISE SET CONT'D

Directions: Use the ROOT/ZERO option to find the **REAL** roots of the following quadratic equations. Sketch the screen display in the indicated viewing window and record the solutions.

12. $-5X^2 + 5X + 8 = 0$

X = _____________ and _____________

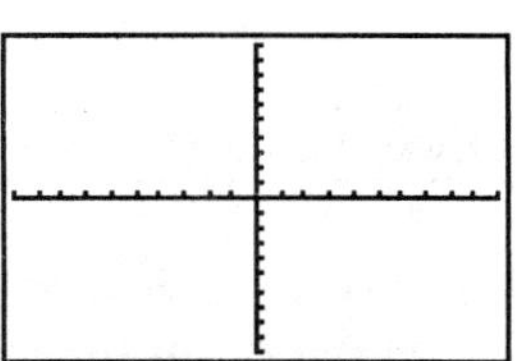

ZStandard

✍13. What happens when you try to convert the roots in #12 to fractions? and why?

14. Solve the quadratic equation in #12 by either completing the square or using the
Quadratic Formula. From the home screen, approximate the solutions and compare
them to the calculator answers recorded. They should be the same.

✍15. $X^2 + 5X + 8 = 0$

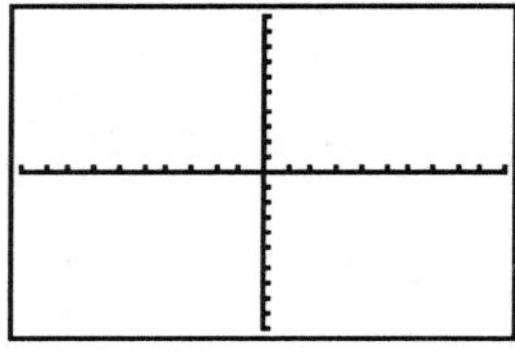

Is it possible to graphically find the roots of this equation using
the calculator?________

Why not?

16. Solve #15 analytically using either the Quadratic Formula or by completing the
square.

Graphs which intersect the X-axis will have real roots because the X-axis represents the set
of real numbers. Roots which are complex numbers will not be represented on the X-axis.
Thus an equation with complex roots will not intersect the X-axis. There will, however, be
two roots because complex roots always occur in conjugate pairs.

17. State the number and type of roots (real or complex) of each of the following equations. DO NOT SOLVE the equations, simply graph the polynomial function in the ZStandard viewing window and check the number of X-intercepts, if any.

a. $6X^2 + 2X - 4 = 0$

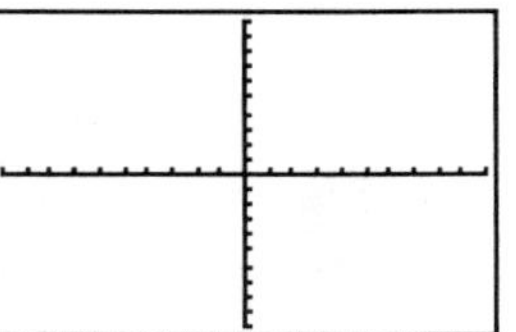

Number of roots:_______

Type of roots:_________

b. $2X^3 - 5X + 5 = 0$

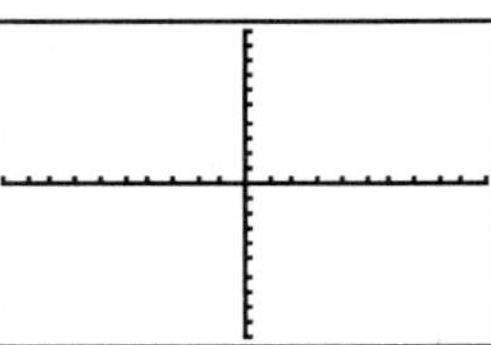

Number of roots:_______

Type of roots:_________

c. $X^4 - 0.5X^3 - 5X^2 + 10 = 0$

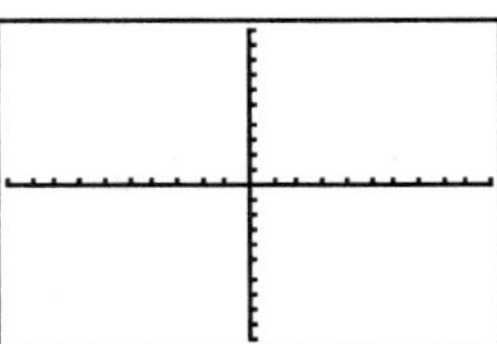

Number of roots:_______

Type of roots:_________

18. $0 = X^6 + 2X^3 - 1$

$X =$ _______________ and _______________

Describe the nature of the other four roots:

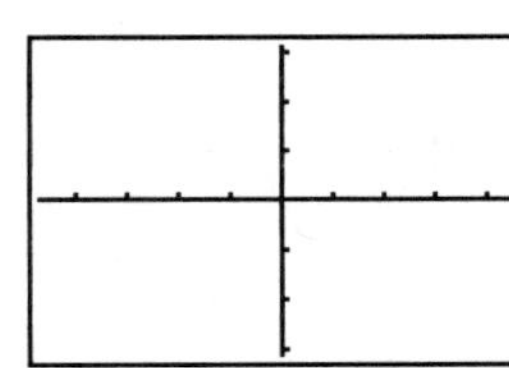

ZDecimal

19. $0 = X^4 - 2X^3 + X - 2$

$X =$ _______________ and _______________

The expression $X^4 - 2X^3 + X - 2$ in completely factored form is $(X - 2)(X + 1)(X^2 - X + 1)$. Use the appropriate algebraic technique to determine the complete solution set.

____

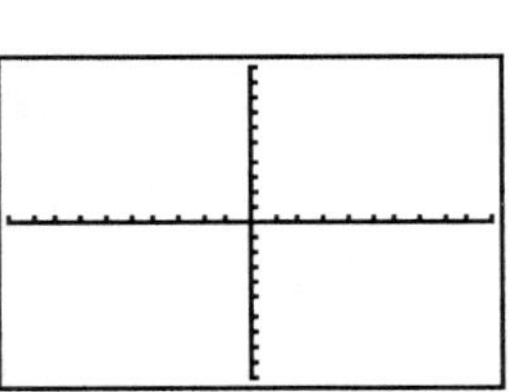

ZStandard

Solutions: 1. $(X+9)(X-1)=0$, $X=-9$ and $X=1$ **2.** $(X-3)(X+2)=0$, $X=-2$ and $X=3$,

3. $(X+3)(X+3)=0$, $X=-3$ **4.** The factor is always the "variable minus the root".

5. $(X-4)(X+2)=0$ **6.** two **7.** #3. We did not see two distinct real roots, but rather two identical roots (often called a double root).

8. $X(X-5)(X-2)=0$, $X=0,2$ and 5   **9.** $(X-2)(X+2)(X-1)(X+1)=0$, $X=2,-2,1$ and -1

10. $X^2(3X+5)(X-1)=0$, $X=0,-5/3$ and 1, Zero is the double root. **11.** The number of

solutions is the same as the degree of the equation. However, not all the solutions are

distinct (different). Some appear as multiple roots. **12.** $X=-.860147$ and $X=1.8601471$

13. These answers will not convert to fractions because they are the decimal approximations of the

irrational numbers $\dfrac{1}{2}\pm\dfrac{\sqrt{185}}{10}$. **15.** No, It does not intersect the X-axis and therefore has no real roots.

16. $X=\dfrac{-5\pm i\sqrt{7}}{2}$ **17. a.** 2, real **b.** 3, 1 real, 2 complex **c.** 0, 4 complex

18. $X = -1.341504$, $X = .74543212$; the remaining four roots consist of 2 pairs of complex

conjugates **19.** $X = -2$, $X = 1$, $\left\{2, -1, \dfrac{1\pm i\sqrt{3}}{2}\right\}$

UNIT 10
APPLICATIONS OF QUADRATIC EQUATIONS

*Unit 9 is a prerequisite for this unit. Answers appear at the end of the unit.

This unit explores the various features of the calculator that can be used to investigate application problems.

1.   A traveling circus has a "human cannonball" act as its grand finale. The equation $Y = -.01X^2 + .64X + 9.76$, where Y = height in feet and X = horizontal distance traveled in feet, represents the flight path of the human cannonball. Display a graphical representation of this equation and use the TRACE feature to answer the questions.

    a. TRACE along the graph in the standard viewing window and examine the numbers displayed at the bottom of the screen.
What do the X values represent?

    What do the Y values represent?

    b. In order to have "friendly" values for the X and Y, change the viewing WINDOW to ZInteger.

TI-85/86   TI-85/86 USERS PRESS [**GRAPH**] [**F3**](**ZOOM**) [**MORE**] [**MORE**] AND THE APPROPRIATE F KEY FOR (**ZINT**).

Press [**ZOOM**], cursor down to highlight 8 for ZInteger and press [**ENTER**]. Be sure your cursor is at X = 0 and Y = 0 (use the arrow keys to move the cursor to the point where the axes intersect to display X = 0 and Y = 0) and press [**ENTER**] again to set the viewing WINDOW to ZInteger. The blinking cursor was positioned at X = 0 and Y = 0 to keep the axes centered on the screen when changing to ZInteger. The cursor may be placed at any position on the screen and the axes will intersect at that point. Now TRACE along the curve and observe the "friendly" values represented at the bottom of the screen.

NOTE: ZInteger yields integer values for X when tracing on the graph. This is particularly valuable when X only has meaning as an integer value.

    c. This curve represents the flight path of the human cannonball. **TRACE** along the path to the right. How far, <u>approximately</u>, has the human cannonball traveled horizontally when he hits the ground?___________

    d. The human cannonball is traveling at speeds up to 65 mph. To land on the ground would mean certain death. If he uses a net for his landing, how far will he have traveled horizontally if the net is 11 feet above the ground?__________

    e. What is the maximum height reached during the course of his flight?_________

    f. How far has he traveled horizontally when he reaches this maximum height?_________

NOTE: The actual distance he has traveled is a length of arc along the curve of the parabola. To calculate this length requires the use of calculus.

g. The human cannonball is shot out of the cannon head first, so all of the distances are measured from his head. TRACE along the curve to X = 0 and Y = 9.76. Explain the meaning of these two values.

2.   Blaire is a pitcher for the Girls Slowpitch softball team at her middle school. The height of the softball X seconds after she releases a pitch is given by the formula $h = -16X^2 + 18X + 3$.

a. Find the length of time it will take the ball to hit the ground if the batter swings and misses.

Solution: When the ball hits the ground the height will be h = _____. Thus the equation we are trying to solve for X is $0 = -16X^2 + 18X + 3$. We want to solve this equation using the root/zero feature, so enter $-16X^2 + 18X + 3$ after Y1 = and set the standard viewing window. You will need to adjust the viewing rectangle!

Press [**TRACE**] to approximate the width of the graph at the points it crosses the X-axis (it appears to cross at about - 0.14 and 1.26) and to determine its approximate height (the highest y value is about 8). Based on this information, press [**WINDOW**] and change the Xmin to - 0.2, Xmax to 1.5, Ymin to - 5 and Ymax to 10. Press [**TRACE**] to simultaneously view the graph and to activate the TRACE feature. TRACE around the curve to see if it is possible to determine the exact time the ball will be 0 feet from the ground. TRACE will not yield the exact answer. Now use the ROOT/ZERO option to determine the two roots/zeroes of the equation.

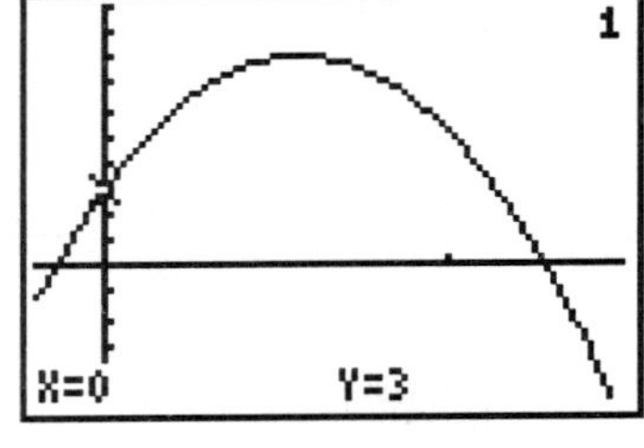

X = __________   or X = __________

One of these roots is not valid. Which one is it and why?

If you are having difficulty answering "why" then ask yourself this question: Can X equal a negative number? Remember, X represents time.

After using the ROOT/ZERO option on the valid root, you should have determined X to be approximately 1.2723635. This means that it will take the ball approximately 1.27 seconds to hit the ground if the batter misses.

b. What is the highest point that the ball reaches during the pitch?

Solution: Again, TRACE can be used to find the highest point, but letting the calculator determine the MAXIMUM point will be more accurate. Press [**2nd**] <**CALC**> [**4:maximum**]. The calculator is now ready to determine the maximum point on the curve. Set upper and lower bounds as in the ROOT/ZERO option.

Remember, the X value displayed represents seconds elapsed since the pitch was thrown and the Y value indicates the height of the ball at that point in time. Y = ___ indicates that the ball reached a maximum height of approximately ______ feet.

c. How **long** does it take the ball to reach its maximum height?_____

d. What is the height of the ball 1.4 seconds after the pitcher releases it? (**TRACE** along the path of the curve to X = 1.4 seconds.) <u>Carefully</u> explain your answer.

e. Use the TRACE cursor to TRACE along the path of the curve from left to right. Explain, in your own words, what information the X and Y values at the bottom of the screen are giving you.

f. Does the curve represent the path of the ball in flight? If you answer yes, then explain which part of the graph display represents the distance the ball travels.

   If you answer no, explain why not.

3.   Bridges are often supported by arches in the shape of a parabola. The equation

$$Y = \frac{10}{7}X - \frac{2}{49}X^2$$, where Y = height and X = distance from the base of the arch,

provides a model for a specific parabolic arch that supports a bridge. Will this arch be tall enough for a road crew to build a county road under?

Solution:
Setting the viewing WINDOW
a. Begin by entering the polynomial at the Y1 = prompt and graph in the standard viewing rectangle.

b. This curve represents the support to a bridge. The entire curve should be visible. TRACE along the curve, recording the following (to the nearest integer): left most X-intercept, maximum Y value, and right most X-intercept.

left most X-intercept:______   maximum Y value:______

right most X-intercept:______

c.   Use the information from part b to set the WINDOW values so that the entire graph is displayed. Press **[WINDOW]** and set the WINDOW values as follows: Xmin = -1, Xmax = 37, Xscl = 1, Ymin = -1, Ymax = 13, Yscl = 1.These values were selected to ensure that the area slightly beyond the perimeter of the graph is displayed.

NOTE: For more information on setting viewing WINDOWS, refer to the unit entitled: "Where Did the Graph Go?".

Solving the problem:

d.   Begin by determining the height (to the nearest tenth) of the highest point under the arch.

Maximum height = ___________

e.   If the average vehicle is no more than 6 feet high, can the vehicle drive under the arch?________ Explain how you determined your answer.

f.   If a two lane road is 20 feet wide, will it fit between the bases of the arch?_________ Explain how you determined your answer.

g.   Can the average vehicle drive in either lane under the arch and not scrape the paint off the roof? (or scrape the roof off the car??) That is to say, if this 20 foot wide road is centered under the arch, is the arch at least 6 feet above the road at all points in its width?

4.   The local community theater is considering increasing the price of its tickets to cover increases in costuming and stage effects. They must be careful because a ticket price that is too low will mean that expenses are not covered and yet a ticket price that is too high will discourage people from attending. They estimate the total profit, Y, by the formula $Y = -X^2 + 35X - 150$, where X is the cost of the ticket.

Before attempting to graphically solve the problem, set the viewing WINDOW by following steps a - c in #3. **MAKE SURE THAT THE ENTIRE CURVE IS DISPLAYED.**

a. What is the maximum amount that can be charged for a ticket to maximize the profit?__________

b. What is the maximum profit?__________

c. If $13 dollars is charged for each ticket, what will the profit be?
What are your solution options here? We could return to the home screen and evaluate the polynomial for X = 13 by using the **STO**re feature or we could scroll through the TABLE in search of X = 13. However, since we have been using the CALC menu to investigate the graph, we will look at value (EVAL X) which is the first entry option under the CALC menu. Access the CALC menu and press **[1:value]** to select value and display the **X** prompt (displayed as EVAL X on the TI-82/85/86). At the prompt, enter 13 and press **[ENTER]**. What will the profit be when $13 is charged for each ticket? ________

TI-85/86 RECALL THAT **EVAL X** IS ACCESSED BY PRESSING **[MORE]** TWICE (THE GRAPH MENU MUST BE DISPLAYED) AND SELECTING **"EVAL."**

BEWARE: The value (EVAL) feature only works when the value selected for X is between the Xmax and Xmin values on the WINDOW screen.

d. If they predict a profit of $150.00 on a play, how much was charged per ticket?
(Scroll through the Y values in the TABLE to answer this question.)__________

e. Use the TABLE display to determine at what point the theater "breaks even", i.e. How much must each ticket cost for there to be no money lost and yet no profit made?

Solutions: **1c.** approx. 76.5 ft. **1d.** 62 ft. **1e.** 20 ft. **1f.** 32 ft. **1g.** It means that his head is 9.76 feet above the ground before he is shot from the cannon. **2a.** h = 0, x = -.1473635, x = 1.2723635 **2b.** Y = 8.0625, 8 feet **2c.** approx. six tenths of a second **2d.** Y = -3.42, the ball is 3.41 feet into the ground. **2e.** The X values represent the time (in seconds) that the ball is in the air; the Y values represent its height. **2f.** Yes: This is obviously an incorrect answer, because nothing represents distance. No: This is the correct response, because the graph is relating the time (X) to the height of the ball (Y). **3d.** 12.5 **3e.** Yes **3f.** Yes, because the supports are 35 feet apart. **3g.** Yes   **4a.** $17.50 **4b.** $156.25 **4c.** $136   **4d.** 15, 20 **4e.** The theater breaks even when tickets are priced at $5 each or $30 each.

*Unit 9 is a prerequisite for this unit. Answers appear at the end of the unit.

This unit will investigate using the INTERSECT and ROOT/ZERO options on the calculator to solve equations that contain radicals. Recall, the solution to an equation is the value(s) for the variable that produce a true arithmetic statement.

Solve $\sqrt{3X + 7} + 2 = 7$ algebraically

Check your solution(s) by substitution:

To graphically solve this equation we will first look at the graphical representation of each side of the equation. Enter $\sqrt{3X + 7} + 2$ at Y1 = and the constant 7 at Y2 =. Graph in the standard viewing window and compare your graph to the screen pictured at the right. Recall, we want to determine graphically where Y1 = Y2. Circle the point of intersection. 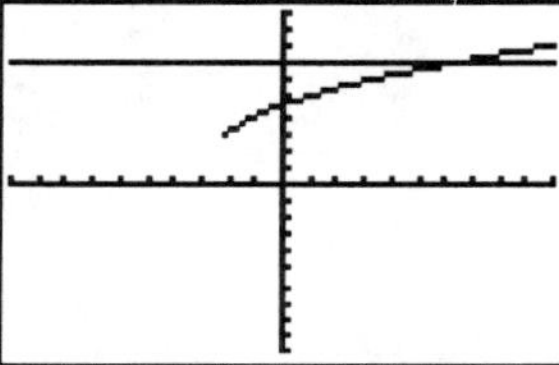

Use the INTERSECT option to graphically find the X-coordinate of the intersection of the two graphs. The X-coordinate of the intersection should be the same as the value found algebraically above: 6. If it is not, recheck both the algebraic solution and the calculator solution.

Now use the ROOT/ZERO option to graphically solve the same radical equation. Remember, you must first rewrite the equation with all terms on one side of the equal sign and the other side equal to 0. Do this in the space below.

If the instructor does not require you to show this algebraic computation, then turn "off" the graphs of Y1 and Y2 and merely enter Y1-Y2 at the Y3 = prompt. (See OPTIONAL NOTE in the middle of page 59.)

Press [GRAPH] and compare your screen to the one pictured at the right. Circle the root/zero, i.e. the X-intercept. 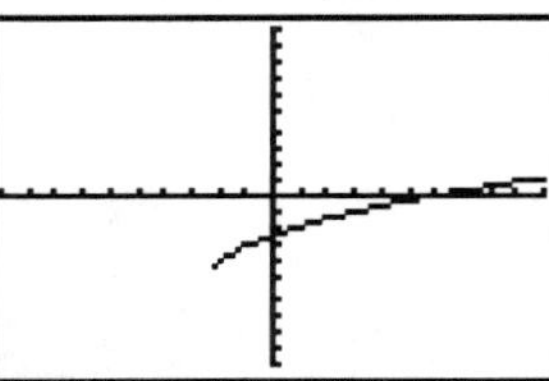

Use the ROOT/ZERO option to find the root of the equation. Your root should, of course, be 6.

Directions: Use either the ROOT/ZERO or the INTERSECT options on the calculator to solve each radical equation below. Sketch the screen display and use the ▸Frac option (under the MATH menu) to convert all decimal results to fractions.

1. $\sqrt{X^2 + 6X + 9} = -X + 6$

   X = _________

   Converted to a fraction, X = _________

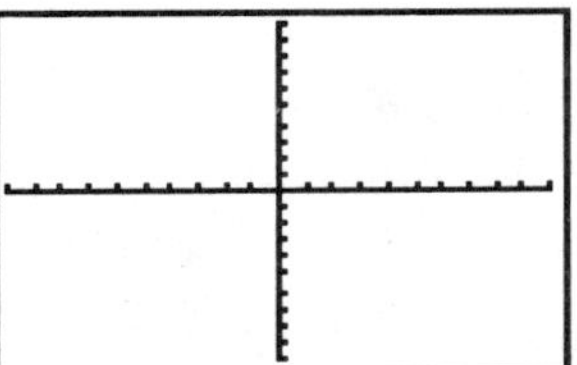

2. $\sqrt{2X + 5} = \sqrt{3 - X}$

   X = _________

   Converted to a fraction, X = _________

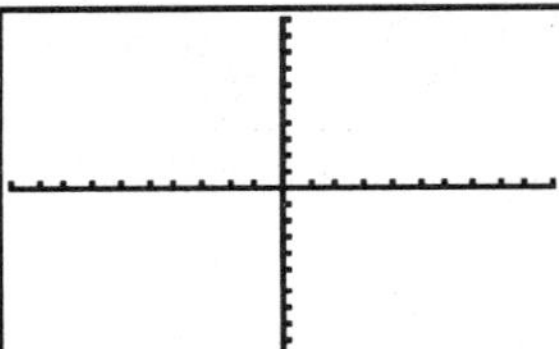

3. $\sqrt[3]{2X + 6} = 2$

   X = _________

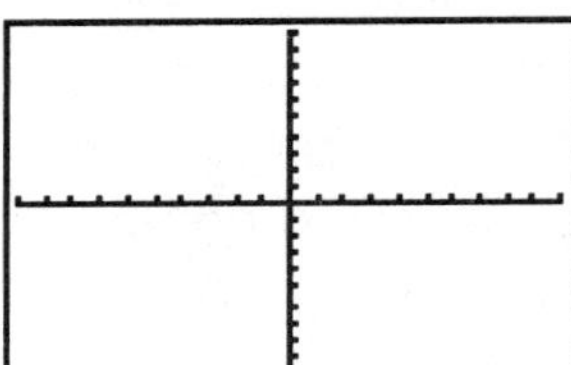

4. $\sqrt{X + 4} + 6 = 3$

   Solution: ___________

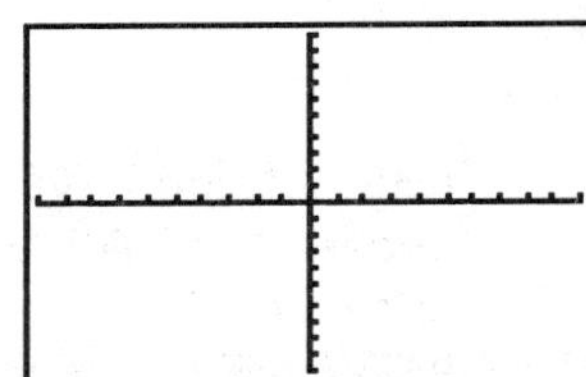

✍ 5. If you solved the equation in #4 algebraically, the first step would be to isolate the radical. Once the radical is isolated, you should realize there are no solutions. Why?

6. Solve the equation $\sqrt{X} + 2 = \sqrt{5 - X} + 3$ algebraically in the space below. Check your solution(s) by substitution.

   Algebraic Solution   Check by substitution

7. Graph the equation in #6 by entering the left side as Y1 = and the right side as Y2 = . Copy the display screen. How many points of intersection do you see? Use the calculator to find the solution.

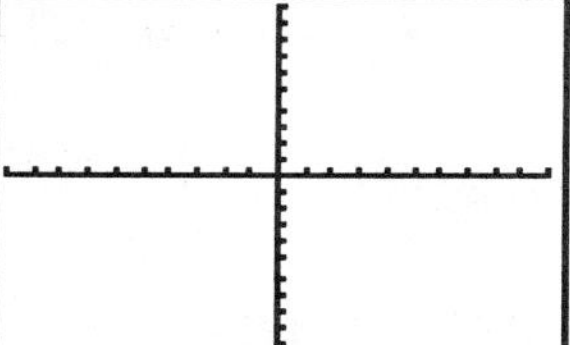

X = _________

8. $\sqrt{2X - 3} = 3 - X$

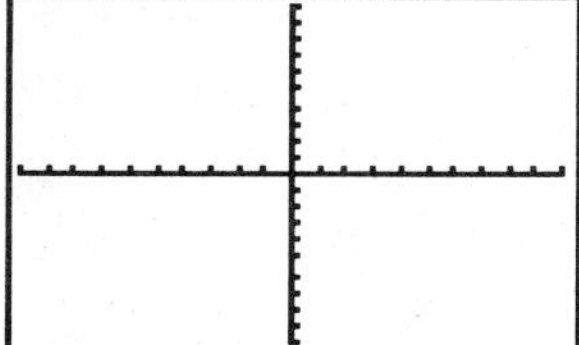

Be Careful! Make sure BOTH the X and Y coordinates are displayed at the bottom of the screen.

X = _______

9. $\sqrt{X^2 - 12X + 36} + 5 = 7$

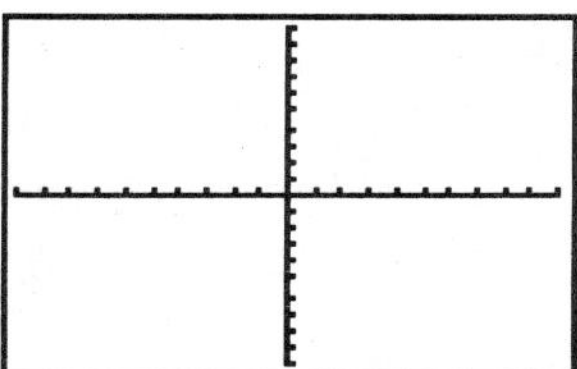

X = _______   X = _______

10. $\sqrt[3]{X^3 + 5X^2 + 9X + 18} = X + 2$

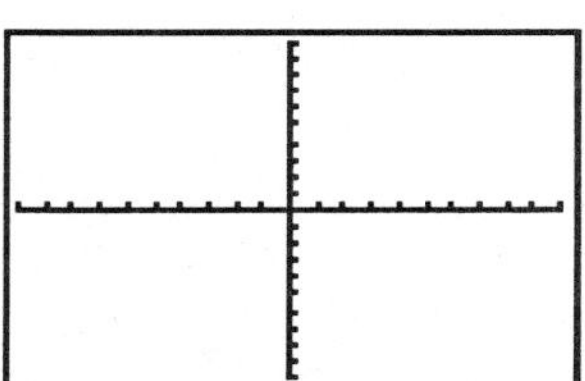

X = _______   X = _______

Hint: If you have difficulty finding the roots using the INTERSECT or ROOT/ZERO option, access the TABLE feature (set the table to begin with X = 1 and increment by 1). Enter the left side of the equation at Y1 = and the right side at Y2 = (as though you were using the INTERSECT option). Access the table, and scroll until you find the X-value(s) for which the Y1 and Y2 values are equal.

11. In the space below, solve $\sqrt{-2X + 6} = 3 - X$ algebraically. Check solutions by substitution.

Algebraic solution: Check by substitution:

12. a. Use the INTERSECT option and have the calculator find the solutions to the equation in #11. Copy your screen display.

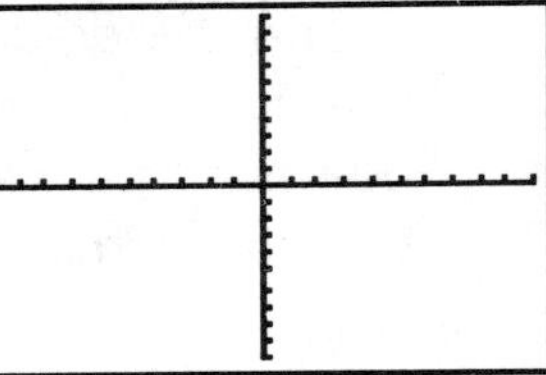

X = ________

b. Use the ROOT/ZERO option and have the calculator find the solutions. Copy your screen display.

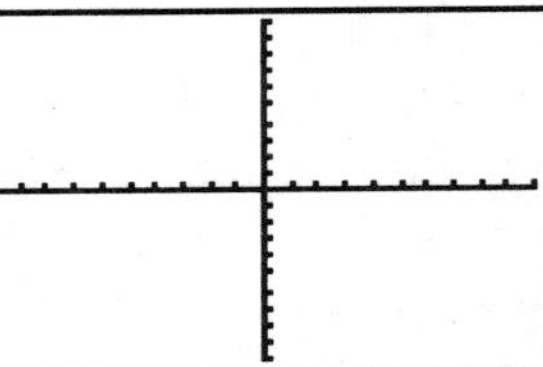

X = ________

✍ c. You found two valid roots algebraically, two roots were displayed graphically, and yet only one root could be computed with the calculator. Which is the correct solution - the algebraic solution in #11 or the calculator solutions above?

d. Set your table to a start value of 1 and increment by 1. Make sure you have the left side of the equation entered at Y1 = and the right side at Y2 =. Access the table. For what values of X are the Y1 and Y2 values equal?

X = ______   X = ______ This confirms your algebraic solution.

✍ e. Explain **why** the calculator was unable to compute both roots in part a.

13. **Application:** The period of a pendulum on a clock is the time required for the pendulum to complete one cycle (one "swing" from a given position back to this initial point). The formula for finding the period of a pendulum is $2\pi\sqrt{\dfrac{X}{32}}$, where T is the time required in seconds and X is the length of the pendulum. A clock company is constructing a clock for a window display. If it takes the pendulum two seconds to complete 1 period, what is the length of the pendulum (to the nearest hundredth of a foot)?

NOTE: Work problems from your text using what you have learned in this unit. Decide what works best for <u>you</u> - algebraic solutions? the ROOT/ZERO option? the INTERSECT option? Then PRACTICE.

Solutions: 1.. X = 1.5, 3/2   **2.** X = -.6666667, -2/3   **3.** 1   **4.** Null Set

5. Because the right side would be equal to -3 and $\sqrt{}$ is defined only for positive roots.

6. X = 4, One is an extraneous root. **7.** X = 4   **8.** X = 2   **9.** X = 4 or X = 8

10. X = -5 or X = 2   **11.** X = 1 or X = 3

12a. X = 1   **12b.** X = 1   **12c.** {1, 3} - Both of these solutions check algebraically.

12d. X = 1 or X = 3

12e. When using the INTERSECT feature, the calculator establishes upper and lower bounds using the domain of the graphed functions. It then searches between these bounds for the point of intersection, **excluding the bounds in its search.**

13. 3.24 feet

UNIT 12
GRAPHICAL SOLUTIONS: LINEAR INEQUALITIES

*Unit 7 is a prerequisite for this unit. Answers appear at the end of the unit.

To solve linear inequalities such as $5X - 1 \geq -3$, we want to find replacement values for X that will produce a true arithmetic sentence. The solution to a first degree inequality in one variable is typically an infinite set of numbers rather than a single number.

> Solve $5X - 1 \geq -3$, algebraically:

The -2/5 you got in your solution is a "critical point." It is so named because it divides the number line into three distinct subsets of numbers - those larger than the number, those smaller than the number, and the number itself.

> In the space below, test the critical point -2/5 in the original inequality.
> (i.e. When X is replaced by -2/5, is the resulting inequality a true statement?)

> In the space below, test a number whose value is larger than that of the critical point.

Because the test number tested true, the implication is that all values to the right of the critical point will also test true. This is also indicated by the mathematical statement $X \geq -2/5$, the algebraic solution.

What *should* happen when a number is tested whose value is smaller than that of the critical point? If you are not sure, choose a value and test it!

The graphing calculator can be used to quickly confirm the solution of $X \geq -0.4$.
Press [**Y =**] and enter 5X - 1 after Y1 = and -3 after Y2 = . Examine the TABLE values where X = -0.4, X > -0.4 and X < -0.4. To do this, press [**2nd**] **<TBLSET>** and set the table to start at -.4 and increment by 1. (TI-86 users press [**TABLE**] and then [**F1**](**TBLST**).)
Press [**2nd**] **<TABLE>** to view the table of values. Compare the Y1 and Y2 columns for X = -0.4, X > -0.4 and X < -0.4. (Recall, Y1 = 5x - 1 and Y2 =-3.) Your table should correspond to the one at the right. When X = -0.4, the expression 5X -1 is equal to -3. When X > -0.4, the expression 5X - 1 has values greater than -3 and when X < -0.4, the expression 5X - 1 has values less than -3.

X	Y1	Y2
-3.4	-18	-3
-2.4	-13	-3
-1.4	-8	-3
-.4	-3	-3
.6	2	-3
1.6	7	-3
2.6	12	-3

X= -.4

Now consider the graphical solution. Since Y1 = 5X - 1 and Y2 = -3 , we want to locate graphically <u>where</u> Y1 $\geq$ Y2.

Press **[ZOOM] [6:ZStandard]** to set the standard viewing WINDOW. The WINDOW values at the right are automatically entered and the graph screen will be displayed. Pressing **[WINDOW]** displays these values.

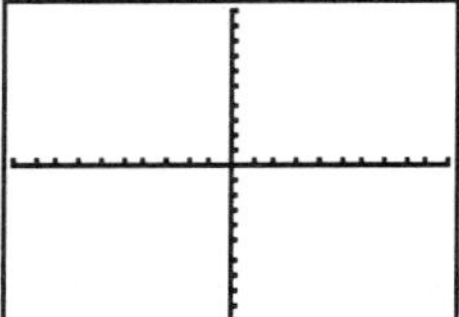

TI-85/86	TI-85/86 USERS PRESS **[GRAPH] [F3](ZOOM)** AND **(ZSTD)** TO SET THE STANDARD VIEWING WINDOW.

Press **[GRAPH]** and sketch the view of the graphical solution to the inequality, being careful to label "Y1" and "Y2" as their graphs are displayed. Circle the point where the two graphs are equal, the point of intersection. Use the INTERSECT option to determine the solution to the equation. You should get X = -0.4.

Graphs should be read from left to right. The graph is a visual representation of numerical information that was previously displayed in the TABLE. For what X values is Y1 greater than Y2? It should be the section that is highlighted on the graph displayed. In general, Y1 > Y2 where the graph of Y1 is <u>above</u> the graph of Y2.

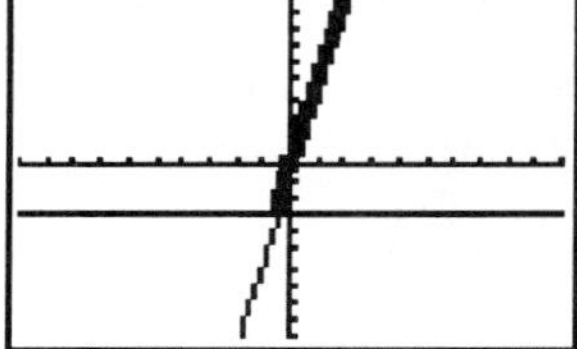

The solution to the inequality is X $\geq$ -0.4.

In set notation this would be written: $\{X | X \geq -0.4\}$

As a number line graph this would be:
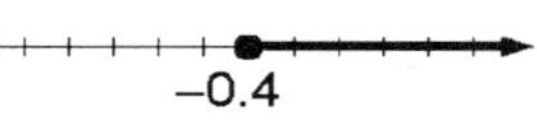

In interval notation this would be written [-0.4,∞).

EXERCISE SET

1.   **Solve** 10 - 3X < 2X + 5 graphically.

Solution Steps:
a. Press **[Y =]** and enter 10 - 3X after Y1 = and 2X + 5 after Y2 =. Sketch the graph displayed. Be observant the first time the graphs are displayed. It will be helpful to label Y1 and Y2. The graph of Y1 will appear first.

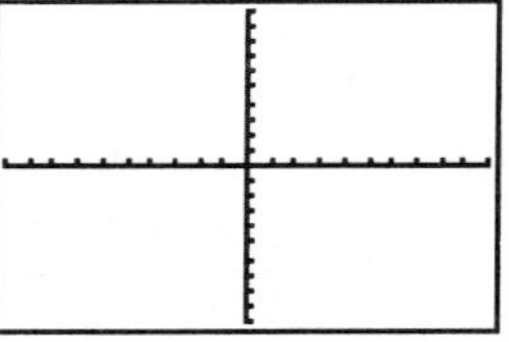

b. Use the INTERSECT feature to find the point of intersection. The intersection is at X = ______. The expressions entered at Y1 and Y2 are equivalent when X = 1. Therefore, X = 1 would be the solution to the *equation* 10 - 3X = 2X + 5.

c. Use a highlighter pen to highlight the section of Y1 that is **less than** Y2. Recall, Y1 < Y2 when the graph of Y1 is <u>below</u> the graph of Y2.

80

d. Determine your solution by setting **TblStart** = *critical point*. Examine the relationship between the values of Y1 and Y2 for X values that are both greater than, equal to, and less than the critical point.

e. Conclusion: Y1 < Y2 when X > 1.
The solution set will be { X | X > 1}.
Translate this solution to a number line graph:

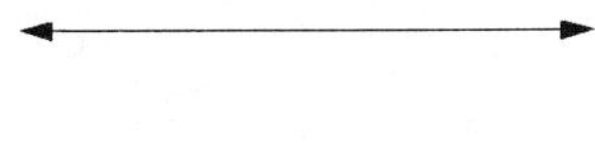

2. **Solve** 6 - 5X $\leq$ -1 graphically, following the steps outlined in 1.

Solution Set:_____________________

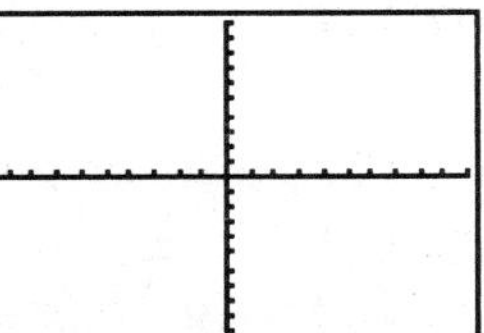

Translate this solution to a number line graph:

3. **Solve** -3 $\geq$ 7 - 2X graphically, following the steps outlined in 1.

Solution Set:_____________________

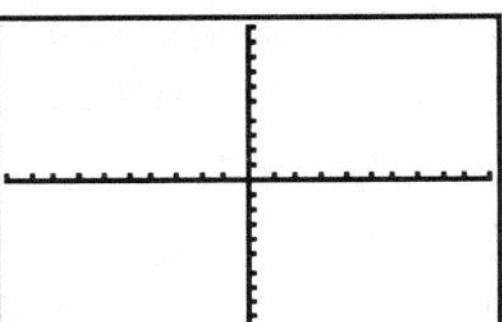

Translate this solution to a number line graph:

4. **Solve** $\dfrac{4X - 2}{6} < \dfrac{2(4 - X)}{3}$ graphically, following the steps outlined in 1.

Solution Set:_______________

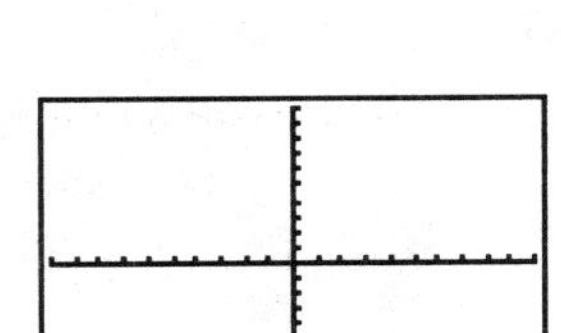

Translate this solution to a number line graph:

5. A linear inequality has been entered on the calculator so that the left side is expressed as Y1 and the right side as Y2.
a. What is the critical point as displayed in the TABLE of values?____

b. What is the solution (expressed as a number line graph) to the inequality Y1 < Y2?

81

The calculator's TEST menu can be used to display a graph that **RESEMBLES** the number line graph of any of the inequalities solved thus far. This provides a means of counterchecking your work. For the example, we will check the original inequality, 5X - 1 $\geq$ -3 with the TEST feature.

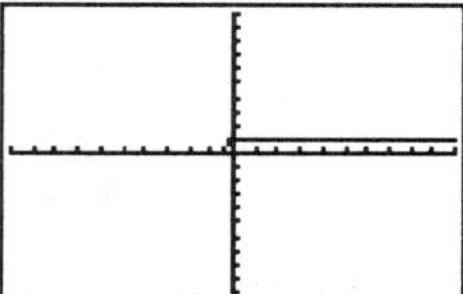

TEST is located above the MATH key (above the "2" key on the TI-85/86). Begin by deleting <u>all</u> entries on the Y = screen. Enter the entire inequality, 5X - 1 $\geq$ -3 on the **Y1** = line. Press **[GRAPH]**.

NOTE: If you TRACE on this graph, the TEST will return a 1 as the Y-value for a true statement and a 0 as the Y-value for a false statement.

Refer to your algebraic solution of the inequality on the first page of the unit and sketch the number line graph. Compare this number line graph to the one sketched on page 80. **BEWARE!** If you TRACE on this number line you will not be able to determine the critical point. This point must still be determined using the procedure outlined in Exercise 1 (graphing each side separately and using the INTERSECT option to locate the critical point). However, the graph illustrated by Y1 = 5x-1$\geq$ -3, displayed above, <u>does</u> show you whether the interval to the right or the interval to the left of the critical point is the solution interval.

EXERCISE SET CONT'D

Directions: Enter the inequality at the Y1 = prompt. Press **[GRAPH]** and sketch the display. Transfer the information displayed to your number line and then convert the information to interval notation. Determine the critical point of the number line by referring back to the corresponding problem in Exercises 1 through 4. Be sure the calculator display agrees with the number line you sketched in each of the Exercises 1 - 4.

6. Use the TEST menu to display a graph that resembles the number line graph of the solution to the inequality
    10 - 3X < 2X + 5.

    Number line graph:
    Interval Notation:___________________

7. Use the TEST menu to display a graph that resembles the number line graph of the solution to the inequality
    6 - 5X $\leq$ -1.
    Be sure the calculator display agrees with the number line you sketched in Exercise 2.

    Number line graph:
    Interval Notation:___________________

8.   Use the TEST menu to display a graph that resembles the
      number line graph of the solution to the inequality $-3 \geq 7 - 2X$.

      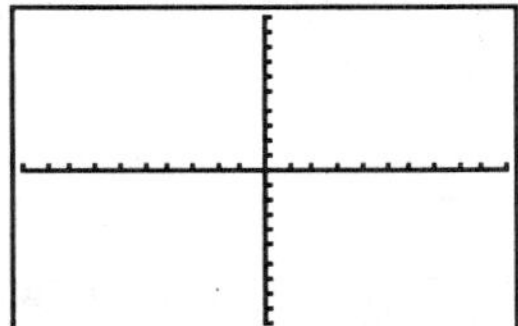

      Be sure the calculator display agrees with the number line you sketched
      in Exercise 3.

      Number line graph:

      Interval Notation:___________________

9.   Use the TEST menu to display a graph that resembles the
      number line graph of the solution to the inequality
      $$\frac{4X - 2}{6} < \frac{2(4 - X)}{3}.$$

      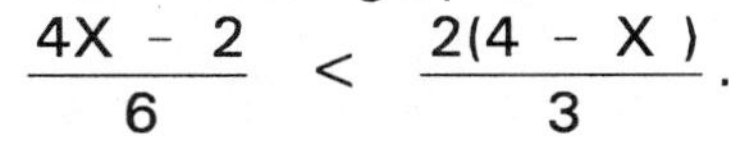

      Be sure your calculator display agrees with the number line you
      sketched in Exercise 4.

      Number line graph:

      Interval Notation:_____________

✐10. In your own words, explain why you cannot use the graphical display in Exercises
      6 - 9 as the only means of displaying the solution to an inequality. What is the one
      major obstacle that this approach has?

Directions: Solve each inequality graphically using the INTERSECT feature. Begin in a
ZStandard WINDOW and adjust the viewing window accordingly. Critical points should be
expressed in fraction form. Remember, you are solving, not applying the TEST menu!

11. $3X - 8 < \frac{4}{5}(3X - 2)$

      Xmin = ______   Xmax = ______

      Ymin = ______   Ymax = ______

      Solution Set:_________________________________   Interval Notation:___________

      Translate this solution to a number line graph:

12. $8 + \dfrac{8}{5}X \geq \dfrac{1 + 9X}{8}$

Xmin = _______   Xmax = _______

Ymin = _______   Ymax = _______

Solution Set:_________________________   Interval Notation:__________

Translate this solution to a number line graph:

13. $-\sqrt{288} + X > 2(2X - 6\sqrt{2})$

Xmin = _______   Xmax = _______

Ymin = _______   Ymax = _______

Solution Set:_________________________   Interval Notation:__________

Translate this solution to a number line graph:

✍14. Solve the combined inequality $-4 < 4 + 2x < 8$
      graphically. Use the steps outlined in Exercise 1 as a model.
      Record any additional steps required to obtain a valid solution.
      Display your graph at the right.

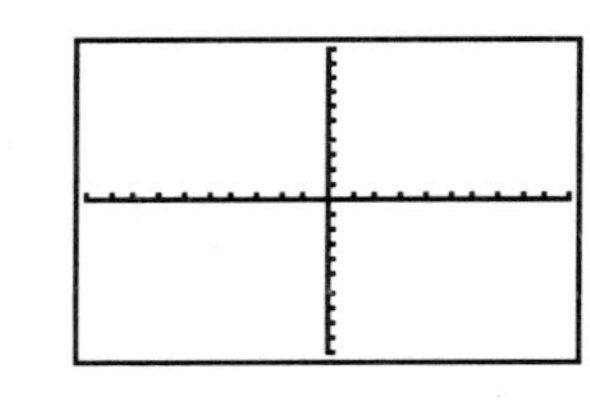

      Solution Set:_____________________

      Translate this solution to a number line graph:

✍15. Use the TEST menu to try to display a graph that resembles the number line graph of
      the solution to the inequality $-4 < 4 + 2x < 8$. If the result produced by the TEST
      menu does not correspond to the solution, explain what is wrong with your entry at
      the **Y1 =** screen.

16. The graph displayed at the right is the graph of the linear equation Y = P where P is a linear polynomial expression. The coordinates displayed represent the point where the line intersects the X-axis (X-intercept). Solve the indicated inequalities using the displayed graph and record your solution on the number line provided.

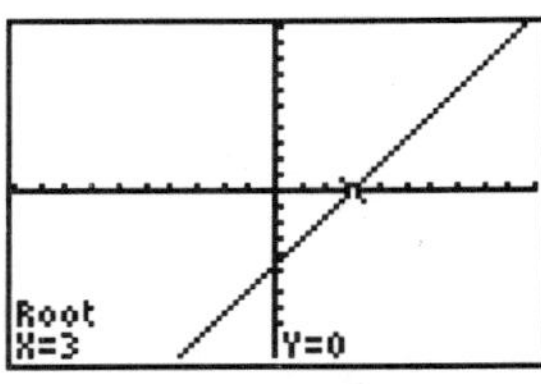

a. $Y = 0$

b. $Y > 0$

c. $Y < 0$

d. $Y \leq 0$

e. $Y \geq 0$

THE FOLLOWING PROVIDES INFORMATION ABOUT THE GRAPH STYLE ICON (APPLICABLE ONLY TO THE TI-83/83PLUS AND TI-86) WHICH YOU MAY WANT TO EXPERIMENT WITH WHEN GRAPHING MORE THAN ONE POLYNOMIAL AT A TIME.

TI-83/83plus

The "graph style icon" is a feature that allows you to distinguish between the graphs of equations. Press **[Y=]** and observe the "\" in front of each Y. Use the left arrow to cursor over to this "\". The diagonal should now be moving up and down. Pressing **[ENTER]** once changes the diagonal from thin to thick. The graph of Y1 will now be displayed as a thick line. Repeatedly pressing **[ENTER]** displays the following:

- ◥: shades above Y1
- ◣: shades below Y1
- -o: traces the leading edge of the graph followed by the graph
- o: traces the path but does not plot
- ⋰: displays graph in dot, not connected MODE

The graphing icon takes precedence over the **MODE** screen. If the icon is set for a solid line and the **MODE** screen is set for DOT and not solid, the graphing icon will determine how the graph is displayed. Pressing **[CLEAR]** to delete an entry at the Y= prompt will automatically reset the graphing icon to default, a solid line.

TI-86

THE "GRAPH STYLE ICON" ALLOWS YOU TO DISTINGUISH BETWEEN THE GRAPHS OF EQUATIONS. PRESS **[GRAPH] [F1](y(x)=)** AND OBSERVE THE "\" IN FRONT OF EACH Y. PRESS **[MORE]** FOLLOWED BY **[F3](STYLE)**. THE DIAGONAL SHOULD NOW HAVE CHANGED FROM THIN TO THICK. THE GRAPH OF Y1 WILL BE DISPLAYED AS A THICK LINE. REPEATEDLY PRESS **[STYLE]** TO DISPLAY THE FOLLOWING:

- ◥: SHADES ABOVE Y1
- ◣: SHADES BELOW Y1
- -O: TRACES THE LEADING EDGE OF THE GRAPH FOLLOWED BY THE GRAPH
- O: TRACES THE PATH BUT DOES NOT PLOT
- ⋰: DISPLAYS GRAPH IN DOT, NOT CONNECTED MODE

IT IS IMPORTANT TO REMEMBER THAT THE GRAPHING ICON TAKES PRECEDENCE OVER THE **MODE** SCREEN. IF THE ICON IS SET FOR A SOLID LINE AND THE **MODE** SCREEN IS SET FOR DOT AND NOT SOLID, THE GRAPHING ICON WILL DETERMINE HOW THE GRAPH IS DISPLAYED. PRESSING **[CLEAR]** TO DELETE AN ENTRY AT THE Y = PROMPT WILL AUTOMATICALLY RESET THE GRAPHING ICON TO DEFAULT, A SOLID LINE.

<u>**Solutions to Exercise Sets:**</u> **1.** $\{x \mid x > 1\}$

2. $\{X \mid X \geq 1.4\}$

3. $\{X \mid X \geq 5\}$

3. $\{X \mid X < 2.25\}$

5. The critical point is 2 and the number line graph of the solution is

6. $(1,\infty)$ **7.** $[1.4,\infty)$ **8.** $[5,\infty)$

9. $(-\infty, 2.25)$

10. The major obstacle is that you are not able to determine the critical point every time. This method IS NOT a graphical solution; it is merely a display of a replica of the number line solution.

11. $\{X \mid X < 32/3\}$, $(-\infty, 32/3)$

12. $\{X \mid X \geq -315/19\}$, $[-315/19, \infty)$

13. $\{X \mid X < 0\}$, $(-\infty, 0)$

*Unit 12 is a prerequisite for this unit. Answers appear at the end of the unit.

Unit #8 examined graphical solutions to absolute value equations. The equation
$|X - (-3)| = 6$ was translated to: *the distance between a number and -3 is 6 units* or *what numbers are 6 units from -3*. In this unit we will examine absolute value inequalities and how they relate to distance.

Consider $|X - (-3)| \leq 6$. This inequality will be translated to: *find all the numbers whose distance from -3 is 6 units __or__ less.* We already know the answer to the first portion of the question: *find all numbers whose distance from -3 is equal to 6 units*. [REVIEW the INTERSECT process from the unit entitled "Graphical Solutions: Linear Equations" if needed.] The points of intersection are X = 3 and X = -9. These are the solutions to the absolute value __equation__ $|X - (-3)| = 6$. Label Y1 and Y2 on the graph at the right. (83/83plus/86 users may want to try the graph style icon and graph Y2 as a thick line instead of using a label.)

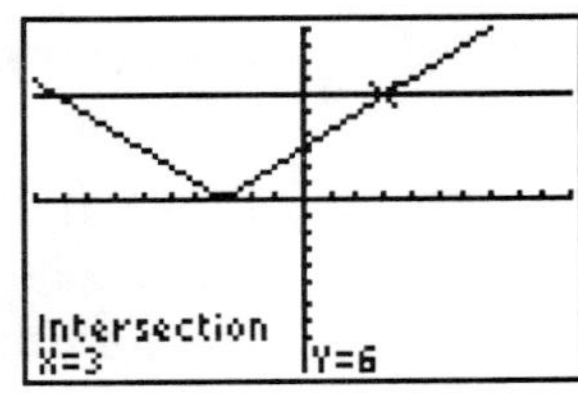

Now examine the second part of the question: *find all numbers whose distance from -3 is less than 6 units*.
Look at the graph displayed above. We know that Y1 < Y2 when the graph of Y1 is __below__ the graph of Y2. To answer the second part of the question, find all X values for which this is true. Press **[TRACE]** and be sure the TRACE cursor is on the graph of Y1.

As you TRACE along the graph of Y1 you will discover that the portion of Y1 that is less than Y2 is the portion of the graph of Y1 that is below the graph of Y2. This portion has been highlighted on the graph at the right.

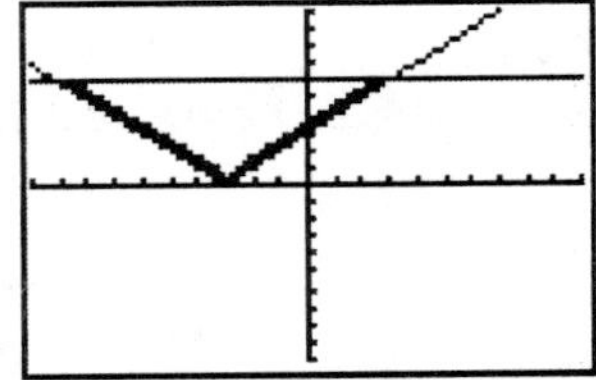

Place the TRACE cursor on the left hand point of intersection and TRACE right along the highlighted portion of the graph of Y1. Observe the X values as you TRACE. What happens to the X values along the highlighted portion of the graph?

On the graph, draw a dotted vertical line from each point of intersection perpendicular to the horizontal axis. Count the tic marks on the horizontal axis and label the points where the perpendicular lines touch the axis.

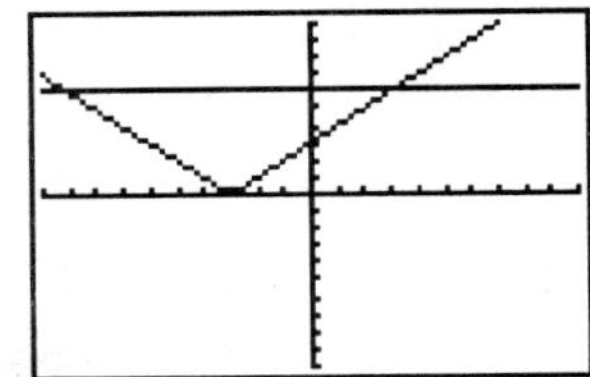

The highlighted portion of the graph of Y1 should be between the two points you labeled.

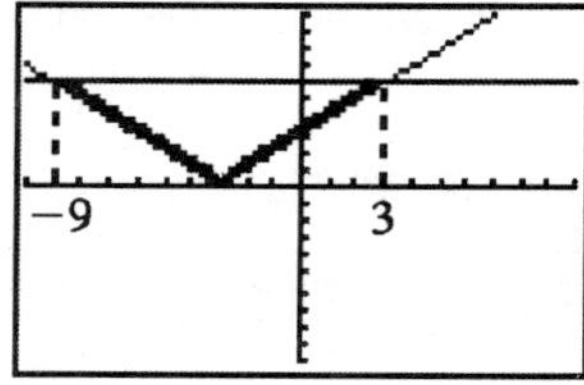

The solutions to the **inequality** $|X - (-3)| < 6$ are all the X values between -9 and 3 (the critical points).

The solutions to the **equation** $|X - (-3)| = 6$ are the two critical points -9 and 3.

Combining this information, we get the solution to $|X - (-3)| \leq 6$ to be $-9 \leq X \leq 3$.

Solution Set: $\{X| -9 \leq X \leq 3\}$, Number line graph: 

Interval Notation: [-9,3]

Graphically solve $|X - (-3)| > 6$ by following the indicated steps.

a.   Press **[Y =]** and enter abs(X + 3) after Y1 = and 6 after Y2 = .

b.   Press **[GRAPH]**. Sketch the display screen and draw in dotted lines from the intersection points perpendicularly to the horizontal axis.

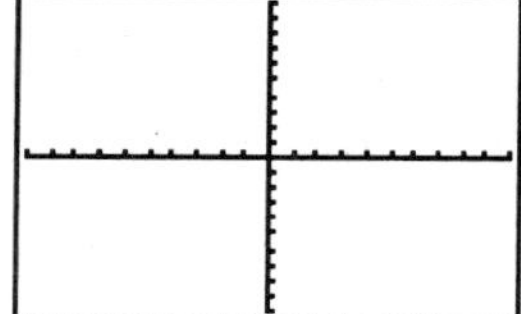

c.   Use the calculator's INTERSECT option to find the points of intersection.
X = _______ and X = _______

d.   Count the tic marks and label the points on the horizontal axis where your perpendicular lines touch the axis.

NOTE: *At this point, the steps listed above are the same steps you used to solve $|X + 3| \leq 6$ in the first part of this unit and to solve $|X + 3| = 6$ in a previous unit.*

e.   The solution is found to be the X values where Y1 > Y2. When the graph of Y1 is above the graph of Y2, as indicated by the highlighted portions, then Y1 > Y2.

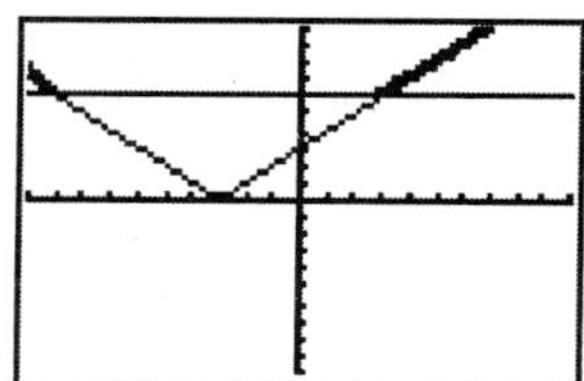

f.   The solutions to the inequality $|X - (-3)| > 6$ are all X values greater than 3 or less than -9: X > 3 or X < -9.

Solution Set: $\{X \mid X < -9 \text{ or } X > 3\}$

Number line graph:

Interval Notation: $(-\infty,-9) \cup (3,\infty)$

NOTE: *-9 and 3 were not included because the inequality states that $|X -(-3)|$ is <u>strictly</u> greater than 6.*

Directions: Graphically solve each of the following inequalities following the steps that were outlined. Record the solution in set notation, as a number line graph and in interval notation.

1. $|2X - 1| \geq 5$

   Solution Set:__________________

   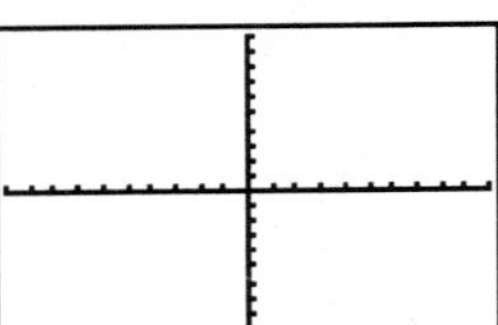

   Number line graph:

   Interval Notation:______________

2. $\left|\dfrac{1}{2}X - 1\right| < 4$

   (Be Careful! Enclose the $\dfrac{1}{2}$ in parentheses.)

   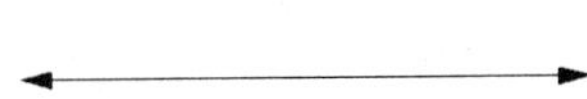

   Solution Set:__________________

   Number line graph:

   Interval Notation:______________

3. $\left|\dfrac{2X + 5}{3}\right| < 4$

   Solution Set:__________________

   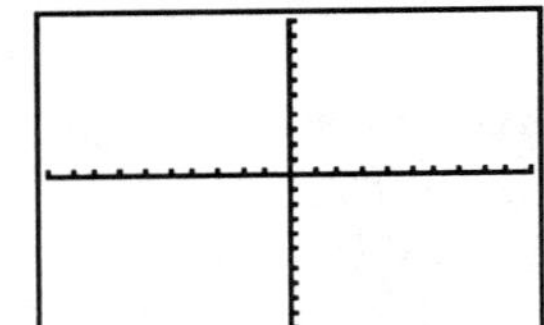

   Number line graph:

   Interval Notation:______________

4. In your own words, explain what type of error could easily be made when graphing the expression $\left|\dfrac{2X + 5}{3}\right|$ or $\left|\dfrac{1}{2}X - 1\right|$.

5. $|4X + 2| > -3$

This particular inequality represents a "special case." Carefully re-TRACE your highlighted portion of the graph before deciding on the solution.

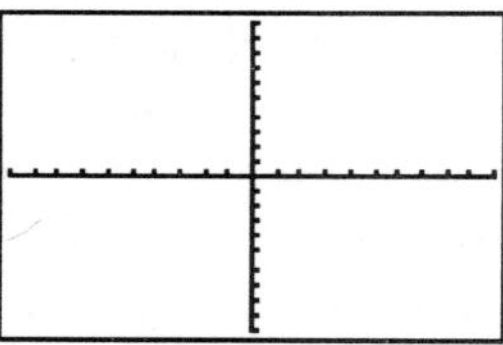

Solution Set:___________________

Number line graph: 

Interval Notation:______________

6. $|4X + 2| < -3$ (another "special case")

Solution Set:__________

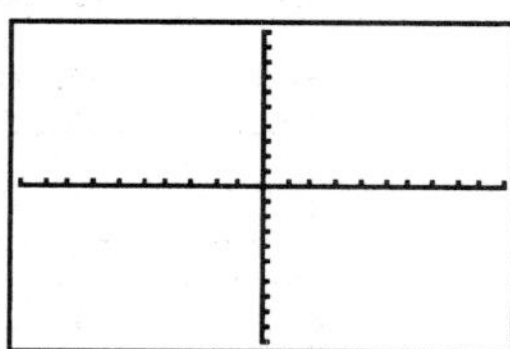

✍ 7. Consider why #5 and #6 are labelled as "special cases." Could #5 and #6 have been solved by merely "looking" at the inequality? Think carefully about the definition of absolute value before formulating your response.

Directions: Use the TEST menu to display a graph that resembles the number line graph of the solution to each of the inequalities solved in this unit. Refer to the unit entitled "Graphical Solutions: Linear Inequalities" for a review of using this menu.

    a. Sketch the display screen for each problem.
    b. The exercise number (where the inequality was originally solved) is displayed in parentheses to the right of the inequality. Return to the referenced problem to determine the critical points.
    c. Label these critical points on the display screen sketch.
    d. If the graph displayed does not agree with the solution previously determined, go back and recheck your work.

8. $|2X - 1| \geq 5$ (Exercise 1)

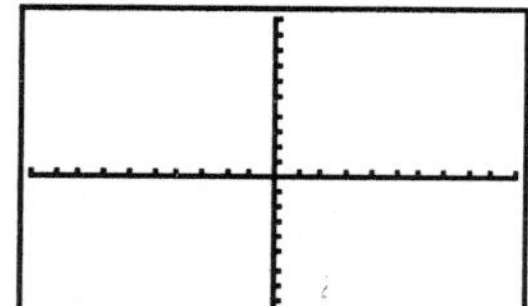

9. $\left|\dfrac{2X + 5}{3}\right| < 4$ (Exercise 3)

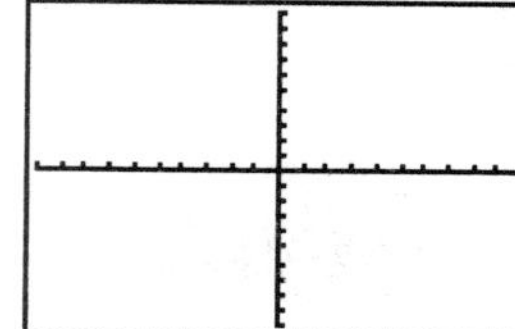

10. $|4X + 2| > -3$ (Exercise 5)

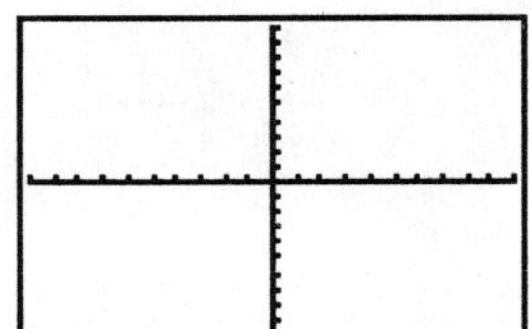

11. $|4X + 2| < -3$ (Exercise 6)

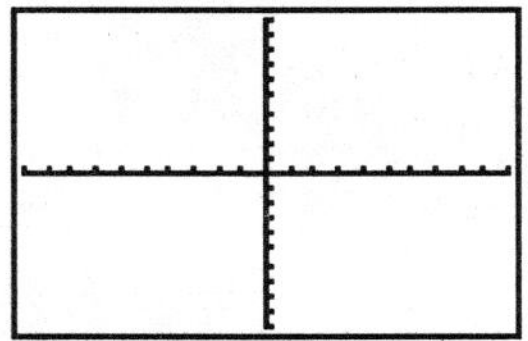

✍12. **Summarizing Results:** Because the left and the right sides of equations and inequalities are graphed as separate expressions, the graphical representations of the solutions to each of the following problems all look alike.

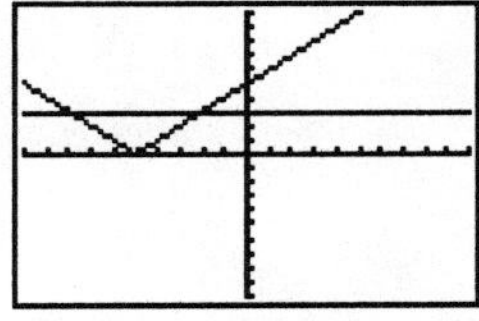

$|X + 5| = 3$

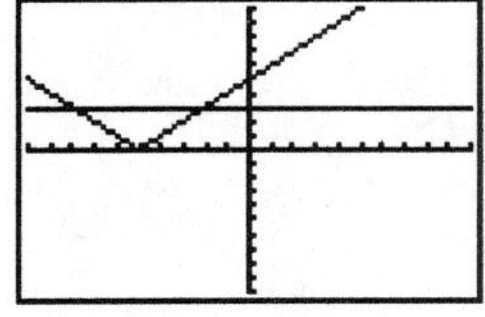

$|X + 5| < 3$

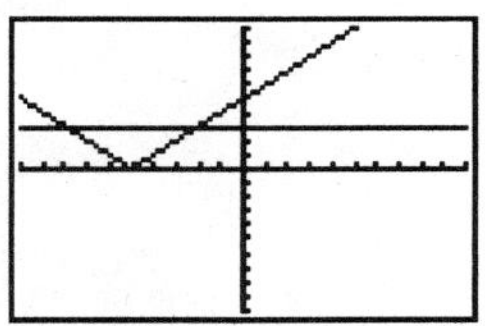

$|X + 5| > 3$

The interpretations of the solutions represented by the graphs above are all different. To summarize your results from this unit, explain how to interpret the solution represented by each graph.

a. Interpretation of $|X + 5| = 3$:

b. Interpretation of $|X + 5| < 3$:

c. Interpretation of $|X + 5| > 3$:

Solutions: **1.** $\{X \mid X \le -2 \text{ or } X \ge 3\}$,   , $(-\infty, -2] \cup [3, \infty)$

2. $\{X \mid -6 < X < 10\}$, $(-6, 10)$,

3. $\{X \mid -8.5 < X < 3.5\}$,   , $(-8.5, 3.5)$

4. You might not put parentheses around the numerator of the fractions or fail to enclose the entire fraction in parentheses when using the absolute value command.

5. $\mathbb{R}$ or $\{X \mid X \text{ is a real number}\}$, $(-\infty, \infty)$

6. null set

7. Remember that an absolute value is at least 0 or larger. Thus, an absolute value is always greater than a negative number for any value of the variable and an absolute is never less than any negative number regardless of the value of the variable.

8.   **9.**   **10.**   **11.**

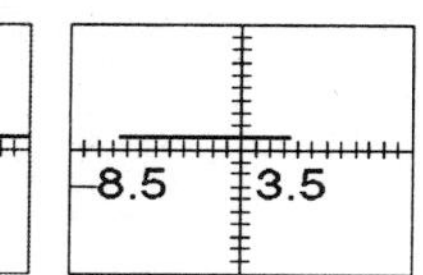
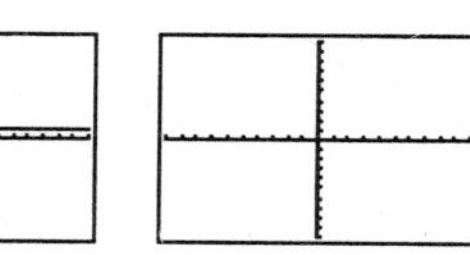

12. The graph displays all look the same because we are always entering the left side of the equation/inequality at Y1 and the right side at Y2. It is the interpretation of these graphs that yields the correct solution.

*Unit 9 is a prerequisite for this unit. Answers appear at the end of the unit.

This unit will graphically examine quadratic inequalities by using the ROOT/ZERO option of the calculator and by interpreting the relationship between the graphical displays of each side of the inequality. REMEMBER: To use the ROOT/ZERO there must be a zero on one side of the inequality.

SPECIAL CASES

The first four examples represent "special cases." They will be the quickest to solve of all the inequalities. All graphs will be displayed in the standard viewing window unless otherwise noted. Set this viewing window now.

Example 1: Graphically solve $2X^2 - X + 1 > 0$.

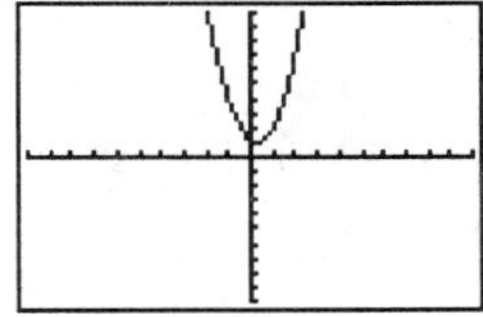

Solution: Let Y1 = represent the left side of the inequality and Y2 = the right side. We want to know <u>where</u> Y1 > Y2. Enter $2X^2 - X + 1$ at the Y1 = prompt and 0 at the Y2 = prompt. Press **[GRAPH]**. Since Y2 = 0 is the X-axis, we do not "see" it as a separate line. It is, however, graphed. Your display should correspond to the display at the right.

If the axes are turned "off" then the display clearly demonstrates that the X-axis and Y2 = 0 are the same line. To turn off the axes, access the **FORMAT** menu. Press **[2nd]** **<FORMAT>** and cursor down and right until **AxesOff** is highlighted. Press **[ENTER]** and then **[GRAPH]**.

 Press **[WINDOW]**, use the right cursor to highlight **FORMAT**, cursor down to **AxesOn**, right to **AxesOff**, and press **[ENTER]**. Now press **[GRAPH]**.

TI-85/86 TO ACCESS THE **FORMAT** MENU, PRESS **[GRAPH]** **[MORE]** **[F3](FORMAT)** AND CURSOR DOWN AND RIGHT TO HIGHLIGHT **AxesOff**. PRESS **[ENTER]** TO CHOOSE THIS OPTION. PRESS **[F5(GRAPH)]**.

From this point on, make a mental note that Y1 is always being compared to the X-axis and do not enter Y2 = 0. **Think** about where Y1 is greater than Y2 (i.e. the X-axis). It is greater than the X-axis where it is **above** the axis. Use your highlighter pen to highlight the portion(s) of Y1 that are above the X-axis. Since all portions of Y1 are greater than the X-axis and since any real number is an acceptable value for X, the solution set is $\mathbb{R}$ (the set of real numbers). ◆

Before proceeding, turn the axes back on.

Example 2: Solve the inequality $2X^2 - X + 1 < 0$ graphically.

Solution: Examine the graph displayed in Example 1. Where is $2X^2 - X + 1 < 0$ (i.e for what values of X is the graph below the X-axis)? Since the graph does not dip below the

X-axis, there are no X-values for which $2X^2 - X + 1 < 0$. The solution is the empty set, ϕ or { }. ◆

Example 3: Solve $X^2 + 4X + 4 \leq 0$ graphically.

Solution: The algebraic statement indicates that the trinomial is less than **OR** equal to zero. Remember, 0 is represented by the X-axis. Enter $X^2 + 4X + 4$ after Y1 and sketch the graph that is displayed. **TRACE** along the path of the curve.

a. At what point(s) is the graph of $X^2 + 4X + 4$ **LESS THAN** 0? While tracing, determine the X-values for which the corresponding Y-values are negative. Clearly, there are none.

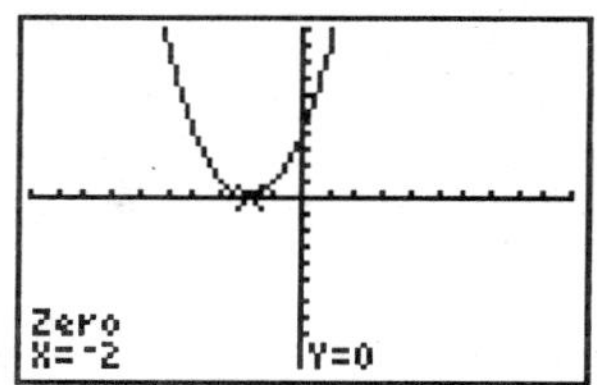

b. Use the ROOT/ZERO option to determine where the graph is **EQUAL TO** zero. Although this is an inequality, there is only **ONE** valid solution: X = -2.

◆

Consider the inequality $X^2 + 4X + 4 \geq 0$. You should be able to interpret the graphical display to determine that the solution set is all real numbers.
Next, consider the inequality $X^2 + 4X + 4 > 0$. Again, correct interpretation of the graphical display confirms the solution of all real numbers except for negative two.

GENERAL INEQUALITIES

In Example 3, $X^2 + 4X + 4 \leq$ 0, the critical point in the solution set was X = -2. At X = -2, $X^2 + 4X + 4 = 0$. Negative two is a root of the equation. When solving inequalities, the critical points (i.e. the roots of the corresponding equation) will be the endpoints of the interval(s) of the solution region(s). The following example illustrates the procedure for solving quadratic inequalities. The way the graphical display is <u>interpreted</u> determines the solution to a given equation or inequality.

The steps for finding the solution set of an inequality will be demonstrated in the next example.

Example 4: Specify the solution set and the number line graph of the solutions of the inequality $X^2 - X - 6 < 0$.

Solution:
    a. Be sure that the right hand side of the inequality is a ZERO.
    b. Enter the polynomial $X^2 - X - 6$ at the Y1 = prompt.
    c. On the display at the right sketch the graphical representation of the solution.
    d. Circle the X-intercepts (i.e. zeroes or roots) of the graph. These are the critical points. The circles will remain OPEN because the strict inequality symbol, "<", indicates that the critical points are not to be included as part of the solution set.
    e. Use the ROOT/ZERO option (under the **CALC** menu) to determine the values of the X-intercepts. Label the values of these two critical points on the display.

<table><tr><td>TI-85/86</td><td>PRESS [GRAPH] [MORE] [F1](MATH) TO LOCATE THE ROOT OPTION.</td></tr></table>

f. Use your highlighter pen to highlight the section of the graph of $X^2 - X - 6$
    that is **LESS THAN** 0 (i.e. below the X-axis).

g. The solution set is the set of all X values that yield the highlighted section of the
    graph. That is, $\{X \mid -2 < X < 3\}$.

Number line graph: 

Alternate Option for TI-82/83/83plus/86 Users: The TABLE feature of the calculator is
helpful in determining solution regions once the critical points have been calculated.

Set the table to start at a value of -2 (-2 was chosen because it is the critical point furthest
to the left) with increments of 1. Press **[2nd]** **<TblSet>** to access this menu. Access the
table by pressing **[2nd] <TABLE>**.

TI-86	PRESS **[TABLE]**, FOLLWED BY **[F2]** (TBLST). THE TABLE IS THEN ACCESSED BY PRESSING **[F1](TABLE)**.

It is clear that Y1<0 when X has values larger than -2 and smaller than 3. Scrolling to X
values smaller than -2 and larger than 3 confirms Y1 values that are positive (Y1 > 0) and
are **not** solutions to the inequality.

We are concerned with solutions to the inequality $X^2 - X - 6 < 0$ and have entered
$X^2 - X - 6$ at the Y1 = prompt. We now know that when X < -2, Y1 > 0 and when
X > 0, Y1 > 0. Therefore, these values should be shaded on the number line graph.

Use the steps outlined in Example 4 (or the TABLE) to solve
$X^2 - X - 6 \geq 0$. Sketch the graph display, label the critical points,
highlight the solution region(s), and interpret the solution.

The interpretation of the graphical display expressed in set notation is
$\{X \mid X \leq -2 \text{ or } X \geq 3\}$ and expressed as a number line graph is

.

EXERCISE SET

Use the steps outlined in a-g previously to graphically solve each of the following
inequalities. For each problem you **must** sketch the graphical display and label the critical
points. Record your solution in the following forms:

    a. solution set   b. number line graph   c. interval notation

1. $2X^2 - X - 10 \leq 0$

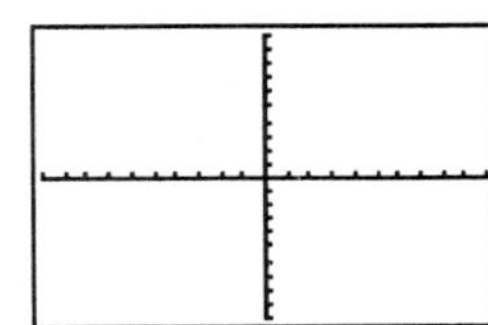

a. Solution Set:_________________________

b. Number line graph:

c. Interval Notation:____________________

2. $3X^2 + X - 4 \geq 0$ (Record critical points as fractions.)

   a. Solution Set:_______________________________

   b. Number line graph: ←————————————→

   c. Interval Notation:___________________

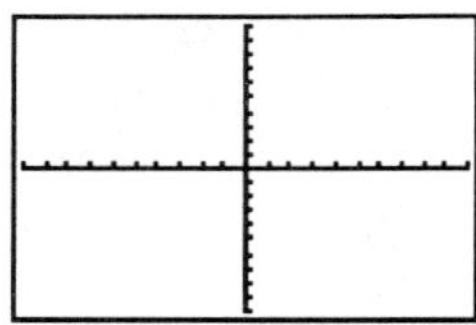

3. $3X^2 + X - 4 < 0$ (Record critical points as fractions)

   a. Solution Set:_______________________________

   b. Number line graph: ←————————————→

   c. Interval Notation:___________________

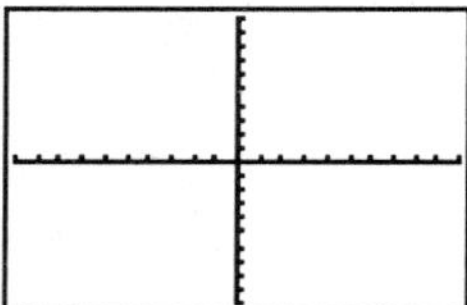

4. $X^2 + 10X + 25 \geq 0$

   a. Solution Set:_______________________________

   b. Number line graph: ←————————————→

   c. Interval Notation:___________________

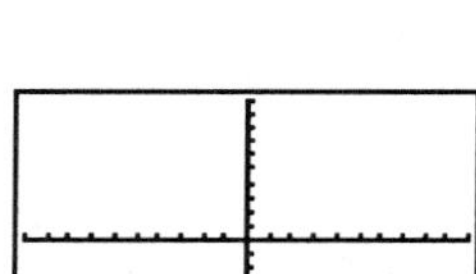

5. $X^2 + 10X + 25 > 0$

   a. Solution Set:_______________________________

   b. Number line graph: ←————————————→

   c. Interval Notation:___________________

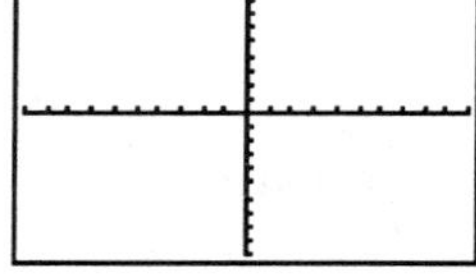

6. $4X^2 - 12X < -9$

   Solution Set:_______________________________

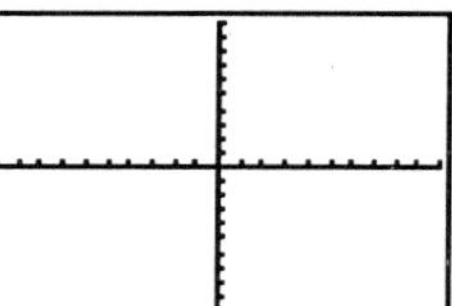

7. $4X^2 - 12X + 9 > 0$

   a. Solution Set:_______________________________

   b. Number line graph: ←————————————→

   c. Interval Notation:___________________

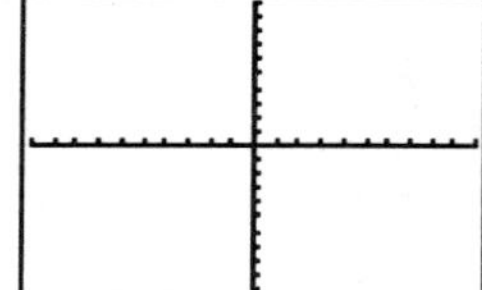

NOTE: Is the critical point actually 1.4999998 or 1.5? Attempting to convert to a fraction would seem to indicate that it is not 1.5. However, since the critical point(s) is the solution to the equation $4X^2-12X+9=0$, then the critical point should be the X value in the TABLE where Y1 = Y2. This can be accomplished in three ways:

a. Set the TABLE to start at X = 1.5 (the table increment is not critical since only one value is being checked). We want to know if Y1 = Y2 when X = 1.5. Press **[2nd]** <**TblSet**>, make this adjustment and then press **[2nd]** <**TABLE**>. At X = 1.5, Y1 = Y2. Thus X = 1.5 is the exact critical point; X = 1.4999998 is an approximation.

b. Use VALUE (**EVAL X**) to determine if Y1 = 0 when X = 1.5

When using the graph screen to solve equations/inequalities, you should be aware that the display coordinate values approximate the actual mathematical coordinates. The accuracy of these display values is determined by the height and width of the pixel space being displayed. The space height/width formulas are discussed in detail in the unit entitled "Preparing to Graph: Calculator Viewing Windows.

8. Explain the similarities and differences between graphically solving equations and inequalities.

9. Explain the significance of the critical points.

<u>**Solutions:**</u> **1.** $\{X \mid -2 \leq X \leq 2.5\}$, , [-2, 2.5]

2. $\{X \mid X \leq -4/3 \text{ or } X \geq 1\}$, , $(-\infty, -4/3] \cup [1, \infty)$

3. $\{X \mid -4/3 < X < 1\}$, , $(-4/3, 1)$

4. $\mathbb{R}$, , $(-\infty, \infty)$ **5.** $\{X \mid X \neq -5\}$, , $(-\infty, -5) \cup (-5, \infty)$

6. Null Set   **7.** $\{X \mid X \neq 1.5\}$, , $(-\infty, 1.5) \cup (1.5, \infty)$

8. Answers may vary.   **9.** Answers may vary.

* Unit 14 is a prerequisite for the unit. Answers appear at the end of the unit.

The previous unit examined solving quadratic inequalities. To solve <u>rational</u> nonlinear inequalities we will use some of the same procedures from the previous unit and investigate necessary modifications.

The following steps were used in solving quadratic inequalities.

SOLVING QUADRATIC INEQUALITIES GRAPHICALLY

a. The inequality was written with one side (usually the left) greater than/less than 0.

b. The left side of the inequality was entered on the calculator at the Y1 = prompt. The standard viewing screen was used to view the graph.

c. Because the right side of the inequality was equal to zero, the X-axis was used as a reference.

d. Once the graph was displayed, the critical points were found (by using the ROOT/ZERO option under the **CALC** menu). Recall these were the values of X that made the equation associated with the given inequality a true statement. Graphically, they were the point(s) where the graph crossed the X-axis, the x-intercepts.

e. The critical points were graphed on a number line. They were enclosed with an open circle if the original inequality was a <u>strict</u> inequality ($>$ or $<$) and by a closed circle if equality was included ($\geq$ or $\leq$).

f. If the inequality was $>$, the solutions were the X-values corresponding to points <u>above</u> the X-axis. Conversely, if the inequality was $<$, solutions were the X-values corresponding to points <u>below</u> the X-axis. Appropriate regions were shaded on the number line graph.

Example 1: Graphically solve $\dfrac{12}{X} \geq 3$.

Solution: To solve $\dfrac{12}{X} \geq 3$, first rewrite the inequality as

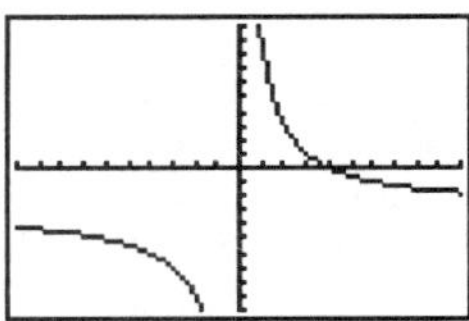

$\dfrac{12}{X} - 3 \geq 0$ (comparing an expression to zero allows the use of the

X-axis as a reference point.) Enter $\dfrac{12}{X} - 3$ at the Y1 = prompt and

view the graph in the standard viewing window. Your display should match the one above.

Find the critical point(s). First, circle the point where the graph crosses the X-axis on the display. Now use the ROOT/ZERO option of the calculator to find the critical point.

To confirm that 4 is a root (critical point), access the TABLE (first set the table to begin at - 1 and increment by 1). When X = 4, Y1 has a value of 0. This confirms that 4 is the root (and thus a critical point). However, notice that when X = 0 that Y1 = ERROR. There is an ERROR message for this value of X. Because the fraction $\frac{12}{X}$ is undefined when X = 0, this value for X IS NOT part of the solution. This is an excluded (or restricted) value.

    a. Locate the excluded value (0) and the root (4) on the graph.

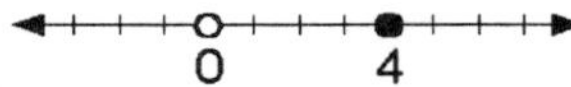

Notice that the coordinate 0 is marked with an open circle, while the coordinate 4 is marked with a closed circle. These are different because 4 is a solution to the inequality, whereas the value 0 cannot be a solution to the inequality because it results in an expression that is undefined.

IT IS IMPERATIVE THAT ALL EXCLUDED VALUES (RESTRICTED VALUES) ARE DETERMINED **BEFORE** GRAPHING. Recall, these are the values that make <u>any</u> denominator equal 0. They must be located on the number line and are <u>ALWAYS</u> marked by an open circle as they are <u>NEVER</u> a part of the solution.

    b. Access the graph and **shade** the part(s) of the pictured number line that correspond to those points of the graph that are <u>above</u> the X-axis.

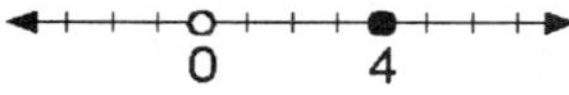

Your graph should correspond to $\{X \mid 0 < X \le 4\}$ which is expressed as (0,4] in interval notation.

    c. Accessing the TABLE feature allows us to check the solution. Notice that for values of X SMALLER than 0, the Y1 values are negative, and are <u>not</u> a part of the desired solution. When X is GREATER than 0 but LESS than 4 the corresponding Y1 values are positive, and thus <u>are</u> solutions to $\frac{12}{X} - 3 \ge 0$. Scrolling past 4 (to X values GREATER than 4) the corresponding Y1 values are negative, and are <u>not</u> a part of the desired solution.

◆

Example 2: Graphically solve the inequality $\dfrac{-8}{X-4} \le \dfrac{5}{4-X}$.

Solution: a. Rewrite the inequality so that one side equals zero: $\dfrac{-8}{X-4} - \dfrac{5}{4-X} \le 0$

Enter the left side at the Y1= prompt and press **[GRAPH]**. Your display should match the one at the right.

    b. Locate the <u>excluded</u> value of 4 by setting denominator factors equal to 0 and solving for X.

    c. Now find any critical point(s). These are the point(s) where the graph crosses the X-axis. Use the ROOT/ZERO option to **try** to find the root(s).

d. To better see what is going on, put the calculator in **DOT** mode. To do this, press **[MODE]** and cursor down to **Connected** and then right to **Dot**. Press **[ENTER]** to highlight this mode. Press **[GRAPH]**. Your display should correspond to the display at the right.

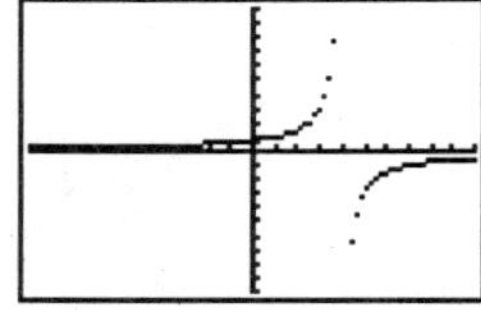

The TI-83/83plus allow the personalization of graphs. Press **[Y=]** and use the arrow key to place the cursor over the graphing icon to the left of Y1. Press **[ENTER]** while observing the change in the graphing icon. When the diagonal pattern of "dots" slanting left to right is displayed, stop. Press **[GRAPH]** and compare your display to the one pictured. If you initially changed **MODE** to DOT then the calculator automatically set the graphing icon to **dot**. TI-83/83plus and TI-86 users should be aware that the graphing icon will override the CONNECTED/DOT **MODE**. Thus, if the **MODE** screen is set to DOT but the graphing icon is set to **solid** then all points will be connected when graphed.

PRESS **[GRAPH] [MORE] [F3](FORMAT)**. CURSOR DOWN AND OVER TO HIGHLIGHT **DRAWDOT** AND PRESS **[ENTER]**.

THIS CALCULATOR ALLOWS THE PERSONALIZATION OF GRAPHS. BE SURE THE GRAPH MENU IS DISPLAYED AT THE BOTTOM OF YOUR SCREEN.   PRESS **[F1](Y(x)=) [MORE] [F3](STYLE)**. CONTINUE PRESSING (STYLE) UNTIL THE DIAGONAL PATTERN OF "DOTS" SLANTING LEFT TO RIGHT IS DISPLAYED. PRESS **[M5](GRAPH)** AND COMPARE YOUR DISPLAY TO THE ONE PICTURED. IF YOU INITIALLY CHANGED **MODE** TO DOT THEN THE CALCULATOR AUTOMATICALLY SET THE GRAPHING ICON TO **DOT**. TI-86 USERS SHOULD BE AWARE THAT THE GRAPHING ICON WILL OVERRIDE THE CONNECTED/DOT **MODE**. THUS, IF THE **MODE** SCREEN IS SET TO **DOT** BUT THE GRAPHING ICON IS SET TO **SOLID** THEN ALL POINTS WILL BE CONNECTED WHEN GRAPHED.

When comparing the two graphical displays (connected mode vs. dot), notice the vertical line (where we believed there was a root) is gone. In CONNECTED MODE this line connected two adjacent pixel points on the graph. However, in DOT MODE it is clear that these two points are at opposite ends of the graph and should not be connected. To connect them would mean that 4 is an acceptable value for X.

The graph <u>never</u> crosses the X-axis but rather jumps from a location above the X-axis to one below it. **TRACE** and observe the X-values to confirm this.

e. Recall, we want the solutions to the inequality $\dfrac{-8}{X-4} - \dfrac{5}{4-X} \le 0$. Place 4 on the number line and circle it (an excluded value) and shade to the right:

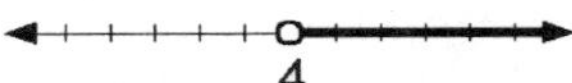

If you are unsure as to why the solution shades to the right, use the VALUE (Eval) feature of your calculator to test points to the right and left of the critical point of 4.

The solution set is $\{X \mid X > 4\}$ or $(4, \infty)$ in interval notation.

f. Accessing the TABLE feature confirms the fact that values of Y1 are positive for X less than 4 and negative for X greater than 4, thus validating the solution above.   ◆

The following steps outline the method of finding the solution(s) to the inequality .

SOLVING RATIONAL INEQUALITIES GRAPHICALLY

a. Rewrite the inequality with one side equal to 0. Enter the non-zero side at the Y1 = prompt. Remember, the calculator should be in DOT MODE or the dot style icon should be activated.

b. Find the excluded values by setting each denominator above equal to 0 and solving for the X values.

Enter these values on the number line and mark them with <u>open circles</u> so that they are not inadvertently included in the solution.

c. Find the roots/zeroes of the equation associated with the inequality by using the ROOT/ZERO option under the **CALC** menu OR by solving the equation associated with the inequality, i.e. change the inequality symbol to an equal sign.

Enter the roots on the number line and mark them with closed circles if the inequality is $\leq$ or $\geq$, and with open circles if the inequality is $<$ or $>$.

d. Shade the regions on the number line that represent the x-values of the ordered pairs on the graph where the y-values are greater than/less than 0 as determined by the inequality. You may also determine the appropriate x-values by testing values in each region of the number line. You can accomplish the same thing by using the TABLE.

e. Write the solution as a solution set and in interval notation.

f. Reset the calculator to CONNECTED MODE and compare the number line graph to the graphical display on the calculator screen. **TRACE** and observe X and Y values. Confirm that the solution above is correct. Notice that when TRACING left the Y values jump from negative values to positive values as the screen scrolls.

EXERCISE SET

Directions: Use the steps outlined above to solve each inequality below. Begin by setting the calculator in Connected MODE with a standard viewing WINDOW. Remember to access the TABLE feature (when appropriate) and convert to dot mode as **YOU** deem appropriate. Use the combination of algebra and calculator that makes you feel comfortable.

1. $\dfrac{X^2 + 6X + 9}{X + 5} > 0$

    a. Number line graph: $\longleftarrow\qquad\longrightarrow$

    b. Solution set: ___________________

    c. Interval notation: ___________________

2. $\dfrac{2}{X-2} < \dfrac{3}{X}$

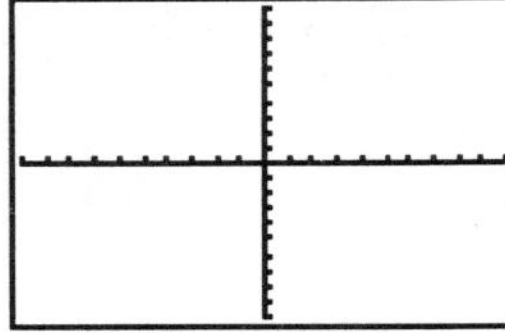

    a. Number line graph:

    b. Solution set: _______________________

    c. Interval notation: _______________________

3. $\dfrac{(2X-1)(X-5)}{X+3} \leq 0$

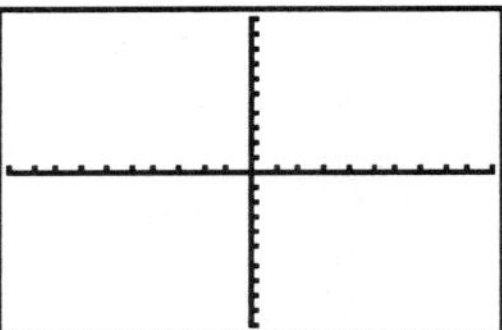

    a. Number line graph:

    b. Solution set: _______________________

    c. Interval notation: _______________________

Note: In #3, did you shade <u>only</u> between the roots of 1/2 and 5? This problem illustrates an important point: ALWAYS look to the left (or right) of the restricted values and/or critical points to see how the graph behaves.

Set the calculator in Connected MODE. TRACE left on the graph of the above expression and go beyond the vertical line that connects the two non-adjacent pixel points. You will not see any more graph (TRACE until your graph shifts <u>at least</u> once). Look at the X and Y coordinates at the bottom of the screen, and then check the window values. We will now scroll down to Ymin and enter -50. Press **[GRAPH]** and compare your display to the one pictured.

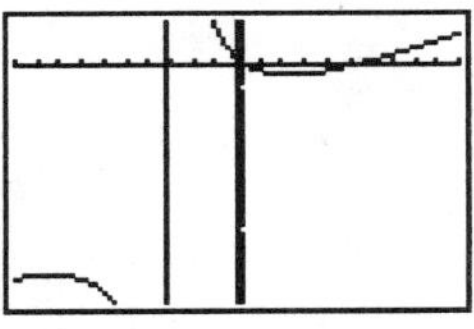

NOW you should understand why the region of the number line less than -3 is shaded.

✍4. Explain what excluded values are, how to find them, and their inclusion or non-inclusion in solutions.

✍5. Discuss the advantages and disadvantages of looking at solutions of non-linear rational inequalities in both Dot and Connected MODE.

Solutions: **1.** $\{X \mid -5 < X < -3 \text{ or } X > -3\}$ $(-5,-3) \cup (-3,\infty)$

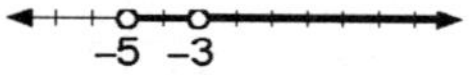

2. $\{X \mid 0 < X < 2 \text{ or } X > 6\}$ $(0,2) \cup (6,\infty)$

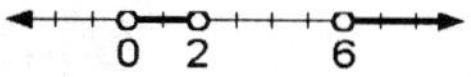

3. $\{X \mid X < -3 \text{ or } \tfrac{1}{2} \leq X \leq 5\}$ $(-\infty,-3) \cup [\tfrac{1}{2},5]$

4. Answers may vary.   **5.** Answers may vary.

CORRELATION CHART

UNITS * Instructors are encouraged to use Units marked by an * to <u>introduce</u> concepts.	PRE-REQ. UNIT(S)	CORRELATING CONCEPT
#16: How Does the Calculator Actually Graph?, p.107	#3	This unit correlates calculator graphing with hand drawn graphs.
#17: Preparing to Graph: Calculator Viewing Windows, p.111	#16	This unit explores pre-set viewing windows on the calculator.
#18 Where Did the Graph Go? , p.123	#17	Adjusting the viewing window to fit your graph.
#19 Functions, p.127	#17	An exploration of functions: notations, domain, range, inverses and evaluation.
#20 Discovering Parabolas, p.135 *	#17	An exploration of the graphs of quadratic functions & the effects of constant values on graphs.
#21 Translating and Stretching Graphs, p.141	#17	Examines the effect of constants on families of graphs
#22 Symmetry of Functions, p.147	#17	Examines the concept of symmetry about the origin and y-axis.
#23 Piecewise Functions, p.153	#18	Uses the TEST menu to graph functions with restricted domains
#24 Rational Functions, p.159	#17	Links algebraic techniques and graphical displays in determining asymptotes
#25 The Algebra of Functions, p.167	#19	Four arithmetic operations of functions and composition are graphically illustrated.
#26 Exponential and Logarithmic Functions, p.171	#19	The graphs and relationships between the exponential function and its inverse are examined
#27 Parabolas Revisited, p. 179	#20	Examination of the graphs of parabolas that are not functions .

Unit Title	TI-82 Keys	TI-83/83plus Keys	TI-85/86 Keys

#16: How Does the Calculator Actually Graph, p.107	GRAPH
	ZOOM
	  6:ZStandard
	WINDOW
		GRAPH
		ZOOM
		  6:ZStandard
		WINDOW
			GRAPH
			GRAPH/ZOOM
			  ZSTD
			RANGE/WIND

#17: Preparing to Graph: Calculator Viewing Windows, p.111	ZOOM
	  4:ZDecimal
	  5:ZSquare
	  8:ZInteger
		ZOOM
		  4:ZDecimal
		  5:ZSquare
		  8:ZInteger
		Graphing Icon
			GRAPH/ZOOM
			  ZDECM
			  ZSQR
			  ZINT
			Graph Icon - 86
			GRAPH/SELCT

#18: Where Did the Graph Go?, p.123	No new keys	No new keys	No new keys

#19: Functions, p.127	DRAW
	  1:ClrDraw
	  3:Horizontal
	  4:Vertical
	  8:DrawInv
	VARS
		DRAW
		  1:ClrDraw
		  3:Horizontal
		  4:Vertical
		  8:DrawInv
		VARS
			GRAPH/MORE
			DRAW
			  CLDRW
			
			  VERT
			  DrInv

#20: Discovering Parabolas, p.135	No new keys	No new keys	No new keys
#21: Translating & Stretching Graphs, p.141	No new keys	No new keys	No new keys
#22: Symmetry of Func. p.147	No new keys	No new keys	No new keys
#23: Piecewise Functions p.153	No new keys	No new keys	No new keys
#24: Rational Functions, p.159	No new keys	No new keys	No new keys
#25: The Algebra of Functions, p.167	No new keys	No new keys	No new keys
#26: Exponential and Logarithmic Functions, p.171	No new keys	No new keys	No new keys
#27: Parabolas Revisited, p.179	No new keys	No new keys	No new keys

UNIT 16
HOW DOES THE CALCULATOR ACTUALLY GRAPH?
(EXPLORING POINTS AND PIXELS)

*Unit 3 is a prerequisite for this unit.

Consider the first degree polynomial expression $\frac{3}{4}X + 6$. Use the STOre feature of your graphing calculator to evaluate the polynomial for each of the given values of X:

X	-8	-4	0	4	8
(3/4)X + 6					

The STOre feature is used to evaluate polynomials for given values of the variable(s). The y = edit screen can quickly accomplish the same task if the polynomial is a one variable polynomial. For calculators with no table, the STOre feature allows the user to generate a table of values. TI-82/83/83plus/86 calculators have a TABLE feature that allows the user to store a polynomial in one variable, and evaluate that polynomial for multiple values. Press [Y=] and enter (store) the expression (3/4)X + 6 after the Y1= prompt. Press [2nd] <TblSet> and set the table to start at -8 to correspond with the chart constructed above.   You may set the table increment (ΔTbl) to 4 or you may leave it at 1 and simply cursor down through the table. Pressing [2nd] <TABLE> will now display the table at the right.

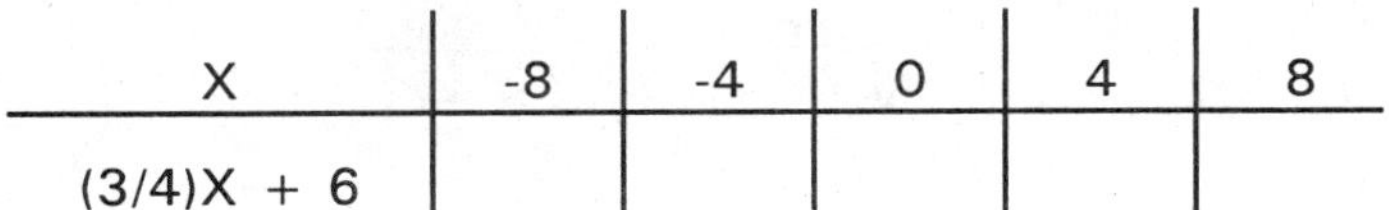

TI-86 PRESS [GRAPH] [F1](Y(X)=) FOR THE Y = EDIT SCREEN AND ENTER THE GIVEN POLYNOMIAL. PRESS [TABLE] [F1](TBLST) TO SET THE INITIAL VALUE OF -8 AND TO INCREMENT THE TABLE. PRESSING [F2](TABLE) WILL DISPLAY THE TABLE..

The expression stored at Y1 has been evaluated for each value of the variable X thus creating ordered pair solutions (x,y) to the equation $Y = \frac{3}{4}X + 6$. Plot each of the ordered

pairs from chart/table above on the graph at the right. Use a ruler to connect the points to graph the line of the equation Y = (3/4)X + 6.

Compare the equation to the calculator graph by pressing **[ZOOM] [6:ZStandard]**. This displays the graph in the standard viewing window. The standard viewing window sets the maximum value of each axis at 10 and the minimum value at -10. Pressing **[WINDOW]** will allow you to see the values of the viewing window.

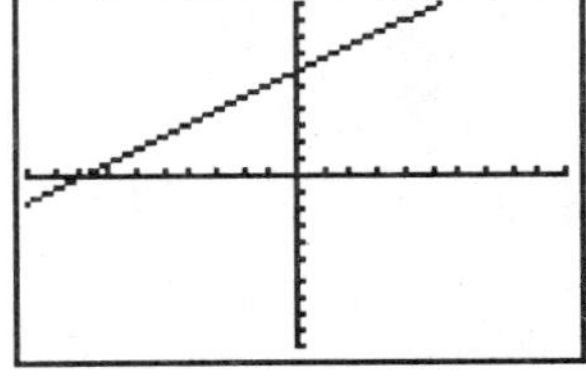

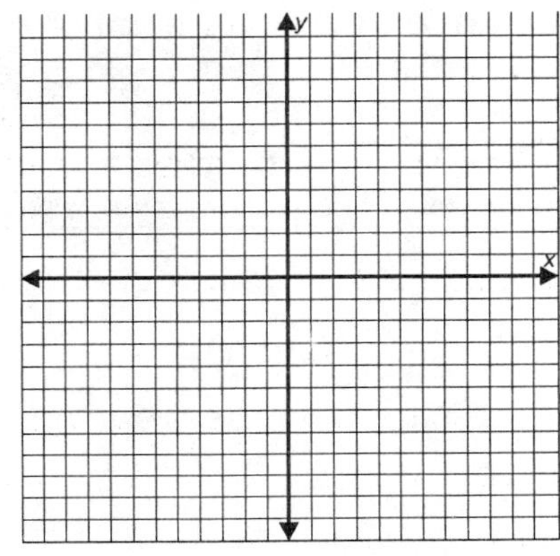

TI-85/86 PRESS [GRAPH] [F3](ZOOM) [F4](ZSTD) TO SET THE STANDARD VIEWING WINDOW. YOU CAN PRESS [2ND] <M2> (RANGE/WIND) TO SEE THE WINDOW VALUES.

Since the calculator drawn graph is not a *straight* line as expected, we will construct a graph in the same manner as the calculator to understand what is happening.

Pretend the graph at the right is the calculator screen and that each box represents a pixel space. "Light up" the X-axis and Y-axis by <u>lightly</u> shading a horizontal line of boxes and a vertical line of boxes where the two axes should be located.

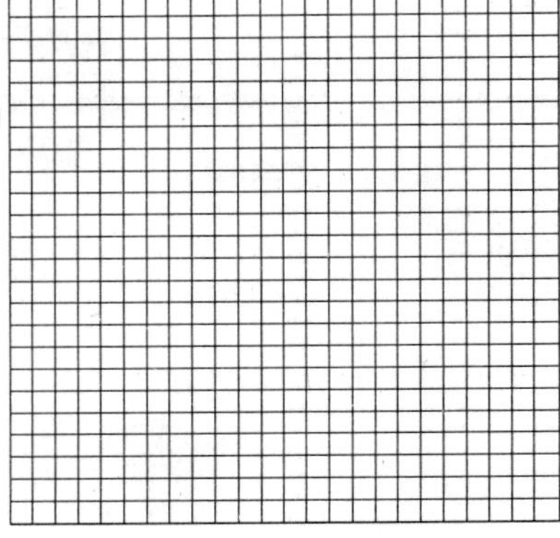

Now *plot* the X-intercept of (-8,0) and the Y-intercept of (0,6) by <u>darkly</u> shading the appropriate pixel space (box).

Place your ruler so that it connects these two pixels. Now <u>darkly</u> shade in a path of pixels along the edge of your ruler. Remember, the entire square representing the pixel must be shaded.

The graphs below show the comparison between your hand drawn graph using the ordered pairs, the actual calculator display and your pixel sketch.

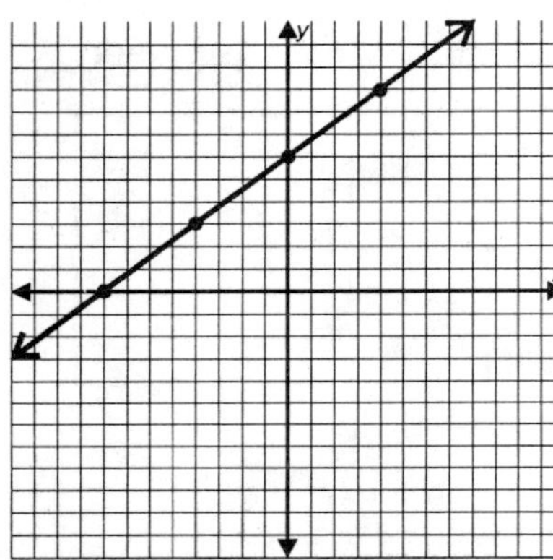

Hand Drawn Graph

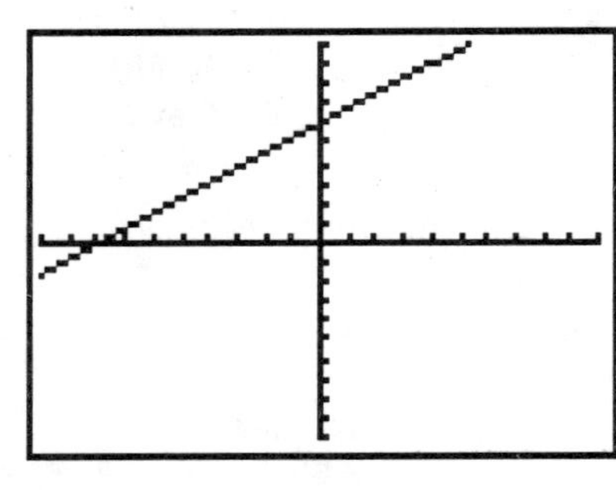

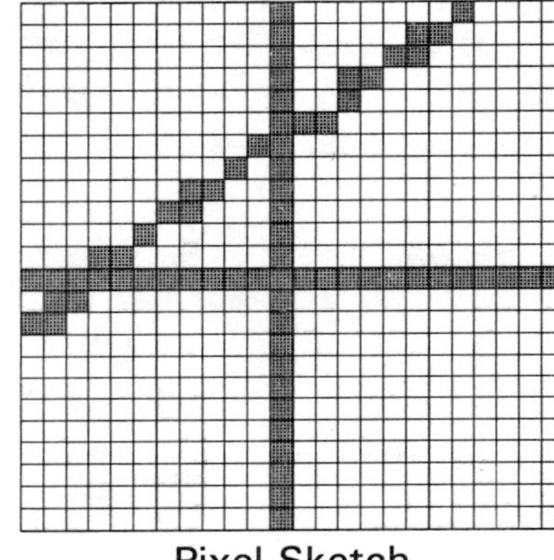

Pixel Sketch

The calculator plots points by lighting up little squares on the screen called pixels. The TI-82/83/83plus screen is 95 pixel points wide (with 94 spaces between the horizontal pixel points) by 63 pixel points high (with 62 spaces between the vertical pixel points). Because there are only a finite number of pixel spaces to light up, the calculator may only be able to "light up" a pixel that is <u>close</u> to the desired point.

TI-85/86	THE TI-85/86 SCREEN IS 127 PIXEL POINTS WIDE (WITH 126 SPACES BETWEEN HORIZONTAL PIXEL POINTS) BY 63 PIXEL POINTS HIGH AS IN THE TI-82/83 SCREEN.

Now that you have seen how the calculator must light up pixels to graph a straight line, we will examine what happens to curves.

A semi-circle with a radius of 5 units has been drawn on the graph at the right. Shade in the squares along the path of the semi-circle to simulate the calculator "lighting up" pixel spaces. The pixels representing the X and Y axes have already been shaded for you.

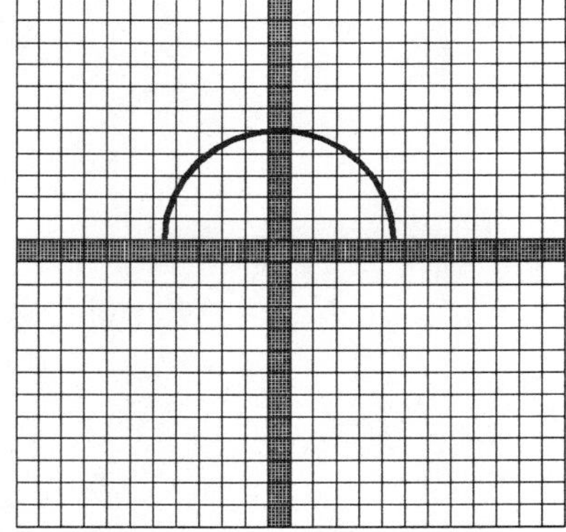

Now graph a semicircle on the calculator by entering the equation Y1 $= \sqrt{25 - X^2}$ on the **Y =** screen. Press [**Y =**] and be sure you enter the expression as: $\sqrt{(25 - X^2)}$. Press [**GRAPH**].

TI-85/86	PRESS [**GRAPH**] [**F1**](Y(x) =) TO ENTER $\sqrt{(100 - X^2)}$ AND TO DISPLAY THE GRAPH, PRESS [**2ND**] <M5> (**GRAPH**).

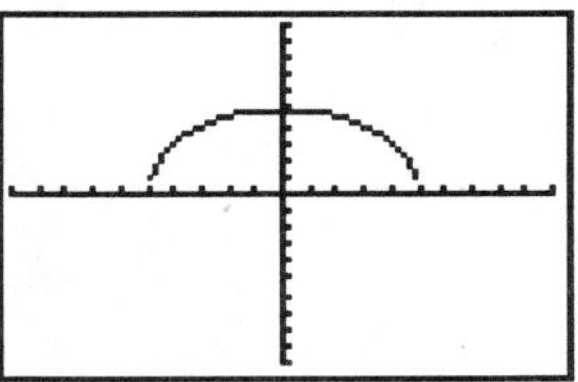

Your graph should look like the one displayed at the right.

Notice how flat the top of the semi-circle appears. All calculator drawn lines and curves will consist of a pattern of boxes, and vertical or horizontal line segments. The **WINDOW** values selected will affect the appearance of the line or curve.

*Unit 16 is a prerequisite for this unit. Answers appear at the end of the unit.

TI-85/86	IF USING THE TI-85/86, GO TO THE GUIDELINES (PG.119).

SETTING UP THE GRAPH DISPLAY

The calculator's display is controlled through the **MODE** and **FORMAT** screens.

Press the **[MODE]** key. **MODE** controls how numbers and graphs are displayed and interpreted. The current settings on each row should be highlighted as displayed. The blinking rectangle can be moved using the 4 **cursor** (arrow) keys. To change the setting on a particular row, move the blinking rectangle to the desired setting and press **[ENTER]**.

NOTE: Items must be highlighted to be activated.

Normal vs. Scientific notation
Floating decimal vs. Fixed to 9 places
Type of angle measurement
Type of graphing: function, parametric, polar, sequence
Graphed points connected or dotted
Functions graphed one by one
Numbers can be viewed as as Real or complex
Screen can be split to view two screens simultaneously

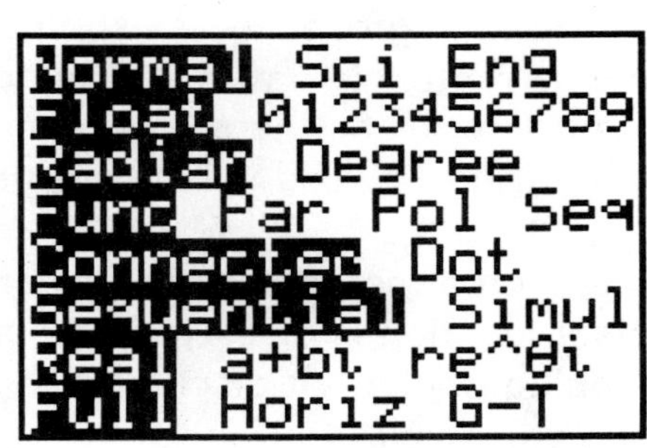

TI-82	The first six options are the same as the TI-83/83plus. The TI-82 will not compute with complex numbers.	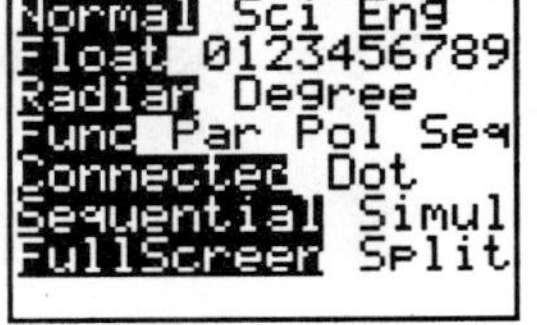

To return to the home screen, at this point, press **[CLEAR]** or **[2nd]** **<QUIT>**.

Press **[Y=]**. The calculator can graph up to 10 different equations at the same time. Because **MODE** is in the sequential setting, the graphs will be displayed sequentially. Note that cursoring down accesses additional Y= prompts. The display of the equations entered on the **Y=** screen is controlled by the size of the viewing window. The dimensions of the viewing window are determined by the values entered on the **WINDOW** screen.

    The TI-83/83plus/86 calculators have a feature called the *graph style icon* that allows you to distinguish between the graphs of equations. To change styles (on the TI-83/83plus), press **[Y=]** and observe the "\" in front of each Y. Use the left arrow to

cursor over to this "\." The diagonal should now be moving up and down. Pressing
[ENTER] once changes the diagonal from thin to thick. The graph of Y1 will now be
displayed as a thick line. Repeatedly pressing **[ENTER]** displays the following:

- ◥: shades above Y1
- ◣: shades below Y1
- -o: traces the leading edge of the graph followed by the graph
- o: traces the path but does not plot
- ∴ : displays graph in dot, not connected MODE

       It is important to remember that the graphing icon takes precedence over the **MODE**
screen. If the icon is set for a solid line and the **MODE** screen is set for DOT and not solid,
the graphing icon will determine how the graph is displayed. Pressing **[CLEAR]** to delete an
entry at the Y= prompt will automatically reset the graphing icon to default, a solid line.

Press **[ZOOM] [6:ZStandard] [WINDOW]**. This is called the standard
viewing window. The information on this screen indicates that in a
rectangular coordinate system the X-values will range from -10 to 10
and the Y-values will range from -10 to 10. The interval notation for
this is [-10,10] by [-10,10]. The Xscl= 1 and Yscl= 1 settings
indicate that the tic marks on the axes are one unit apart. The values
entered on this screen may be changed by using the cursor arrows to move to the desired
line and typing over the existing entry. When drawing a graph, you may set the desired
viewing rectangle on the calculator as well as scale the X-axis and Y-axis. The row labeled
Xres=determines the screen resolution. It should be set equal to 1, which means that each
pixel on the   X-axis will be evaluated and graphed.

Graphs on a rectangular coordinate system
Cursor location is displayed on screen
Graphing grid is not displayed
Axes are visible
Axes are not labeled with an X and Y
The expression entered at Y1 is displayed on the graph
when trace is activated.

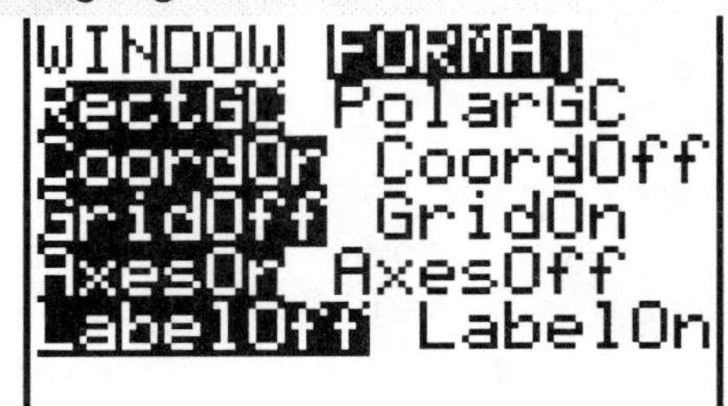

(**TI-82**) Be sure the cursor is on the word **WINDOW** and use the right arrow key to cursor
over to **FORMAT**. The following settings should be highlighted:

Graphs on a rectangular coordinate system
Cursor location is displayed on screen
Graphing grid is not displayed
Axes are visible
Axes are not labeled with an X and Y

Directions: Before proceeding further, press [Y=] and clear all entries.

1. Press **[WINDOW]** and enter Xmin = -5, Xmax = 5, Xscl = 1,
    Ymin = -12, Ymax = 7, Yscl = 1. Be sure to use the gray [(-)]
    key for negative signs. Press **[GRAPH]** to view the coordinate
    axes. Count the tic marks on the axes and see how these
    marks correspond to the max and min values. Label the last tic
    mark on each axis (i.e. farthest tic mark left, right, up and
    down) with the appropriate integral value.

TI-85/86	TI-85/86 USERS MUST PRESS [CLEAR] TO DELETE MENU DISPLAY BEFORE COUNTING TIC MARKS.

2. Change the viewing window to Xmin = -20, Xmax = 70, Xscl = 10, Ymin = -5,
    Ymax = 15, Yscl = 3 and press **[GRAPH]** to view the axes. How many tic marks
    are on the positive portion of the X-axis?_____   How many units does each of the tic
    marks on this axis represent?_____   Based on your last two answers, how many units
    long is the positive portion of the X-axis?_____   Does this number correspond to the
    Xmax value given in the problem? _______

    In your own words, explain what is happening.

3. To help "de-bug" errors in graphing set ups later on, describe what you think would
    happen if Xmin = 10 and Xmax = -5. You might want to draw your own set of
    coordinate axes and <u>try</u> to label them in this manner. Enter these values on the
    WINDOW screen and press **[GRAPH]**. What did happen?

4. Reset Xmin = -10, Xmax = 10 and describe what you think will happen if you set
    Ymin = 5, Ymax = 5. Again, enter the values and press **[GRAPH]**. What did
    happen? (Try drawing your own set of axes and labeling them as indicated.)

5. What should the relationship between Max and Min be? (i.e. Min > Max,
    Min < Max, or Min = Max)

Reset the viewing window to ZStandard, by pressing **[ZOOM] [6:ZStandard]**.

TI-85/86	TI-85/86 USERS PRESS **[GRAPH] [F3](ZOOM) [F4](ZSTD)**.

Press **[Y=]**. On the screen Y1= is followed by a blinking cursor. Anything else can be cleared by pressing **[CLEAR]**. Enter Y = -2X+ 6, and press **[ENTER]**. The cursor is now on the second line following Y2=. At this prompt, enter the equation Y = (1/2)X - 4. Note that the equal signs beside both Y1 and Y2 are highlighted. This means that both equations will be graphed. Press **[GRAPH]** to display the graph screen. See display.

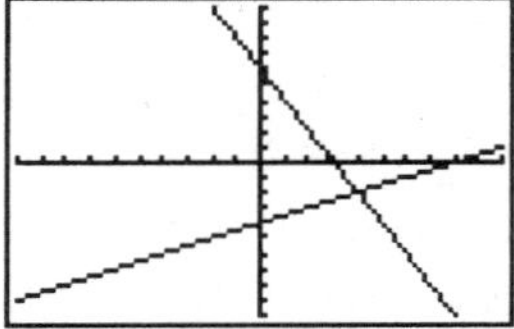

Note: Y = (1/2)X - 4 would be displayed in your text as $y = \dfrac{1}{2}x - 4$.

GRAPHING

To graph Y = -2X + 6 only, press **[Y=]** and use the arrow key to move the cursor over the equal sign beside Y2. Press **[ENTER]**. Notice that the equal sign beside Y2 is *not* highlighted, whereas the equal sign beside Y1 *is* highlighted. Press **[GRAPH]**; only the highlighted equation, Y1, is graphed.

TI-85/86	TO GRAPH Y = -2X + 6 ONLY, PRESS **[F1](Y(X)=)**, PLACE THE CURSOR ON THE Y2 EQUATION AND PRESS **[F5](SELCT)**. THE EQUAL SIGN IS NO LONGER HIGHLIGHTED, INDICATING THAT THE GRAPH OF Y2 WILL NOT BE DISPLAYED. THE Y2 EQUATION CAN BE RESELECTED FOR GRAPHING BY PLACING THE CURSOR ON THE EQUATION AND PRESSING **[F5](SELCT)** AGAIN.

On the viewing screen at right, the calculator draws a set of axes whose minimum and maximum values and scale match the choices under **WINDOW**. The graph of Y1 is drawn from left to right. Return to the Y= menu by pressing **[Y=]**. Cursor down to Y2 and *turn on* this graph by highlighting the equal sign. Press **[GRAPH]** and notice that the two graphs are drawn in sequence. **SEQUENTIAL** was chosen from the **MODE** menu earlier.

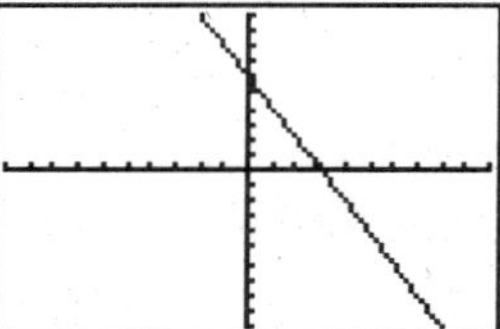

TI-85/86	TI-85/86 USERS GO TO THE GUIDELINES (PG.120), AND READ THE SECTION ENTITLED ALTERING THE VIEWING WINDOW.

ALTERING THE VIEWING WINDOW

The last unit addressed the size of the calculator screen (viewing window). Because the screen is 95 pixel points wide by 63 pixel points high, there are 94 horizontal spaces and 62 vertical spaces to light up. When tracing on a graph, the readout changes according to the size of the space. The size of the space can be controlled by the following formulas:

$$\frac{Xmax - Xmin}{94} = \text{horizontal space width}, \quad \frac{Ymax - Ymin}{62} = \text{vertical space height}.$$

We will examine some preset viewing windows and how they affect the pixel space size.

Press **[ZOOM]**. There are ten entries on this screen. (The TI-82 has nine entries.) The down arrow key can be used to view remaining entries.

1: Boxes in and enlarges a designated area.
2: Acts like a telephoto lens and "zooms in."
3: Acts like a wide-angle lens and "zooms out."
4: Cursor moves are ONE tenth of a unit per move.
5: "Squares up" the previously used viewing window.
6: Sets axes to [-10,10] by [-10,10].
7: Used for graphing trigonometric functions.
8: Cursor moves are ONE integer unit per move.
9: Used when graphing statistics.
0: Replots function, recalculating Ymin and Ymax.

ZDecimal is useful for graphs that require the use of the calculator's TRACE feature. Applying the horizontal space width formula,

$$\frac{Xmax - Xmin}{94} = \frac{4.7 - (-4.7)}{94} = 0.1,$$ changes the X-values by one-tenth of a unit each time the cursor is moved. This is why this screen yields "friendly" values when TRACING. (In general, Xmax - Xmin needs to be a multiple of 94 to produce a "friendly" screen.)

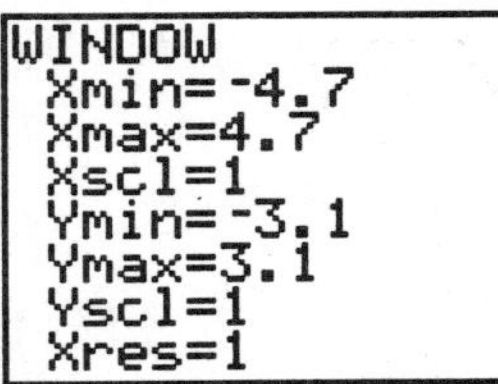

| **TI-85/86** | TI-85/86 USERS ARE REMINDED THAT THE DENOMINATOR OF THE HORIZONTAL SPACE WIDTH FORMULA SHOULD BE 126. IN GENERAL, xMAX - xMIN NEEDS TO BE A MULTIPLE OF 126 TO PRODUCE A "FRIENDLY" SCREEN. |

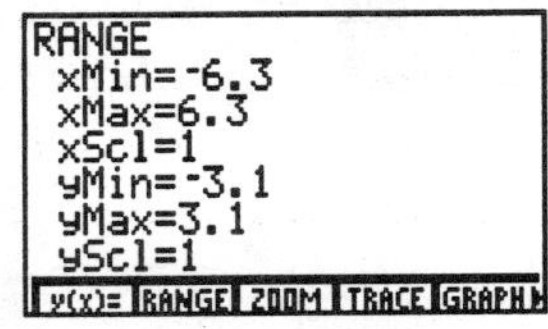

Enter Y1 = -2X + 3 and Y2 = ½X - 2 on the Y= screen. Press **[ZOOM] [4:ZDecimal]** (TI-85/86 users press **[GRAPH] [F3] [MORE] [F4] (ZDECM)**)to display the graph of these two lines in the ZDecimal viewing window. TRACE along the graph of one of the lines and observe the changes in X values. The change should be one-tenth of a unit. (Remember, the Y-values are dependent on the values selected for X.)

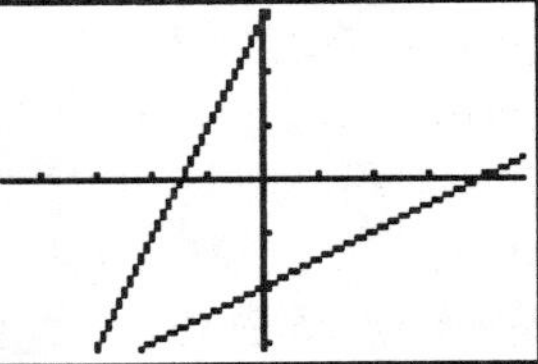

ZInteger is useful for application problems where the X-value is valid only if represented as an integer (such as when X equals the number of tickets sold, number of passengers in a vehicle, etc.). The horizontal space width formula, $\frac{Xmax - Xmin}{94} = \frac{47 - (-47)}{94} = 1$, changes the X-values by one unit each time the cursor is moved.

| **TI-85/86** | REMEMBER, THE TI-85/86 SCREEN IS 127 PIXELS WIDE THUS THE HORIZONTAL SPACE WIDTH FORMULA STILL NEEDS A DIVISOR OF 126. |

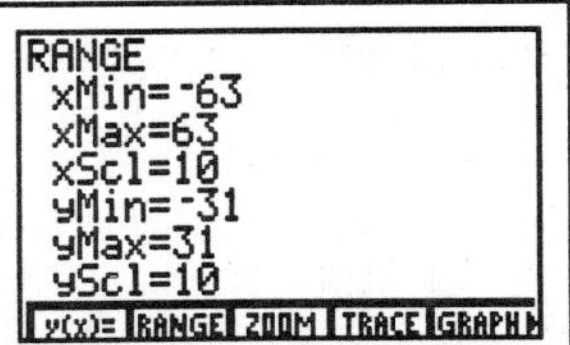

Press **[ZOOM] [8:ZInteger]** , move the cursor to the origin of the graph (x = 0 and y = 0), and press **[ENTER]** to display the graph of these two lines as indicated at the right.

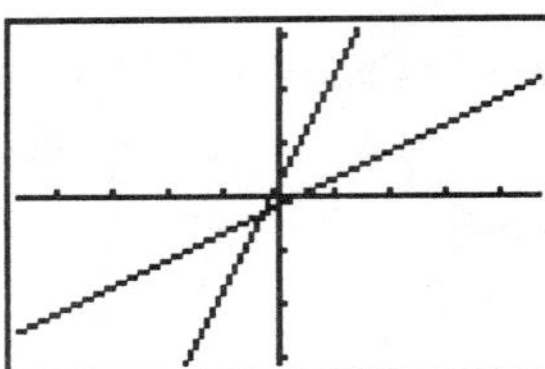

TI-85/86	PRESS **[GRAPH] [ZOOM] [MORE] [MORE] (ZINT)** AND MOVE THE CURSOR AS DIRECTED ABOVE.

TRACE along the graph of one of the lines and observe the changes in X values. The change should be one unit.

ZStandard provides a good visual comparison between hand sketched graphs (or textbook graphs) that are approximately [-10,10] by [-10,10]. Applying the horizontal space width formula,

$$\frac{Xmax - Xmin}{94} = \frac{10 - (-10)}{94} \approx 0.212765974,$$ the X-values will

```
WINDOW
Xmin=-10
Xmax=10
Xscl=1
Ymin=-10
Ymax=10
Yscl=1
Xres=1
```

change by .212765974 each time the TRACE cursor is moved. If you are using the TRACE feature, you will usually want a screen with "friendlier" X-values than this one provides.

Press **[ZOOM] [6:ZStandard]** to display the graph of these two lines as indicated at the right. TRACE along the graph of one of the lines and observe the changes in x values. If you select two consecutive x-values and find the difference, it should be 0.212765974.

ZDecimal x n: ZDecimal frequently does not provide a large enough viewing window. When this is the case, you may multiply the Xmin and Xmax by the same constant and the Ymin and Ymax by the same constant to produce a larger viewing rectangle which still provides cursor moves in tenths of units. Multiplying Xmin and Xmax by 2 would mean cursor moves of two-tenths of a unit:

```
WINDOW
Xmin=-9.4
Xmax=9.4
Xscl=1
Ymin=-6.2
Ymax=6.2
Yscl=1
Xres=1
```

$$\frac{Xmax - Xmin}{94} = \frac{2(4.7) - 2(-4.7)}{94} = 0.2,$$ whereas multiplying by 3 would mean cursor

moves of three-tenths of a unit. The screen above is the ZDecimal screen with the max and min values multiplied by 2. This WINDOW will be referred to in the future as ZDecimal x 2.

| TI-85/86 | REMEMBER, THE TI-85/86 SCREEN IS 127 PIXELS WIDE THUS THE HORIZONTAL SPACE WIDTH FORMULA STILL NEEDS A DIVISOR OF 126. | |

```
RANGE
xMin=-12.6
xMax=12.6
xScl=1
yMin=-6.2
yMax=6.2
yScl=1
y(x)= RANGE ZOOM TRACE GRAPH
```

Press **[Y =]** and clear all entries. Enter Y1 $= \frac{2}{3}$X + 6 and Y2 $= -\frac{3}{2}$X - 5 . (Did you remember to put parentheses around the fractions?)

$Y1 = \dfrac{2}{3}X + 6$ and $Y2 = -\dfrac{3}{2}X - 5$ are perpendicular lines whose intersection forms a 90° angle. Press **[ZOOM] [6:ZStandard]** for the standard viewing rectangle. Notice that the lines do not appear to be perpendicular. This is because the screen is rectangular - not square.

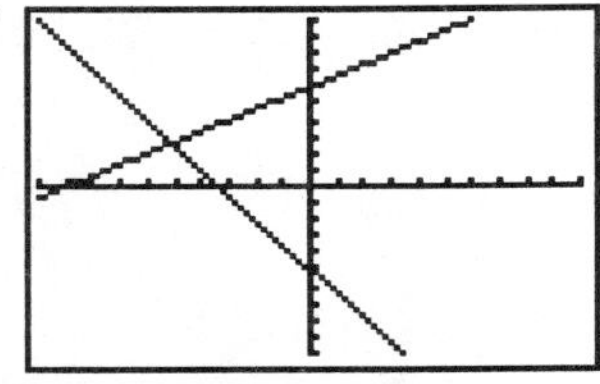

TI-85/86	TI-85/86 USERS SHOULD PRESS **[GRAPH] [F3](ZOOM) [F4](ZSTD)** TO AUTOMATICALLY SET THE STANDARD VIEWING RECTANGLE.

Press **[ZOOM]** again and this time select **[5:ZSquare]**. ZSquare *squares up* the viewing screen based on the previous viewing window. The lines should now appear to be perpendicular. Notice that the tic marks are all evenly spaced now.

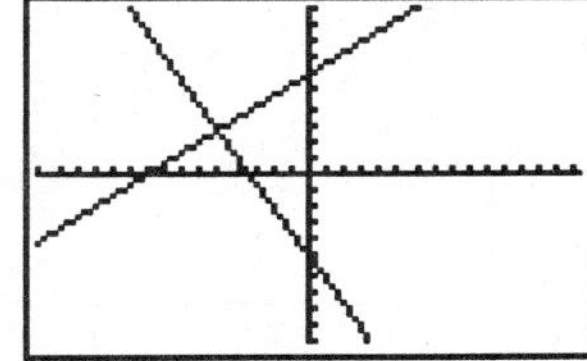

EXERCISE SET

✍6. Press **[WINDOW]** to see how the Max and Min values were affected by the Zsquare command applied above. Enter the WINDOW values displayed. Explain how the viewing window is different from the ZStandard viewing window.

NOTE: The ZDecimal screen (or any multiplicity of this screen) will provide a "squared up" graph screen on the TI-82/83/83plus. TI-85/86 users will need to use ZSquare .

7. **CLEAR** all entries on the **Y=** screen. Using
$Y1 = \dfrac{1}{2}X - 4$, press **[ZOOM] [4:ZDecimal]**
and at the right, sketch the screen as displayed.
Write the appropriate value for the "endpoint" of each
axis on the graph.

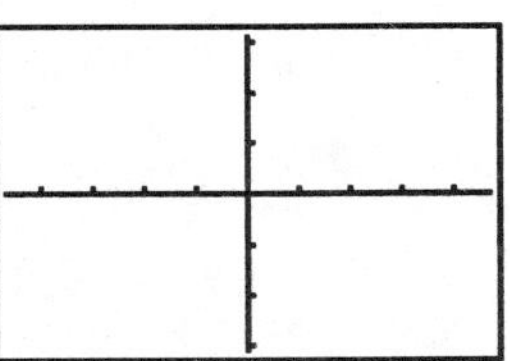

8. Now, press **[ZOOM] [8:ZInteger]** (pause for
your graph to be displayed) **[ENTER]** and at
the right, sketch the screen as displayed.
Write the appropriate value for the "endpoint" of each
axis on the graph.

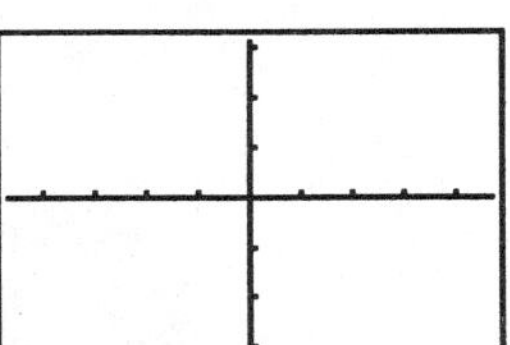

NOTE: ZDecimal and ZStandard do not require you to press [ENTER] to activate the viewing WINDOW, but ZInteger demands it.

✎9. In your own words, explain the differences in the displays in #7 and #8. What accounts for these differences?

10. Enter the equation Y = 12X^6 - 58X^4 + 84X^2 + 8 at the **Y=** prompt. Try to view the graph in each of the pre-set viewing windows discussed in this unit - ZDecimal, ZStandard, ZInteger, then sketch the graph as displayed in each of the indicated viewing WINDOWS. Both the Xscl and Yscl should be equal to zero.

<table>
<tr><td></td><td></td><td></td></tr>
</table>

     [0,2] by [15,35]   [-2,2] by [-5,30]   [-5,5] by [-5,75]

Pre-set viewing WINDOWS can provide a "starting point" for displaying a complete graph but frequently do not display all of the critical features of the graph. The next unit examines in detail the approach necessary for setting a good viewing WINDOW for individual graphs.

Solutions: **1.** left: -5, right: 5, top: 7, bottom: -12 **2** 7, 10, 70, yes; By increasing the value of the scale, the axes are increased in size without physically extending the length.

3. Because the maximum does not exceed the minimum (i.e. -5 is NOT greater than 10) the calculator displayed an error message. Moreover, that error message tells you that <u>you</u> made an error in setting the values.

4. The y-axis has not been given a defined length by setting the max and min at the same value. Again, an error message is displayed.

5. Max > Min   **6.** To have a "square" screen, tic marks must be evenly spaced. This was accomplished by adding tic marks to the X-axis.

7.   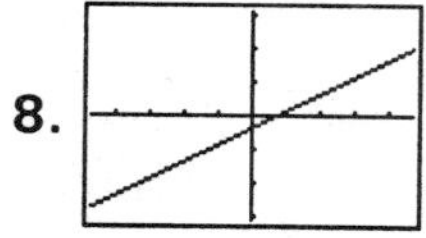 **8.**

Xmin: -4.7, Xmax: 4.7, Ymin: -3.1, Ymax: 3.1   Xmin: -47, Xmax: 47, Ymin: -31, Ymax: 31
TI-85/86: xMin: -6.3, xMax: 6.3,   TI-85/86: xMin: -63, xMax: 63,
        yMin: -3.1, yMax: 3.1   yMin: -31, yMax: 31

9. The difference was the amount and position of graph displayed. More of the graph was displayed on the ZInteger screen. This was because the <u>scales</u> were different on the two screens.

SETTING UP THE GRAPH DISPLAY

The calculator's display is controlled through the **MODE** and **GRAPH/FORMAT** screens.

Press **[2nd]** **<MODE>**. The current settings on each row should be highlighted as displayed. The blinking rectangle can be moved using the 4 **cursor** (arrow) keys. To change the setting on a particular row, move the blinking rectangle to the desired setting and press **[ENTER]**.

> **NOTE:** Items must be highlighted to be activated.

Normal vs. Scientific notation
Floating decimal vs. Fixed to 11 places
Type of angle measurement
Complex number display
Type of graphing: function, polar, parametric, differential eq.
Performs computations in bases other than base 10
Format of vector display
Type of differentiation

The **FORMAT** screen is accessed by pressing **[GRAPH]** **[MORE]** **[F3](FORMT)**. The following settings should be highlighted.

Graphs on both rectangular and polar coordinate system
Cursor location is displayed on the screen
Graphed points are connected or discrete
Functions displayed sequentially or simultaneously
Graphing grid is not displayed
Axes are visible
Axes are not labeled with "x" and "y"

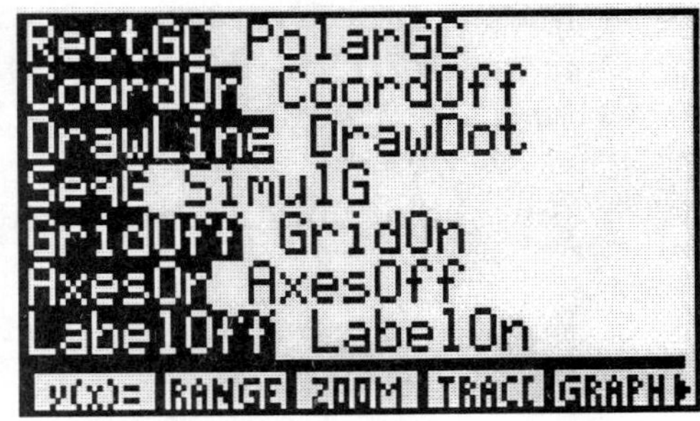

To return to the home screen, at this point, press **[CLEAR]** or **[EXIT]**.

Press **[GRAPH]** **[F1](y(x)=)**. The calculator can graph up to 99 different equations at the same time. Because **MODE** is in the sequential setting, the graphs will be displayed sequentially. The display of the equation(s) entered on the y(x)= screen is controlled by the size of the viewing window. The dimensions of the viewing window are determined by the values entered on **RANGE** screen, which is referred to as the **WINDOW** screen in the core units.

THIS CALCULATOR HAS A FEATURE CALLED THE "GRAPH STYLE ICON" THAT ALLOWS YOU TO DISTINGUISH BETWEEN THE GRAPHS OF EQUATIONS. PRESS **[GRAPH] [F1]**(Y(x) =) AND OBSERVE THE "\" IN FRONT OF EACH Y. PRESS **[MORE]** FOLLOWED BY **[F3](STYLE)**. THE DIAGONAL SHOULD NOW HAVE CHANGED FROM THIN TO THICK. THE GRAPH OF Y1 WILL NOW BE DISPLAYED AS A THICK LINE. REPEATEDLY PRESSING **[STYLE]** DISPLAYS THE FOLLOWING:

       ◥: SHADES ABOVE Y1

       ◣: SHADES BELOW Y1

       -O: TRACES THE LEADING EDGE OF THE GRAPH FOLLOWED BY THE GRAPH

       O: TRACES THE PATH BUT DOES NOT PLOT

       ⋱: DISPLAYS GRAPH IN DOT, NOT CONNECTED MODE

IT IS IMPORTANT TO REMEMBER THAT THE GRAPHING ICON TAKES PRECEDENCE OVER THE **MODE** SCREEN. IF THE ICON IS SET FOR A SOLID LINE AND THE **MODE** SCREEN IS SET FOR DOT AND NOT SOLID, THE GRAPHING ICON WILL DETERMINE HOW THE GRAPH IS DISPLAYED. PRESSING **[CLEAR]** TO DELETE AN ENTRY AT THE Y = PROMPT WILL AUTOMATICALLY RESET THE GRAPHING ICON TO DEFAULT, A SOLID LINE.

Press **[GRAPH] [F3](ZOOM) [F4](ZSTD) [2nd] <M2>(RANGE).** (WIND on the TI-86.) This is called the standard viewing window. The information on this screen indicates that in a rectangular coordinate system the x-values will range from -10 to 10 and the y-values will range from -10 to 10. The interval notation for this is [-10,10] by [-10,10]. The xScl = 1 and yScl = 1 settings indicate that the tic marks on the axes are one unit apart. The values entered on this screen may be changed by using the cursor arrows to move to the desired line and typing over the existing entry. When drawing a graph, you may set the desired viewing window on the calculator as well as scale the X-axis and Y-axis.

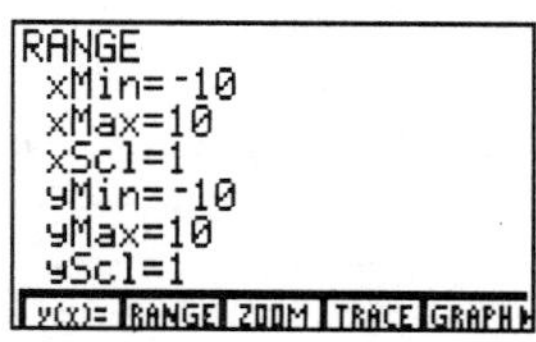

     REMEMBER: Every time the core unit indicates "press **[WINDOW]**" you will need to press **[GRAPH] [F2](RANGE)/ (WIND).** When the core unit indicates "press **[GRAPH]**", you should press **[F5](GRAPH)** from the **GRAPH** menu.

Before proceeding further, press [GRAPH]   [F1](y(x) =) and clear all entries.

☞   RETURN TO THE CORE UNIT AND COMPLETE THE EXERCISES ON PAGE 113.

ALTERING THE VIEWING WINDOW

The last unit addressed the size of the calculator screen (viewing window). Because the screen is 127 pixel points wide by 63 pixel points high, there are 126 horizontal spaces and 62 vertical spaces to light up. When tracing on a graph, the readout changes according to the size of the space. The size of the space can be controlled by the following formulas:

$$\frac{Xmax - Xmin}{126} = \text{horizontal space width},\qquad \frac{Ymax - Ymin}{62} = \text{vertical space height}.$$ We

will now examine some preset viewing windows and how they affect the pixel space size.

Press **[GRAPH]** **[F3]**(ZOOM). Pressing **[MORE]** repeatedly will display the remaining menu selections and then return the display to the original screen.

ZBOX	Boxes in and enlarges a designated area.
ZIN	Acts like a telephoto lens and "zooms in".
ZOUT	Acts like a wide-angle lens and "zooms out".
ZSTD	Automatically sets standard viewing window to [-10,10] by [-10,10].
ZPREV	Resets **RANGE** values to values used prior to the previous ZOOM operation.
ZFIT	Resets yMin and yMax on the **RANGE** screen to include the minimum and maximum y-values that occur between the current xMin and xMax settings.
ZSQR	"Squares up" the previously used viewing window.
ZTRIG	Sets the **RANGE** to built-in trig values.
ZDECM	Sets cursor moves to ONE tenth of a unit per move.
ZDATA	*TI-86 only: Automatically sets the viewing window to accomodate statistical data.*
ZRCL	Sets **RANGE** values to those stored by the user (see ZSTO).
ZFACT	Sets the zoom factors used in **ZIN** and **ZOUT**.
ZOOMX	Graph display is based on xFact only when zooming in or out.
ZOOMY	Graph display is based on yFact only when zooming in or out.
ZINT	Cursor moves are ONE integer unit per move.
ZSTO	Stores current **RANGE** values for future use. Values are recalled by **ZRCL**.

☞   Return to core unit pg.115 and begin reading at **ZDecimal**.

*Unit 17 is a prerequisite for this unit. Answers appear at the end of the unit.

Many times students are frustrated when the equation they have carefully keystroked into the **Y =** screen does not appear when **GRAPH** is pressed. What actually happens to the graph? Suppose you graphed $Y = 2X^2 + 4X + 12$ on graph paper and then graphed this same equation on the calculator with the viewing window set to ZStandard. The figure at the right illustrates the handsketched graph with the section displayed on the ZStandard screen outlined in a bold black line. The viewing window selected is not large enough to display the graph. This unit will give you the practice necessary to feel confident about setting the viewing window correctly to display a complete graph. A complete graph is defined to be one in which all x- and y-intercepts are visible as well as any peaks/maximums and valleys/minimums.

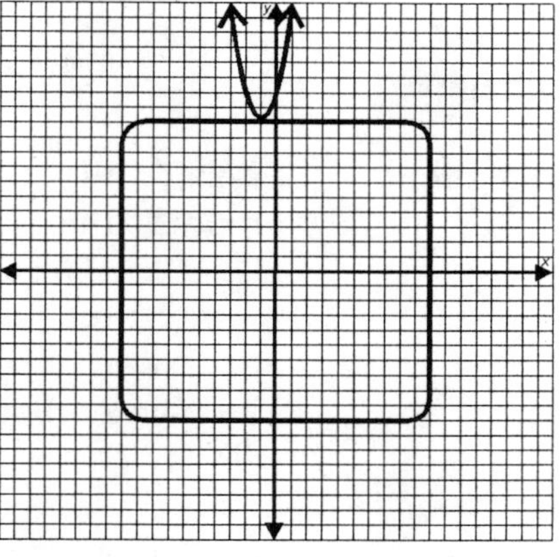

Before proceeding, press **[ZOOM] [6:ZStandard]** (TI-85/86 USERS PRESS **[GRAPH] [F3](ZOOM) [F4](ZSTD)**) to set the standard viewing WINDOW.) Enter Y1 = 4X - 18 on the **Y =** screen and press **[GRAPH]**. The graph at the right should be displayed. The X-intercept is visible, but the Y-intercept is not.

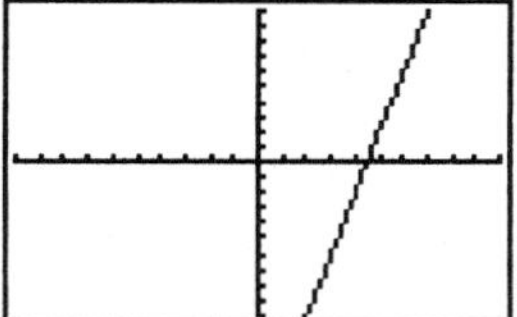

If the equation entered is not displayed on the graphing screen, the first item to be checked is the entry of the equation on the **Y =** screen. Is the equation *SELECTED* to be graphed? That is, is the equal symbol highlighted? If it is, proceed. If it is not, move the cursor over the equal sign and press **[ENTER]** to highlight the equal sign, thus activating the equation.

TI-85/86	TI-85/86 USERS SHOULD REMEMBER THAT **[F5](SELCT)** ON THE Y(X) = MENU WILL BE USED TO ACTIVATE AND DEACTIVATE EQUATIONS.

If the equation is activated, then begin the process of adjusting the WINDOW by locating the X and Y-intercepts of the graph.

Press **< 2nd >[Calc] [1:value]** to access the value option. Enter 0 at the X = prompt (because y-intercepts have an x-value of 0) and press **[ENTER]**. At the bottom of the screen the cursor's location (the y-intercept) of X = 0 and Y = -18 is displayed.

TI-85/86	WITH THE **GRAPH** MENU DISPLAYED, PRESS **[MORE] [MORE]** AND THE APPROPRIATE F KEY FOR **EVAL**.

Exit the GRAPH screen and enter the WINDOW screen by pressing **[WINDOW]**.

TI-85/86	PRESS **[F2](RANGE)**, WHICH IS **(WIND)** ON THE TI-86.

Use the down arrow key to move the cursor to Ymin and replace the Ymin with -20, a value

Use the down arrow key to move the cursor to Ymin and replace the Ymin with -20, a value
smaller than -18. This smaller value allows a clear view of the Y-intercept.
Pressing [**GRAPH**] displays the screen at the right. This a satisfactory
graph because both the X and Y-intercepts are displayed.

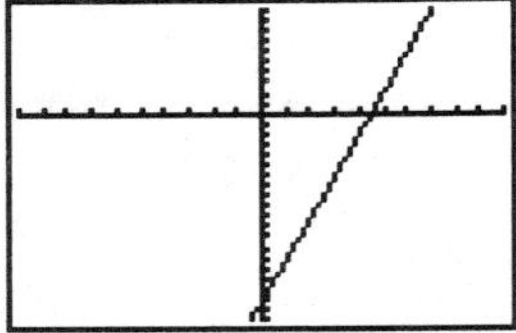

Reset the WINDOW to the standard viewing window.

Example 1:   Graph $y = x^3 - 15x^2 + 26x$ in an appropriate viewing window.

Solution: Press [**Y =**], enter Y1 = $X^3 - 15X^2 + 26X$ and press [**GRAPH**]. A satisfactory
graph of this equation should display all of its interesting features. Press
[**TRACE**] (TI-85/86 users press [**F4**] (**TRACE**)) and TRACE along the graph to
the right (using the right arrow key) and record the X and Y-intercepts as
encountered. Remember, you are not on the ZInteger or ZDecimal screens.
The X-intercepts may only be close approximations with the TRACE feature.
(HINT: This is a third degree equation. There could be three X-intercepts.)
Each of the screens below indicate the points closest to the X-intercepts that
you should be able to locate.

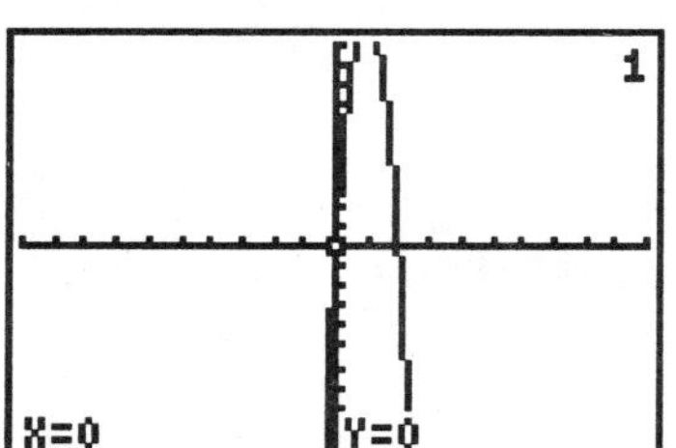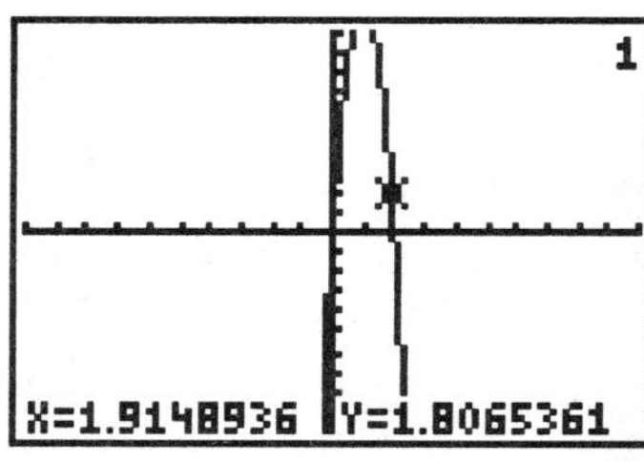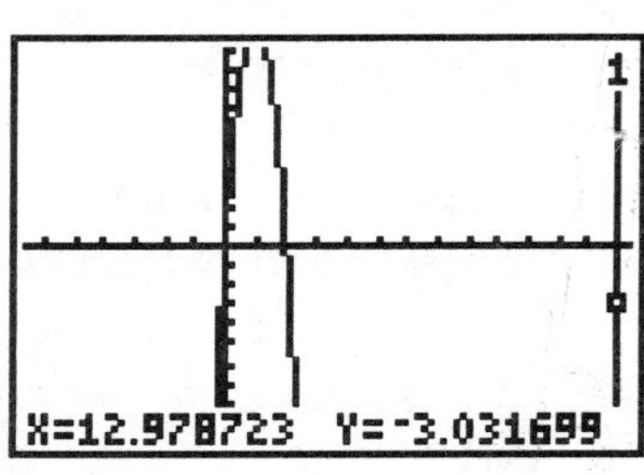

    X and Y-intercept   X-intercept ≈ 2   X-intercept ≈ 13

Are all the peaks (maximums) and valleys (minimums) of the graph displayed? A
satisfactory viewing window is a window that includes the X and Y-intercepts as well
as all the peaks and valleys of the graph. TRACE the curve again, going left this
time, to determine the lowest **Y value** (valley/minimum) and the highest **Y value**
(peak/maximum) displayed (to the nearest whole number value).

valley/minimum = _______   peak/maximum = _______

Press [**WINDOW**] and adjust the Max and Min values for both X and Y to include the
intercept points on the X-axis and the maximum (y = 12) and minimum (y = -252)
Y values. It is suggested that you enter values that are a few units larger (or smaller)
than the intercepts, maximum, and minimum you recorded. Suggested values are:

Xmin = -2   Ymin = -255
Xmax = 15   Ymax = 15

The graph is displayed using the above WINDOW values. The
tic marks have been adjusted by setting the Xscl = 1 and the
Yscl = 20. You may prefer to set both scales equal to zero
and eliminate all tic marks.

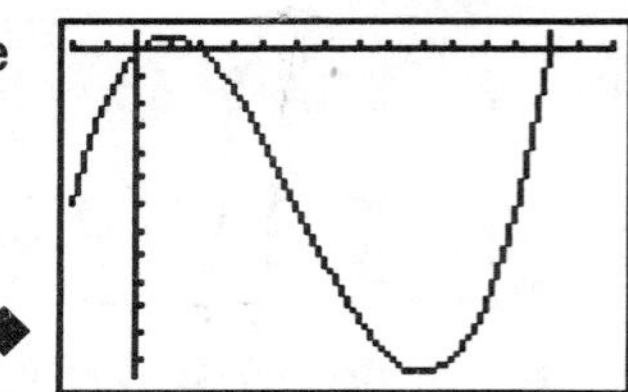

As your algebraic knowledge increases, you will be able to blend this knowledge with calculator expertise to expedite the process of determining a good viewing window.

EXERCISE SET

Directions: Begin each problem by viewing the graph in the ZStandard viewing window. Sketch a graph of the equation once a good viewing window has been established.
   a. Record the values used to determine your WINDOW.
   b. Sketch the graph that is displayed for these WINDOW values.
   c. Record the WINDOW in interval notation.

1. $Y = -2X^2 + 19$

Y-intercept = _________   Ymax = _______

X-intercepts = _________, and _________

[_________,_________] by [_________,_________]
  Xmin   Xmax   Ymin   Ymax

2. $Y = (1/2)X^4 - 10X^2 + 25$

Y-intercept = _______

X-intercepts = _________, _________, _________, _________

Ymax = _______   Ymin = _______, _______

[_________,_________] by [_________,_________]
  Xmin   Xmax   Ymin   Ymax

3. $Y = \sqrt[3]{X^2 - 5X - 300}$

(If you do not remember how to enter radicals other than square
roots, refer to Unit 4.)

Y-intercept = _______

X-intercepts = _________, _________,

Ymin = _______

[_______,_______] by [_______,_______]
  Xmin   Xmax   Ymin   Ymax

4.   a. Establish an appropriate viewing window for the equation
$Y = 0.1X^4 - 8X^2 + 5X + 1$ and sketch the result.

[_______ , _______] by [_______ , _______]

b. Using the space width formulas discussed in the unit entitled "Preparing to
Graph", determine the missing window values if each move of the TRACE
cursor is to be **ONE** unit.

[-45, X] by [-200,10]   X = _________

Sketch the display provided by this window.

✍ c. Space width formulas can be used for both coordinates and yet when tracing,
the Y-coordinates displayed are not consistently integers. Explain.

Solutions: **1.** Y-intercept:19, Ymax: at least 20, X-intercepts: approximately -3 and 3;
[-4, 4] by [-10, 20]

2. Y-intercept:25, X-intercepts: approximately -4, -1.7, 1.7, 4, Ymax:25, Ymin:-26, -26;
[-5, 5] by [-30, 30]

3. Y-intercept: approximately -7, X-intercepts:-15,20, Ymin: -7; [-25, 30] by [-10, 10].

4. a. [-10, 10] by [-200, 10]   b. [-45, 49] by [-200,10]   c. Y-values are determined by
the x-values that are selected to
replace the variables in the
equation.

UNIT 19
FUNCTIONS

*Unit17 is a prerequisite for this unit. Answers appear at the end of this unit.

A function is a collection of ordered pairs in which each *x*-value is paired with a unique *y*-value.

Domain and Range

The set of all *acceptable* *x*-values of the ordered pairs in a function is called the **domain**. The set of all resultant y-values is called the **range**. When determining domains and ranges of functions, bear in mind that there are three tools: the algebraic definition of the function, the table feature of the graphing calculator (82/83/83plus/86 users) and the graph.

TI-85 USERS SHOULD BE CAREFUL NOT TO CONFUSE THE RANGE OF A FUNCTION (THE SET OF ALL Y VALUES) WITH THE **RANGE** SCREEN WHERE PARAMETERS ARE DETERMINED FOR THE VIEWING WINDOW.

EXERCISE SET

Directions: Use the **TRACE** feature to determine the domain and range of each of the following functions. TRACE along the path of the graph and examine the X and Y values displayed at the bottom of the screen. These will be of assistance in determining the domain and range. You may use any viewing WINDOW you desire, however you may discover that some viewing WINDOWS are more "informative" than others. Suggestion: view and TRACE on each graph in each of the following WINDOWS - ZStandard, ZDecimal, ZInteger before determining the domain. (TI-85/86 users recall that these preset windows are listed under the **ZOOM** menu as ZSTD, ZDECM, and ZINT.)

1. $Y = X^2 + 1$ Domain = _________________________

     Range = _________________________

2. $Y = 3X + 5$ Domain = _________________________

     Range = _________________________

3. $Y = \sqrt{X + 5}$ Domain = _________________________

     Range = _________________________

4. $Y = X^3 + 4X^2 + 2$ Domain = _________________________

     Range = _________________________

✍ 5. Consider the two functions Y1 = X and Y2 = $\sqrt{X} \cdot \sqrt{X}$ and respond to each of the
following questions.
a. Would you expect the two domains to be the same? Why or why not?

b. Considering only the algebraic equations (no calculator allowed), what should the
domains should be?

c. What does the table of each function indicate the domain should be?

d. What does the graph of each function indicate the domain should be? Is one
viewing window more informative than another?

✍ 6. Consider the function $Y = \dfrac{(2X + 3)(X + 2)}{2X + 3}$ and respond to each of the following
questions.
a. What do you expect the domain to be?

b. Considering only the algebraic equation (no calculator allowed), what should the
domain should be?

c. What does the table of the function indicate the domain should be?

d. What does the graph of the function indicate the domain should be? Is one
viewing window more informative than another?

When an equation is graphed, the VERTICAL LINE TEST can be used to determine if the equation is a function. Remember, to be a function there must be only one Y value for each X value. Thus, when a vertical line is passed across the graph (from left to right) the line will intersect the graph in only one point at a time if the graph represents a function.

Enter the equation $Y = X^2 + 2X + 2$ to be graphed and display the graph on the standard viewing WINDOW. To "draw" a vertical line with the calculator, press **[2nd] <DRAW>**, **[4:Vertical]** (i.e. vertical line).

TI-85/86	TO ACCESS THE VERTICAL LINE, PRESS **[GRAPH] [MORE] [F2](DRAW) [F3](VERT)**.

The vertical line is actually displayed on the Y-axis initially. Press the **[▶]** and **[◀]** arrow keys to move the vertical line right and/or left across the graph. Since the vertical line does not intersect the graph in more than one place at a time the equation represents a function. Using the **[▶] or [◀]** keys, return the vertical line to the Y-axis so that it is no longer visible. Each time you graph a new equation you must go back to the **DRAW** menu to access the Vertical Line option (as well as other options in this menu). Note: Anything that is DRAWN on the graph via the **DRAW** menu can only be cleared by pressing **[1:ClrDraw]**.

TI-85/86	ACCESS THE **DRAW** MENU, PRESS **[MORE]** UNTIL **(CLDRW)** FOR CLEAR DRAW IS DISPLAYED, THEN PRESS THE APPROPRIATE **F** KEY.

EXERCISE SET CONTINUED

Directions: Use the Vertical Line option from the **DRAW** menu on the graph of each of the following equations to decide if the graph of the equation represents a function. Sketch the graph displayed **AND** sketch the vertical line at some point on the graph.

7. $Y = -X^2 - 8X - 10$

    Function? (yes or no)________

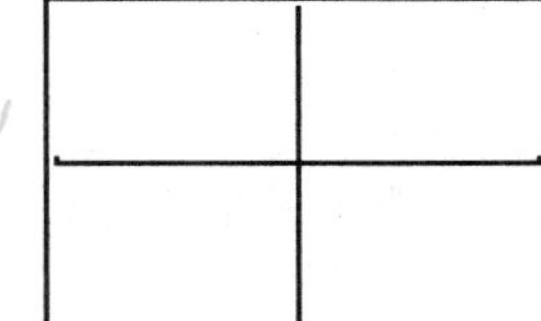

8. $Y = \dfrac{1}{2}X^3 + X^2 - 2X + 1$

    Function? (yes or no)________

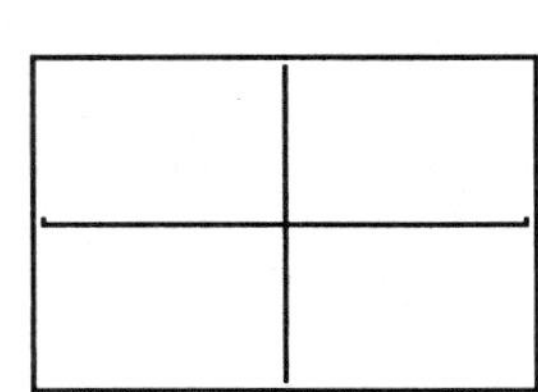

9. $Y = 2\sqrt{X + 4}$

    Function? (yes or no)________

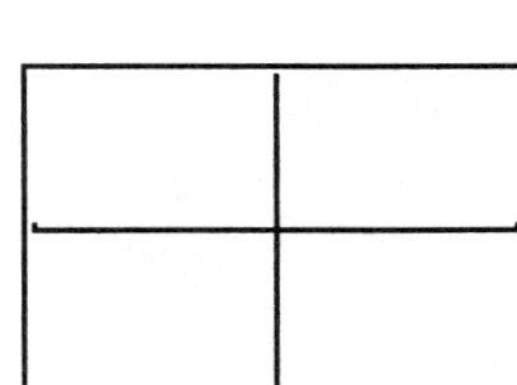

The horizontal line test is used on the graph of a function to determine if it is one-to-one. A function is one-to-one if for each Y-value in the range there is one and only one X-value in the domain. A horizontal line passed across the graph will not intersect the graph in more than one place at a time if the graph is that of a one-to-one function.

TI-85/86	THE TI-85 DOES NOT HAVE A HORIZONTAL LINE OPTION. USE A STRAIGHT EDGE TO PERFORM THE HORIZONTAL LINE TEST ON THE EXAMPLE AND THE EXERCISES. THE TI-86 HAS THE HORIZONTAL LINE OPTION. IT IS LOCATED NEXT TO THE VERT ON THE DRAW MENU.

Re-enter $Y = X^2 + 2X + 2$ and display the graph. Press **[2nd]** <**DRAW**> **[3:Horizontal]**. Use the **[▲]** and **[▼]** arrow keys to move the horizontal line up and down the graph. The horizontal line is originally positioned on the X-axis. When you are finished with this option, return the horizontal line to its beginning position on the X-axis. This is not the graph of a one-to-one function because the horizontal line intersects the graph in <u>two places</u> everywhere except at the vertex.

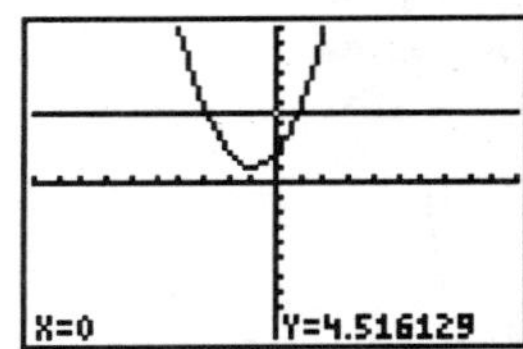

EXERCISE SET CONTINUED

Directions: Use the Horizontal Line option from the **DRAW** menu on the graph of each of the functions to decide if the graph represents a one-to-one function. Sketch the graph displayed and the horizontal line at a location where it intersects the function in more than one place. If the function is one-to-one do not draw in the horizontal line.

10. $Y = -X^2 - 8X - 10$

      1-1 Function? (yes or no)________

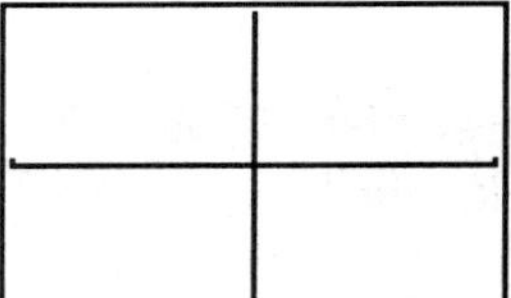

11. $Y = \dfrac{1}{2}X^3 + X^2 - 2X + 1$

      1-1 Function? (yes or no)________

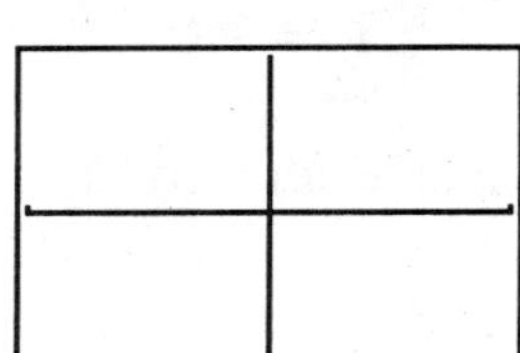

12. $Y = 2\sqrt{X + 4}$

      1-1 Function? (yes or no)________

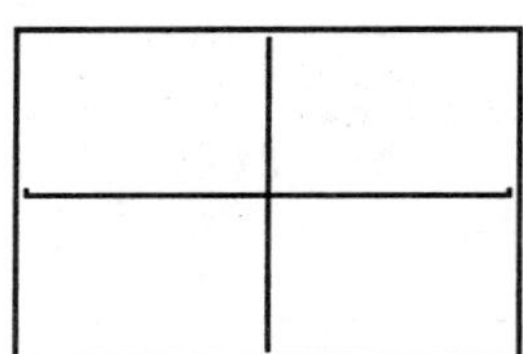

INVERSES

If a function is one-to-one, i.e. passes both the vertical and horizontal line tests, it will have an inverse function.

Consider the function Y = 2X + 3. Enter 2X + 3 at the Y1 = prompt on the calculator. Press **[ZOOM] [4:ZDecimal] [WINDOW]** and double all window values except for Xscl and Yscl. This will be referred to in the future as the **ZDecimal x 2** viewing window. Copy the display. This is a one-to-one function because the graph passes both the vertical and horizontal line tests.

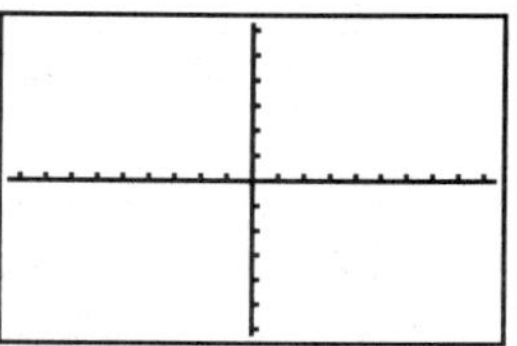

Algebraically find the inverse below by (a) interchanging the X and Y variables and (b) solving for Y.

The inverse equation should be $Y = \dfrac{1}{2}X - \dfrac{3}{2}$. Enter this equation at the Y2 = prompt. At the Y3 = prompt, enter X to graph Y = X. Your display of all three functions should look like the one at the right.

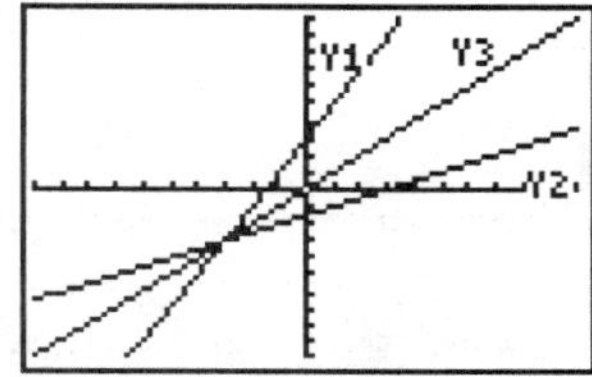

Y1 and Y2 are symmetric across the line Y = X (the Y3 line). This will always be true of a function and its inverse.

To use the calculator to __DRAW__ the inverse function, the function must be entered on the **Y =** screen. **Since the function is entered at Y1, go back and delete Y2 and Y3 before proceeding.** Press **[2nd]** <QUIT> to return to the home screen.

Instruct the calculator to __DRAW__ the inverse of Y1 (DrawInv): Press **[2nd]** <DRAW> **[8:DrawInv] [Vars] [▶]**(to highlight Y-Vars) **[1:Function…] [1:Y1] [ENTER]**.

Y1 will be <u>graphed</u> first and the INVERSE of Y1 will be <u>drawn</u> second. The inverse that is drawn is the same line as the one graphed at Y2, however, because it is drawn (and not graphed from the Y = screen) you will not be able to interact with the graph. That is to say, you will not be able to TRACE, use INTERSECT, ROOT/ZERO, VALUE, TABLE, etc.

Standard function notation is **f(X)**, where f denotes the function, X is the independent variable and f(X) represents the function's value at X. If an equation represents the graph of a function, then the Y variable in the equation may be replaced by the f(X) notation.

In Exercise 7, $Y = -X^2 - 8X - 10$ was determined to be a function. The equation can now be written as $f(X) = -X^2 - 8X - 10$. Since the calculator will only graph functions, Y1 (or y1(x) on the TI-85/86) is equivalent to the function denoted as **f**.

To evaluate the function $f(X) = -X^2 - 8X - 10$ at X = 2, write: $f(2) = -(2)^2 - 8(2) - 10 = -30$. Thus when X = 2, f(X) = -30 (i.e. Y = -30). This would yield the ordered pair (2,-30) on the graph of f(X). Verify with the TABLE. Be sure it is incremented by 1 and enter $- X^2 - 8X - 10$ at the Y1 = prompt. Find X = 2 in the table. When X = 2, the table indicates that Y1 = - 30. You could also use the VALUE (EVAL) feature as discussed in Unit 10.

SINCE THERE IS **NO TABLE** FEATURE, YOU SHOULD CONSIDER USING THE FUNCTION EVALUATION FEATURE. PRESS **[GRAPH] [F1](y(x)=)** AND ENTER $-X^2 - 8X - 10$ AT THE y1(x) = PROMPT. PRESS **[GRAPH] [MORE] [MORE] [F1]EVAL)** AND ENTER THE DESIRED X VALUE AT THE PROMPT. THE **EVAL** OPTION IS RESTRICTED TO X VALUES BETWEEN xMIN AND xMAX ON THE **RANGE** SCREEN. ADJUST THE **RANGE** AS NECESSARY. IF MORE THAN ONE EQUATION IS ENTERED ON THE y(x) = SCREEN, EACH EQUATION WILL BE EVALUATED AND THE RESULTS FOR EACH CAN BE DISPLAYED BY PRESSING THE UP OR DOWN CURSOR ARROWS. THE EQUATION NUMBER IS DISPLAYED IN THE TOP RIGHT HAND CORNER OF THE SCREEN.

The TI-82/83/83plus/86 uses the equivalent function notation of Y(X) instead of f(X). To evaluate $f(X) = -X^2 - 8X - 10$ at X = 2, the f(X) expression must be entered on the **Y=** screen. Enter the polynomial at the Y1 = prompt. Return to the home screen by pressing **[2nd] <QUIT>**. To compute the value for f(X) when X = 2, press **[VARS] [▶] [1:function] [1:Y1] [(] [2] [)]** and then **[ENTER]** to compute. (TI-86 users press **[2nd] <alpha> <y> [1] [(] [2] [)])**.

```
Y1(2)
                          -30
```


Find Y1 by pressing **[2nd] <VARS>   [1:Function]
[1:Y1]**.

The value of the function will be displayed for the X-value of 2. In ordered pair form this would be (2,-30).

TI-85 USERS MUST RELY ON THE **EVAL** OR **evalf** FEATURES. FOR A REVIEW OF THE USE OF **EVALF**, SEE THE TI-85 BOX AT THE END OF UNIT 10. IF y1(2) IS ENTERED ON THE TI-85, THE CALCULATOR COMPUTES THE VALUE OF y1 WITH THE CURRENT VALUE STORED IN X-VAR AND THEN MULTIPLIES THE RESULT BY 2.

EXERCISE SET CONTINUED

Directions: Enter each of the following functions on the **Y=** screen. Evaluate for the indicated value of X using the Y-VARS capability of the calculator (i.e. Y(X) notation). Copy the screen that displays the answer. Record your final information as an ordered pair.

13.   Evaluate $Y = 3X^3 - 2X^2 + X - 5$ for $X = -\dfrac{3}{20}$.

    Screen display:

    Ordered pair:_________________________ (in decimal form)

14.   Evaluate $Y = \sqrt{2X - 5}$ for $X = 3.5$.

Screen display:

Ordered pair:______________

15.   Evaluate $Y = \dfrac{2X + 3}{X^2 + 4X - 5}$ for $X = -\dfrac{1}{2}$

Screen display:

Ordered pair:______________ (in fraction form)

✍16. Evaluate $Y = \sqrt{3X + 5}$ for $X = -8$.

Screen display:

Why did you get this display? Explain carefully.

17.   The profit or loss for a publishing company on a textbook supplement can be
       represented by the function $f(X) = 10X - 15000$ (X is the number of supplements
       sold and f(X) is the resulting profit or loss).
       a. Use the Y-VARS capability to determine the amount of profit (or loss) if 2000
       supplements are sold. Copy the screen display to justify your work.

       ✍ b. How can you tell if the $5000 is profit or loss?

       c. What would be the profit (or loss) if 1000 supplements are sold?
          Copy the screen display to justify your work.

✍18. In your own words, explain the various approaches for determining domain and range of functions. Include both algebraic and calculator methods in your discussion.

✍19. In your own words, explain the similarities and differences between a function and a one to one function. Your discussion should not be limited to the vertical and horizontal line tests.

Solutions: 1. domain $= \mathbb{R}$, range $= \{Y\,|\,Y\geq 1\}$ **2.** domain $=\mathbb{R}$, range $= \mathbb{R}$

3. domain $= \{X\,|\,X\geq$-5$\}$, range $= \{Y\,|\,Y\geq 0\}$ **4.** domain $= \mathbb{R}$, range $= \mathbb{R}$

5. a. answers may vary   **b.** $Y=x$ should have a domain of all real numbers and $y = \sqrt{x}\sqrt{x}$ should have a domain of all non-negative real numbers.   **c.** The table agrees with the statement made in part b.   **d.** The graph agrees with the statement made in part b. With respect to viewing windows, answers may vary.

6. a. The domain should be $\{x\,|\,x \neq$ -3/2$\}$.   **b.** The domain is $\{x\,|\,x \neq$ -3/2$\}$.   **c.** The table will not support the declared domain unless you know to increment the table by $\frac{1}{2}$. The graph supports the domain stated in part b. With respect to viewing windows, answers may vary.

7. Yes   **8.** Yes   **9.** Yes   **10.** No   **11.** No   **12.** Yes

13. (.15, -5.205125), (-3/20, -41641/8000)   **14.** (3.5, 1.414213562)

15. (-1/2, -8/27)   **16.** $\sqrt{3X + 5}$ is equivalent to $\sqrt{-19}$ when $X=$-8. The square root function is undefined for negative radicands.

17. a. $Y_1(2000)$ 5000   **b.** The $5,000 is positive, and therefore a profit.
c. $Y_1(1000)$ -5000   The negative indicates that the $5,000 would be a loss.
18. answers may vary

UNIT 20
DISCOVERING PARABOLAS

*Unit 17 is a prerequisite for this unit. Answers appear at the end of the unit.

This unit explores the graphs of quadratic equations, specifically quadratic functions. A quadratic function is an equation in the form $y = ax^2 + bx + c$, where a, b, and c are real numbers. These values (a,b,c) will affect the size and location of the curve. This unit is an exercise in discovery. As each equation is graphed, study the size and location of the parabola and compare this information to the coefficients in the equation.

1.　　Set the viewing window to ZDecimal x 2, [-9.4,9.4] by [-6.2,6.2], with both scales equal to 1. (The TI-85/86 RANGE/WIND values will be [-12.6,12.6] by [-6.2,6.2].) The TRACE feature will be used to help discover some of the characteristics of these parabolas. This viewing window was selected because the cursor moves will be in tenths of units.

2.　　The vertex is the minimum point (or maximum point) on the graph of a parabolic curve. GRAPH and TRACE to find the vertex of each of the parabolas graphed by the given equation. Sketch the graph and record the coordinates of the vertex as ordered pairs.

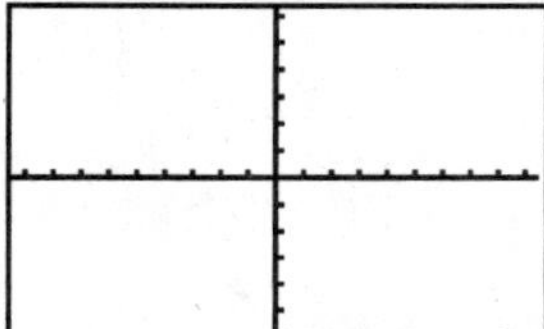

Y = X²

vertex:_______

Y = 3X²

vertex:_______

Y = (¼)X²

vertex:_______

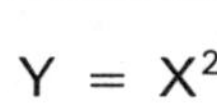3.　　Observation: What effect does the coefficient on the X² term have on the graph of the equation?

4.　　GRAPH and TRACE to find the vertex on each of the equations. Sketch the graph and record the coordinates of the vertex.

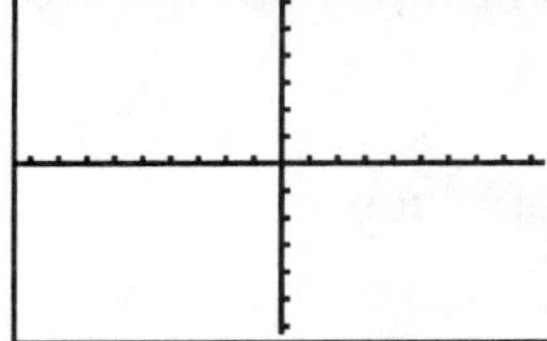 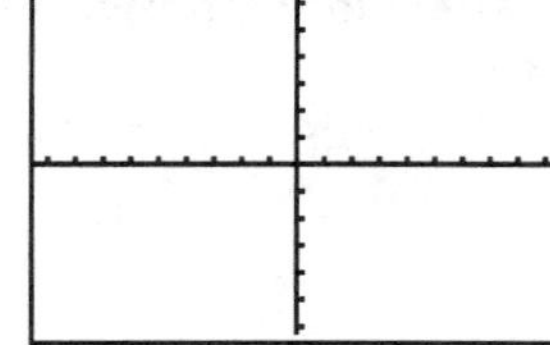 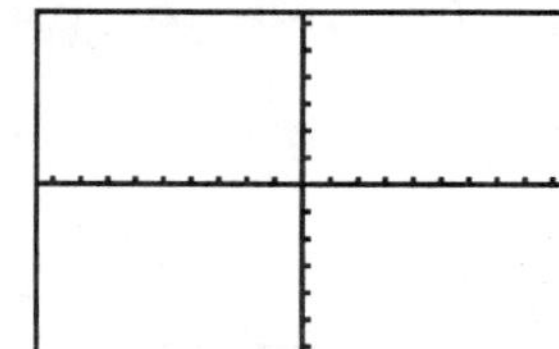

Y = X² + 3

vertex:________

Y = X² + 1

vertex:________

Y = X² - 2

vertex:______

5.　　Observation: What effect does the constant term have on the graph of the equation?

6. GRAPH and TRACE to find the vertex on each of the equations. Sketch the graph and record the coordinates of the vertex.

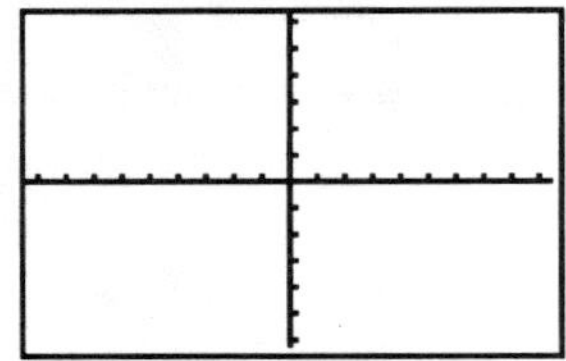

$Y = (X + 6)^2$

vertex:________

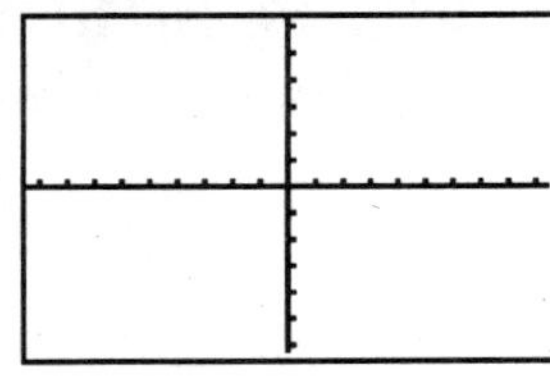

$Y = (X + 1)^2$

vertex:_________

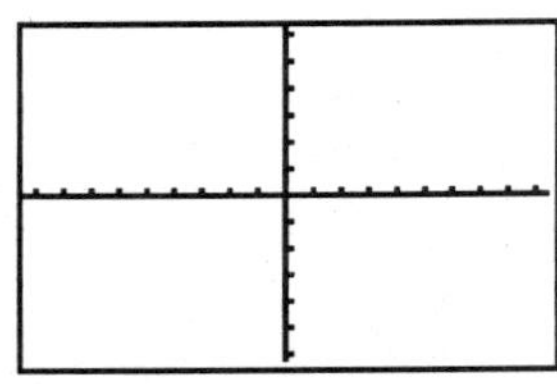

$Y = (X - 3)^2$

vertex:______

7. Observation: When a value is added or subtracted to the X <u>before</u> the quantity is squared, what effect does it have on the graph?

8. GRAPH and TRACE to find the vertex on each of the equations. Sketch the graph and record the coordinates of the vertex.

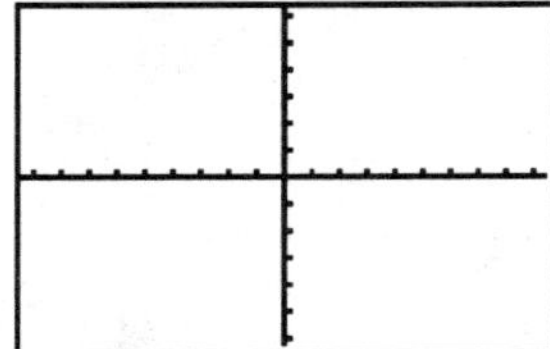

$Y = (X + 3)^2 + 2$

vertex:________

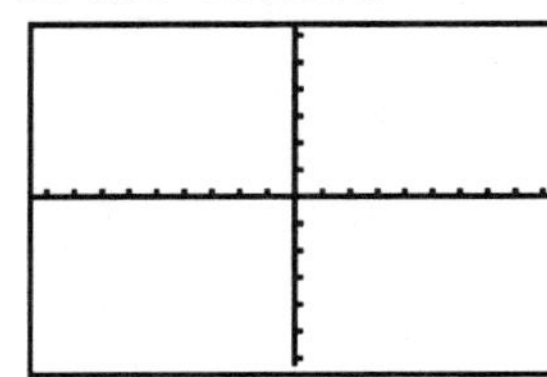

$Y = (X + 3)^2 - 2$

vertex:________

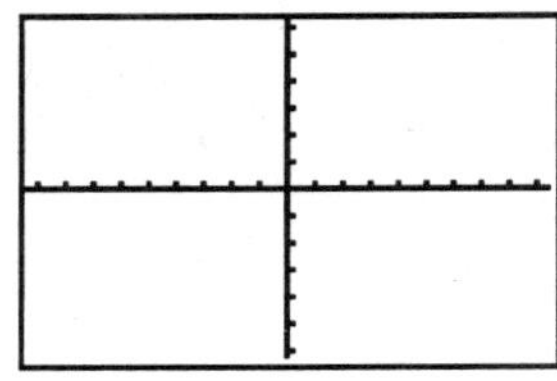

$Y = (X - 3)^2 + 2$

vertex:______

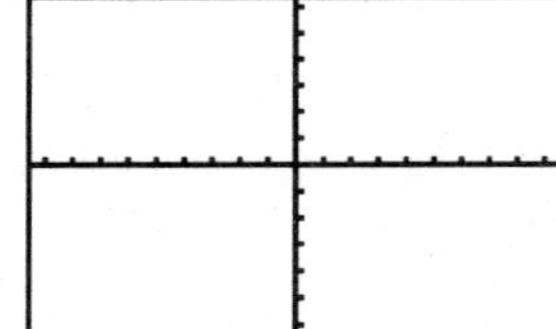

$Y = (X - 3)^2 - 2$

vertex:________

9. Based on your observations from the previous problems, what should the vertex of the parabola $Y = (X + 115)^2 - 38$ be? **DO NOT ATTEMPT TO ANSWER THIS QUESTION BY GRAPHING THE EQUATION!** Use the information gathered in the previous problems to determine the vertex.

____

10. Compare each pair of equations by graphing on the same graph screen, ZDecimal x 2.

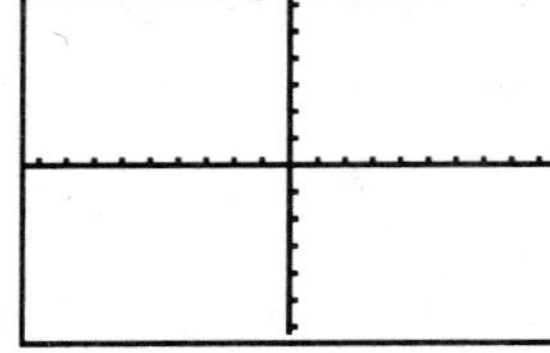

a. $Y = 3(X + 2)^2 + 2$

b. $Y = (X + 2)^2 + 2$

What effect did the "3" have on the shape of the graph?

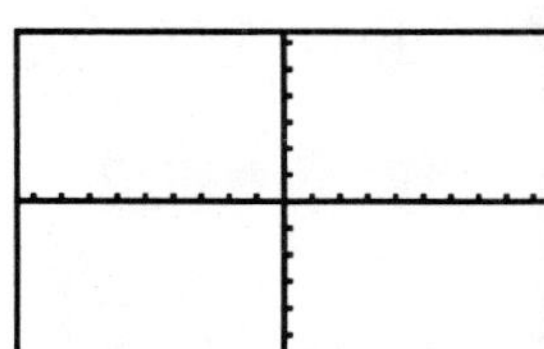

a. $Y = (¼)(X + 2)^2 + 2$

b. $Y = (X + 2)^2 + 2$

What effect did the "¼" have on the shape of the graph?

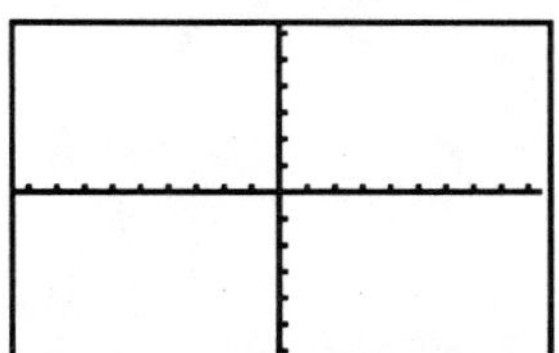

a. $Y = 3(X + 2)^2 + 2$

b. $Y = -3(X + 2)^2 + 2$

What effect did the negative sign on the 3 have on the shape/orientation of the graph?

Using what you have learned in this unit, match each graph below with one of the equations #11-14. DO NOT ENTER ANY EXPRESSIONS ON THE CALCULATOR.

_____11. $Y = (X + 5)^2 - 4$ _____13. $Y = (X + 4)^2 + 5$

_____12. $Y = (X - 5)^2 + 4$ _____14. $Y = (X - 5)^2 - 4$

A. 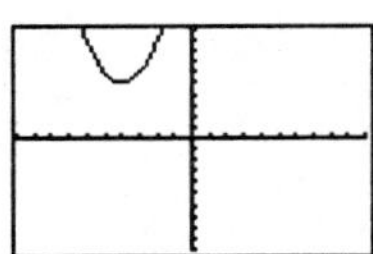   B. 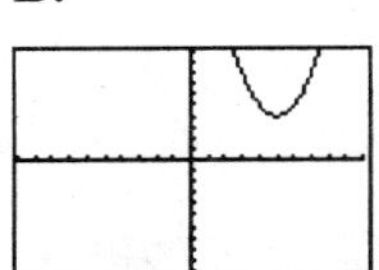   C. 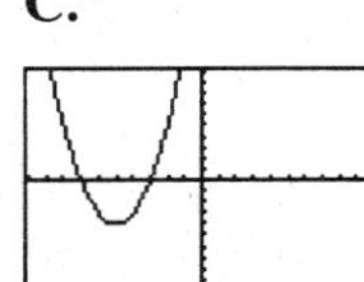   D. 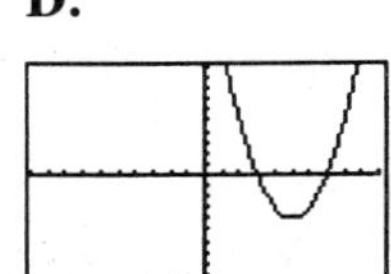

CONCLUSIONS: Answer the following questions without graphing the equation on the graphing calculator. You may then go back and check your answers with the calculator. Some questions may have more than one correct response.

_____15.   Which of these graphs will have its vertex at the origin?
a. $Y = (X - 5)^2$
b. $Y = X^2$
c. $Y = (4/5)X^2$
d. $Y = 2X^2 + 7$
e. $Y = 4(X + 3)^2 - 7$

_____16.   Which of these is the graph of $Y = X^2$ translated (shifted) two units to the left of the Y-axis?
a. $Y = 2X^2$
b. $Y = (X + 2)^2$
c. $Y = X^2 - 2$
d. $Y = (X + 2)^2 - 2$
e. $Y = (X - 2)^2$

_____17.   Which of these is the graph of $Y = X^2$ translated 2 units down from the
           X-axis?
           a. $Y = 2X^2$
           b. $Y = (X + 2)^2$
           c. $Y = X^2 - 2$
           d. $Y = (X + 2)^2 - 2$
           e. $Y = (X - 2)^2$

_____18.   Which of these graphs has a maximum point?
           a. $Y = 2X^2$
           b. $Y = -2(X - 2)^2$
           c. $Y = X^2 - 2$
           d. $Y = (X + 2)^2 - 2$
           e. $Y = 2 - X^2$

✍19. a. Based on your observations, for the parabola $y = a(x - h)^2 + k$, what effect does
         the sign of "a" have on the orientation of the parabola?

     b. As $|a|$ increases, what affect does it have on the size of the parabola?

     c. The value of h will shift (translate) the parabola which direction?

     d. The value of k will shift (translate) the parabola which direction?

20. Consider the quadratic equation $Y = 2X^2 - 12X + 19$. Enter
     this equation on the calculator and sketch the graph on the
     display at the right.

     a. Access the **CALC** menu and select **[3:minimum]** (or
     4:maximum as appropriate) to find the vertex of the parabola.
     The vertex appears to be (3,1). Confirm this by setting the
     TABLE start/min to 0, and accessing the TABLE. Notice that coordinates are
     symmetric on either side of (3,1).

TI-85/86	FMIN AND FMAX ARE LOCATED UNDER THE MATH SUBMENU OF THE GRAPH MENU OF BOTH OF THESE CALCULATORS.
	TI-85 USERS ONLY: SINCE THERE IS **NO TABLE** FEATURE, YOU SHOULD CONSIDER USING THE **EVAL** OR **EVALF** FEATURES. FOR A REVIEW OF THE USE OF **EVAL OR EVALF**, SEE THE TI-85 NOTE AT THE END OF UNIT 10.

NOTE: When using the graph screen to determine coordinates, you should be aware that the display coordinate values approximate the actual mathematical coordinates. The accuracy of these display values is determined by the height and width of the pixel space being displayed. The space height/width formulas are discussed in detail in Unit 17.

   b. Previously the vertex of the parabola could be read from the equation that was in the form $y = a(x - h)^2 + k$. However, when the equation is in the form $y = ax^2 + bx + c$ you must complete the square on x to be able to read the vertex.

   c. Since the vertex is at (3,1), then we can expect the equation to have the form $y = a(x - 3)^2 + 1$. The only value we do not know is "a". Can you predict the value of **a**?_______

   d. To complete the square on X of $Y = 2X^2 - 12X + 19$, the first step would be to factor out the coefficient of X^2, i.e. $Y = 2(X^2 - 6X + \underline{\quad}) + 19$.

   Continuing this process produces:
   $Y = 2(X^2 - 6X + \underline{9}) + 19 - \underline{2(9)}$
   $Y = 2(X - 3)^2 + 1$

   Notice **a** = 2. Was your prediction correct? Will **a** always be equal to the coefficient of the x^2 term? Why or why not?

21.   a. Graph the equation $Y = 4X^2 + 8X - 2$. Sketch the display.

   b. Use the "minimum" option of the CALC menu to find the vertex of the parabola. Record the vertex coordinates:

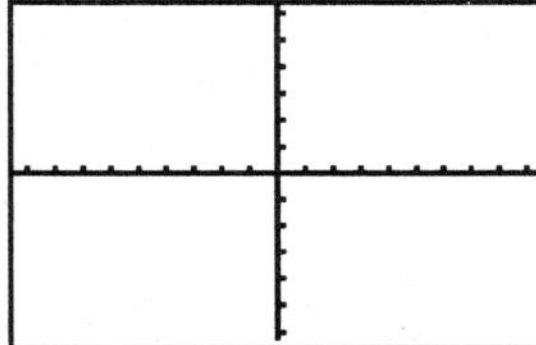

   (_____ , _____)

   c. Using what you learned in #20, write the equation of the parabola $Y = 4X^2 + 8X - 2$ in $y = a(x - h)^2 + k$ form:_______________________

   d. Complete the square on $Y = 4X^2 + 8X - 2$ to write the equation in $y = a(x - h)^2 + k$ form to be sure that the result agrees with your response in letter **c**.

<u>**Solutions:**</u> **2.** (0,0), (0,0), (0,0) **3.** It affects the width of the parabola. **4.** (0,3), (0,1), (0,-2) **5.** It moves the parabola up or down the Y-axis. **6.** (-6,0), (-1,0), (3,0) **7.** It shifts the parabola left or right along the X-axis. **8.** (-3,2), (-3,-2), (3,2), (3,-2) **9.** (-115,-38) **10.** The "3" made the graph "steeper". The "¼" made the graph "flatter". The negative sign turned the graph "upside down".

11. c **12.** b **13.** a **14.** d **15.** b,c **16.** b,d **17.** c,d **18.** b,e **19a.** If "a" is positive, the parabola opens up and has a minimum. If "a" is negative , the parabola opens down and has a maximum.

19b. $|a|$ determines the width of the parabola. **19c.** "h" shifts the parabola right or left.

19d. "k" shifts the parabola up or down. **21b.** (-1,-6)   **21c.** $y = 4(x + 1)^2 - 6$

UNIT 21
TRANSLATING AND STRETCHING GRAPHS

*Unit 17 is a prerequisite for this unit. Answers appear at the end of the unit.

Translations

When the graphs of two curves are identical except for their location on the coordinate plane, then a translation has occurred. These translations may be either horizontal, vertical or both. The unit "Discovering Parabolas" was an in-depth look at the graphs of parabolic functions and their corresponding equations. You may want to complete the unit "Discovering Parabolas" before working on this unit; however, it is not a prerequisite. The exercises in this unit explore the translations of various graphs.

Before proceeding, set the viewing WINDOW to a standard viewing window.

EXERCISE SET

1. a. Each of the following equations is in the form
$y = f(x) + c$. The reference graph $f(x) = X^2$ is displayed.
Graph $Y = X^2 + 4$ and $Y = X^2 - 4$ on this same set of axes, labeling each graph.
The range of $Y = X^2 + 4$ is __________________.
The range of $Y = X^2 - 4$ is __________________.
The domain for both graphs is $\Re$.

   b. The next group of graphs is in the form $y = f(x + b)$.
The reference graph of $f(x) = X^2$ is displayed.
Graph $Y = (X + 5)^2$ and $Y = (X - 5)^2$ on this same set of axes, labeling each graph.
The domain for both graphs is ______________ and the range for both graphs is __________________.

   ✍ c. CONCLUSION: In the general form
$y = f(x + b) + c$, horizontal shifts result from changes in the variable ________ and vertical shifts result from changes in the variable ________ .

   d. In the case of a parabola, does the horizontal shift ever affect the domain or range?

   e. Does the vertical shift affect the domain or range?

f. Based on the above conclusions, the graph of y = (x - 28)² + 12 should translate
vertically _______ units _______ (up/down) and horizontally _______ units _______
(left/right) when compared to f(x) = x².
This would locate the vertex of the parabola at the coordinates (_____ , _____).

2.   a. Using the same viewing WINDOW, consider the
following absolute value equations in the form
y = f(x) + c. The graph of f(x) = |X| has been
graphed as the reference graph. Graph Y = |X| + 4 and
Y = |X| - 4 on this same set of axes, labeling each
graph.
The range of Y = |X| + 4 is _________________.
The range of Y = |X| - 4 is _________________.
The domain for both graphs is ℜ.

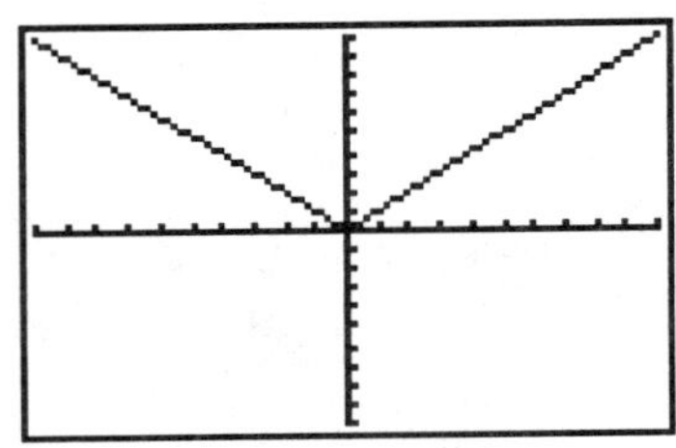

b. The next group of graphs is in the form y = f(x + b).
The reference graph of f(x) = |X| is already displayed.
Graph Y = |X + 5| and Y = |X - 5| on this same set of
axes, labeling each graph.
The domain for both graphs is _____________ and the
range for both graphs is _________________.

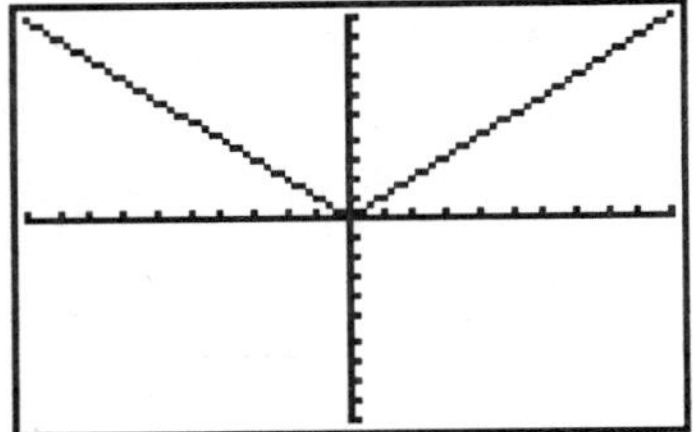

✎ c. CONCLUSION: In the general form
y = f(x + b) + c, horizontal shifts result from changes in the variable _______ and
vertical shifts result from changes in the variable _______ .

d. In the case of an absolute value function, does the horizontal shift ever affect the
domain or range?

e. Does the vertical shift affect the domain or range?

f. Based on your conclusions, the graph of y = |x + 32| - 42 should translate
vertically _______ units _______ (up/down) and horizontally _______ units _______
(left/right) when compared to the graph of f(x) = |x|.
This would locate the vertex of the absolute value function at the coordinates
(_____ , _____).

3.   a. Using the same viewing WINDOW, consider the
following square root functions in the form
y = f(x) + c. The graph of f(x) = √X has been
graphed as the reference graph. Graph Y = √X + 4
and Y = √X - 4 on this same set of axes, labeling each
graph.
The range of Y = √X + 4 is _________________.
The range of Y = √X - 4 is _________________.
The domain of both graphs is _________________.

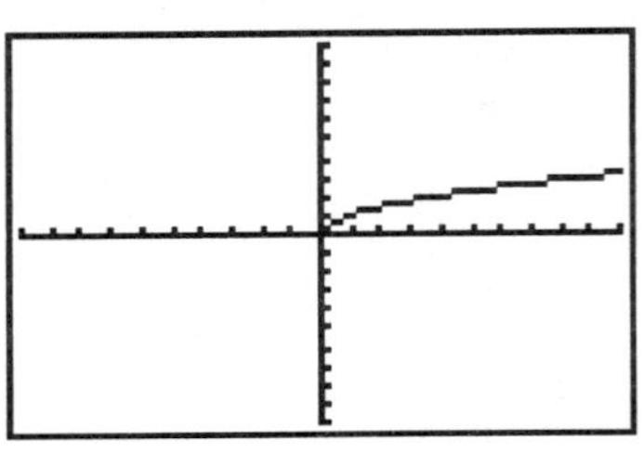

b. The next group of graphs is in the form $y = f(x + b)$.
The reference graph of $f(x) = \sqrt{X}$ is displayed.
Graph $Y = \sqrt{X + 5}$ and $Y = \sqrt{X - 5}$ on this same set of
axes, labeling each graph.

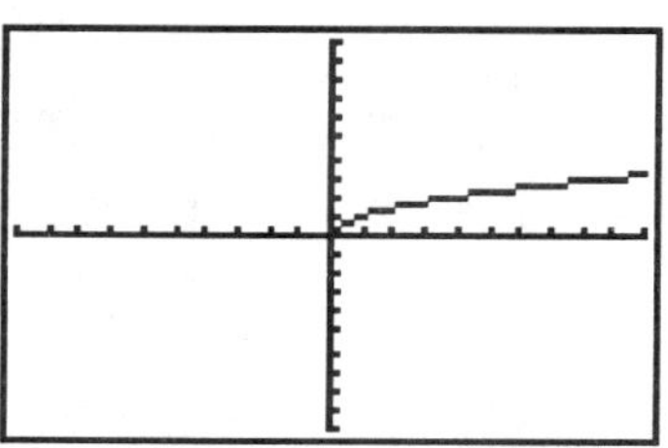

The domain of $Y = \sqrt{X + 5}$ is _________________ .

The domain of $Y = \sqrt{X - 5}$ is _________________ .

The range for both graphs is _________________ .

c. CONCLUSION: In the general form $y = f(x + b) + c$, horizontal shifts result from changes in the variable _______ and vertical shifts result from changes in the variable

_______ .

d. In the case of a square root function, does the horizontal shift ever affect the domain or range?

e. Does the vertical shift affect the domain or range?

f. Based on your conclusions, the graph of $y = \sqrt{x + 23} - 31$ should translate
horizontally _______ units _________ (left/right) and vertically _______ units _______
(up/down) when compared to $f(x) = \sqrt{x}$.
This would locate the initial point of the square root curve at the coordinates
(_____ , _____).

Unlike translations, when a graph is stretched its location (and hence domain and range) are not affected. (Note: This does not apply to trigonometric functions.) Stretching affects only the size of the graph. We will continue to use a standard viewing WINDOW.

4.   Graph each of the following groups of graphs that are in the form $y = a \cdot f(x)$. The reference graph, listed first in the series, is graphed for you. Label each graph on the display.

$Y = X^2, \; Y = 4X^2, \; Y = (\tfrac{1}{2})X^2$ $\qquad\qquad\qquad$ $Y = |X|, \; Y = 4|X|, \; Y = (\tfrac{1}{2})|X|$

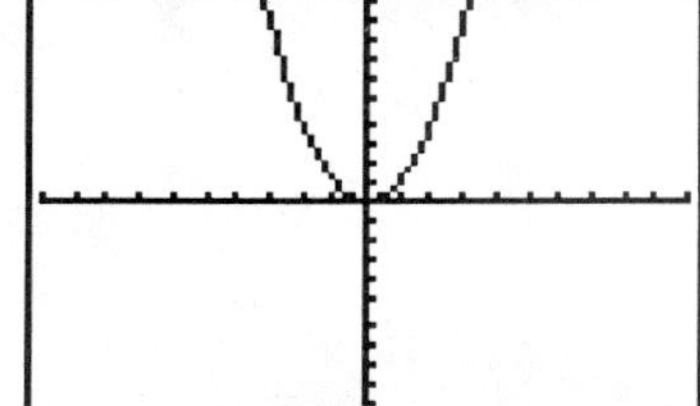 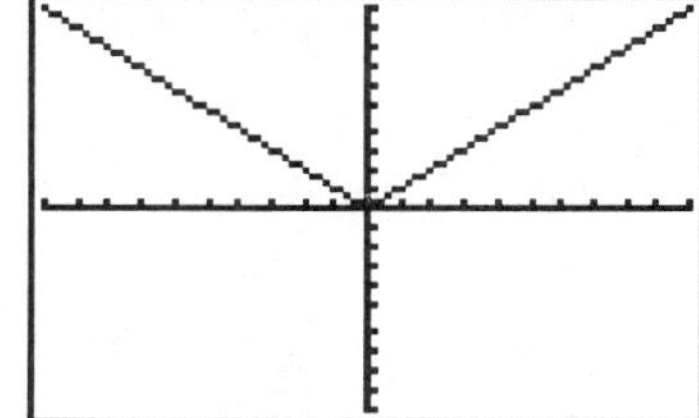

$Y = \sqrt{X}$, $Y = 4\sqrt{X}$, $Y = (\tfrac{1}{2})\sqrt{X}$

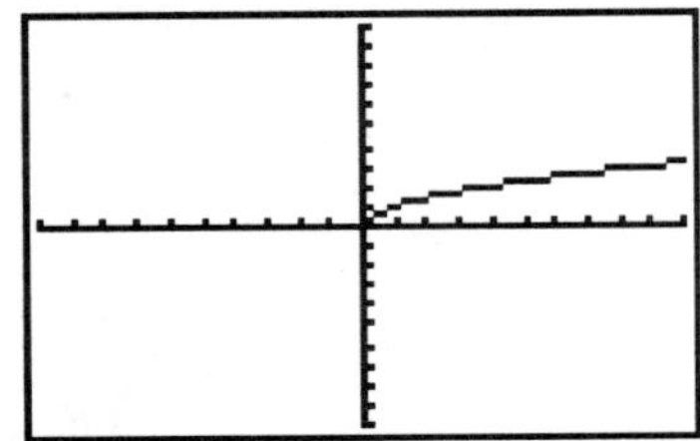

a. In general, if a > 0, what effect does **a** have on the graph of y = a · f(x)?

b. What will happen to the graph of y = a · f(x) if a < 0? (If need be, go back to the above problems and graph when a < 0 instead of a > 0.)

5. The graph of f(x) = $\sqrt{x}$ swings right and up, f(x) = $-\sqrt{x}$ swings right and down. What must be done to f(x) so that the graph of the square root function swings left?

6. Write the equation of a square root function whose domain is {X| X ≤ 1} and whose range is {Y| Y ≥ -3}.

7. Write the equation of a square root function whose domain is {X| X ≤ -1} and whose range is {Y| Y ≤ 3}.

8. Write the equation of a square root function whose domain is {X| X ≥ 4}, whose range is {Y| Y ≤ 5}.

9. Write the equation of an absolute value function whose domain is ℜ, whose range is {Y| Y ≤ 5}.

<u>**SOLUTIONS:**</u> **1. a.** range of Y = X^2+4: Y≥4, range of Y = X^2-4: Y≥-4   **b.** In the next group of graphs the domain is $\Re$ and the range is Y≥0 for both graphs. **c.** Conclusion: "b" affects horizontal shift and "c" affects vertical shift. **d.** No   **e.** The vertical shift affects only the range.
f. Thus y = $(x-28)^2$ +12 translates vertically 12 units and horizontally 28 units with the vertex at (28,12).

2. a. range of Y = $|X|$ +4: Y≥4, range of Y = $|X|$ - 4: Y≥-4   **b.** In the next group of graphs the domain is $\Re$ and the range is Y≥0 for both graphs. **c.** Conclusion: "b" affects horizontal shift and "c" affects vertical shift. **d.** No   **e.** The vertical shift affects only the range.
f. Thus y = $|x+32|$ - 42 translates vertically 42 units down and horizontally 32 units left with the vertex at (-32, - 42).

3. a. range of Y = $\sqrt{X}$ + 4 : Y≥4, range of Y = $\sqrt{X}$ - 4: Y≥-4. The domain of both graphs is X≥0.
b. In the next group of graphs the domain of Y = $\sqrt{X + 5}$ is X≥-5 and the domain of Y = $\sqrt{X - 5}$ is X≥5. The range is Y≥0 for both graphs. **c.** Conclusion: "b" affects horizontal shift and "c" affects vertical shift. **d.** Only the domain is affected. **e.** Only the range is affected.
f. Thus y = $\sqrt{X + 23}$ - 31 translates horizontally 23 units left and vertically 31 units down with the initial point of the curve at (- 23,- 31).

4. a. "a" affects the "width" of the graph. **b.** If a<0 the graph will be symetric across the X-axis to the graph whose "a" is positive.

5. The opposite of the radicand must be graphed.   **6.** f(x) = $\sqrt{1 - x}$ - 3

7. f(x) = $- \sqrt{-1 - x}$ - 3   **8.** f(x) = $- \sqrt{x - 4}$ + 5   **9.** f(x) = $-|x|$ + 5

145

UNIT 22
SYMMETRY OF FUNCTIONS

*Unit 17 is a prerequisite for this unit. Answers appear at the end of the unit.

Finding the X and Y-intercepts is helpful in adjusting the WINDOW values on the calculator so that a complete graph is displayed. Also helpful in sketching (and determining a good viewing window) is the concept of symmetry.

> **TI-85** THROUGHOUT THIS UNIT IT IS SUGGESTED THAT YOU USE THE **EVALF** FEATURE ON THE **CALC** MENU TO COMPLETE TABLES OR THAT YOU **TRACE** ON THE GRAPH, IN A ZDECM X 2 SCREEN, TO LOCATE DESIRED COORDINATES.

SYMMETRY ABOUT THE Y-AXIS

Graph the equation $Y = X^2 - 4$ (use a ZDecimal x 2 size screen).

Is the graph symmetric with respect to the Y-axis? In other words, if the graph could be folded along the Y-axis, does the left-side <u>exactly</u> match the right side? To confirm that the left side exactly matches the right side (the curve has y-axis symmetry) then points of the curve will have ordered pairs of the form (X, Y) and (-X, Y). Notice the coordinates of the points marked on the graphs below:

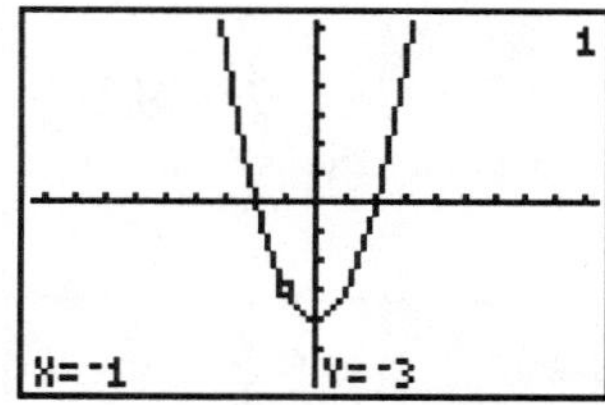

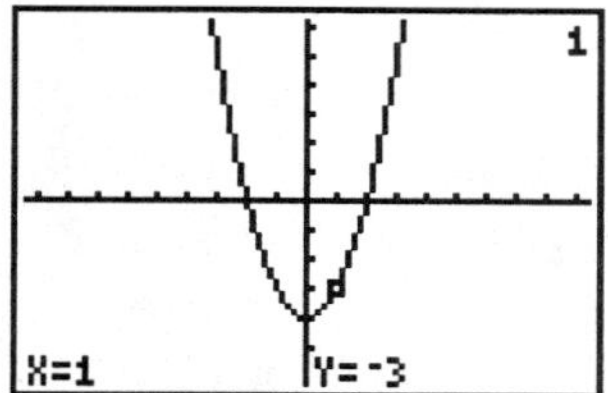

When the **TABLE** feature of the calculator is accessed (beginning at X = -3 and incremented by 1), the symmetry becomes numerically apparent.

X	Y1	
-3	5	
-2	0	
-1	-3	
0	-4	
1	-3	
2	0	
3	5	
X=-3		

This confirms, but *does not prove*, that the graph of $Y = X^2 - 4$ is symmetric with respect to the Y-axis.

> **TI-85** USE THE **EVALF(** FEATURE TO CONSTRUCT THE TABLE ENTRIES.

Graph the equation $Y = \dfrac{3}{X}$.

Is the graph symmetric with respect to the origin? Imagine rotating the graph 180° about the origin. The "top" part of the graph would line up with the "bottom" part if there is symmetry about the origin. Points on the curve will have ordered pairs of the form (X, Y) and (-X, -Y). Notice the coordinates on the graphs below:

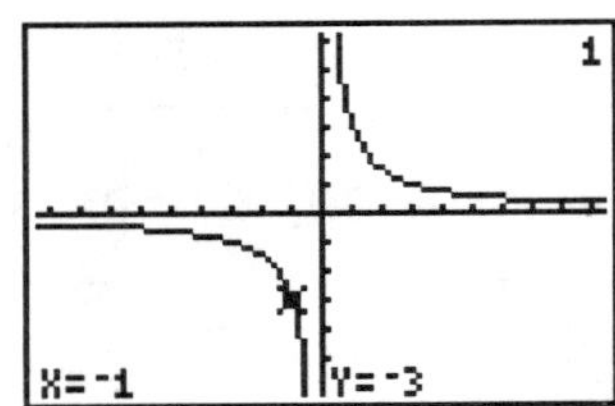

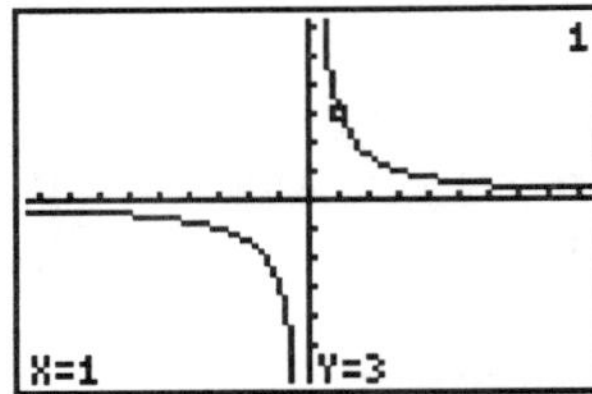

Set the TABLE to begin at X = -3 and to be incremented by 1.

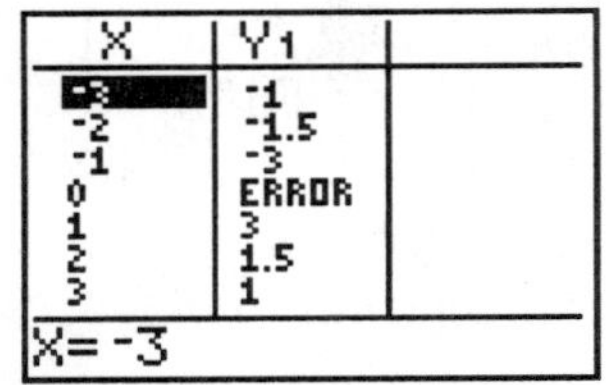

Because coordinates of the form (X, Y) and (-X, -Y) are on the curve, the graph is symmetric with respect to the origin.

EXERCISE SET

1. Each pictured graph below is symmetric with respect to the Y-axis. Using the information displayed at the bottom of the screen, state the coordinates of another point which also lies on the graph.

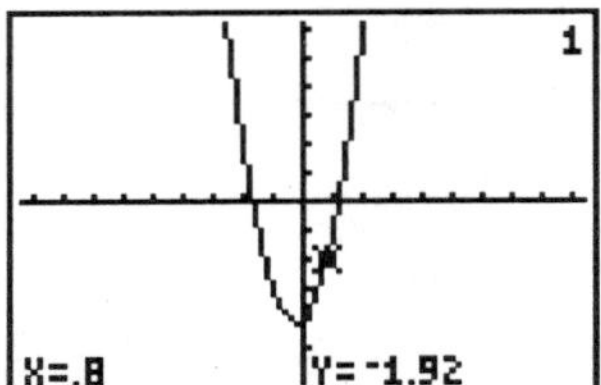

Coordinates:______________

Coordinates:______________

2. Each pictured graph below is symmetric with respect to the origin. Using the information displayed at the bottom of the screen, state the coordinates of another point which also lies on the graph.

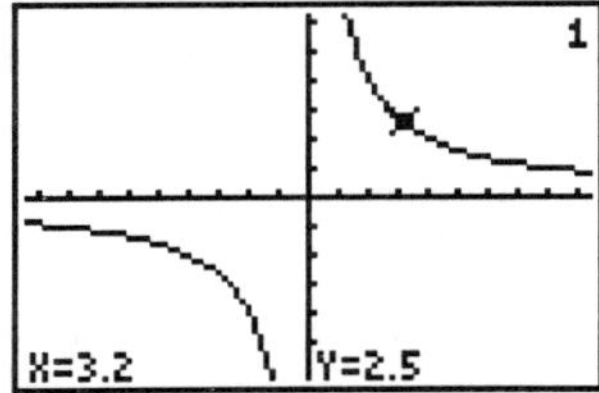

Coordinates:______________

Coordinates:______________

DIRECTIONS: Sketch the graph of each of the following polynomial functions on the grid provided. Use a ZDecimal x 2 screen. State whether the function is symmetric with respect to the Y-axis, and/or with respect to the origin, or has no symmetry. Justify the symmetry (or lack of) by stating the coordinates of three <u>pairs</u> of points whose coordinates confirm the Y-axis symmetry or symmetry about the origin.

3. $Y = |X|$ 

Type(s) of symmetry:

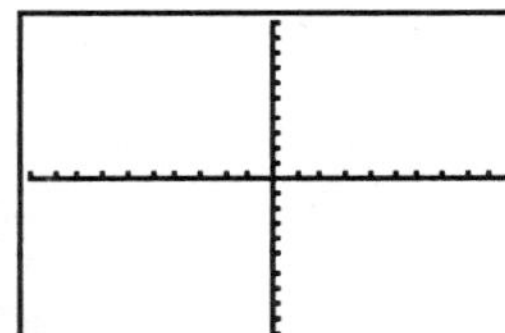

Justification:

4. $Y = X^4 - 2X^2$

Type(s) of symmetry:

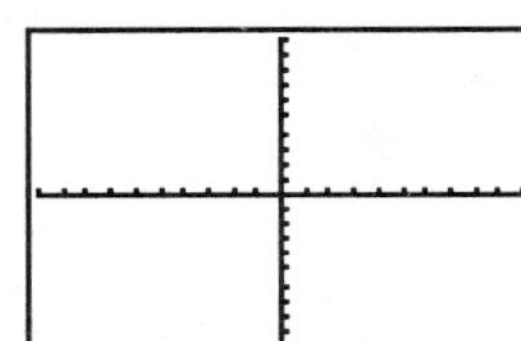

Justification:

5. $Y = X^4 + X^3 - 2X^2$

Type(s) of symmetry:

Justification:

6. $XY = 8$

Type(s) of symmetry:

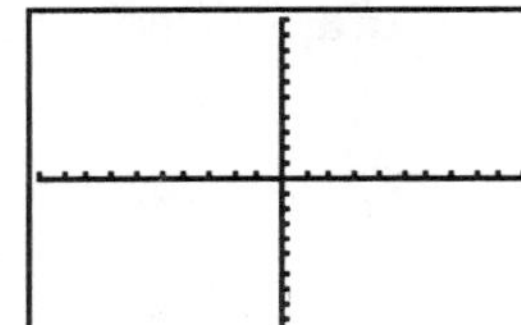

Justification:

7. $Y = X^3 - X$

Type(s) of symmetry:

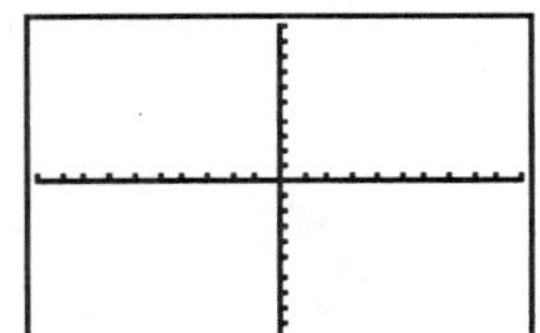

Justification:

8. $Y = X^3 + 3$

Type(s) of symmetry:

Justification:

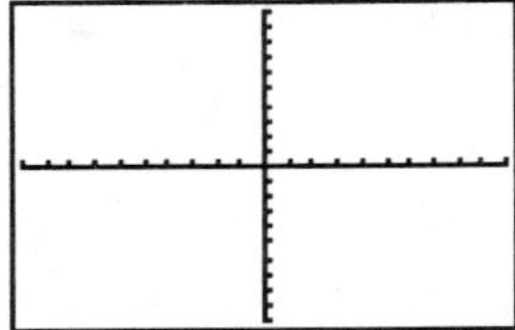

✍9.   A function that is symmetric about the Y-axis is called an <u>even</u> function whereas one whose graph is symmetric about the origin is called an <u>odd</u> function. Use the calculator to explore polynomial functions. Consider graphs of polynomial functions whose powers of the variable X are (a) even, (b) odd, (c) a combination of even and odd. What conclusions (if any) can be made about these functions and symmetry?

✍10. This unit has only examined symmetry about the Y-axis and the origin. Another type of symmetry is symmetry about the X-axis. The graphs of these equations are <u>NOT</u> functions. Explore symmetry about the X-axis by graphing $Y^2 = X$. (Hint: You must graph TWO curves that together **represent** the graph of $Y^2 = X$.) In conclusion, specify the form taken by the ordered pairs that satisfy this relation.

<u>Solutions:</u>

1. (-.8,-1.92) and (-1.8,4.0176)

2. (-3.2,-2.5) and (1.2,1.728)

Type of Symmetry	Justification (points may vary)
3. Y-axis	(-3,3) (-2,2) (-1,1) (3,3) (2,2) (1,1)
4. Y-axis	(-3,63) (-2,8) (-1,-1) (3,63) (2,8) (1,-1)
5. None	(-3,36) (-2,0) (-1,-2) (3,90) (2,16) (1,0)
6. Origin	(-3,-2.667) (-2,-4) (-1,-8) (3, 2.667) (2, 4) (1, 8)
7. Origin	(-3,-24) (-2,-6) (-1,0) (3, 24) (2, 6) (1,0)
8. None	(-3,-24) (-2,-5) (-1,2) (3, 30) (2,11) (1,4)

9. Answers may vary.   10. Answers may vary.

UNIT 23
PIECEWISE FUNCTIONS

*Unit 18 is a prerequisite for this unit. Answers appear at the end of the unit.

This unit examines piecewise functions; functions in which f(x) varies for different intervals of the domain. In these functions, the domain is segmented into a finite number of pieces. The **TEST** menu will be used to graph the different pieces of the function. Because we <u>do not</u> want the pieces connected, set the calculator mode to dot at this time.

TI-83/83plus and TI-86 users have the option of changing to **DOT** on the **MODE** screen or simply setting the graphing style icon on the **Y =** menu to DOT for each equation graphed. In either case, the graphing style icon should reflect the dot graphing format and should be verified each time a new graph is displayed.

> **TI-85**   CHANGE TO **DOT** MODE VIA THE **GRAPH/FORMAT** MENU BY PRESSING [GRAPH] [MORE] [F3](FORMT).
> SELECT **DRAWDOT** AND PRESS [ENTER] TO ACTIVATE.

TEST MENU

How does the TEST menu work? When an inequality symbol from the TEST menu is designated in the function, the calculator evaluates the function for all real numbers in the domain. For each value tested that **IS** in the designated interval, the calculator returns a **1** and for each value tested that **IS NOT** in the designated interval the calculator returns a **0**. Thus, the **1** turns the point **on** and allows it to be 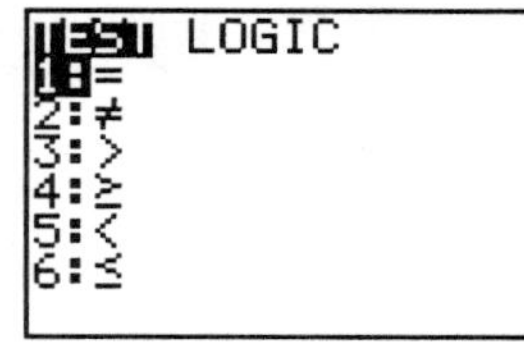
displayed, whereas, the **0** turns the point **off** so it is not displayed as part of the function. The following examples demonstrate the calculator's response.

Example 1:   Graph the piecewise function f(X) = X + 2 when X ≤ 5 on the grid provided.

Solution: Because the domain is X ≤ 5, the graph of the equation will <u>**begin**</u> at X = 5 and pass through all points whose X value is less than or equal to five. Plot the ordered pairs and draw the **ray** which represents the graph of f(X) = X + 2 when X ≤ 5.

X	Y
5	7
4	6
3	5

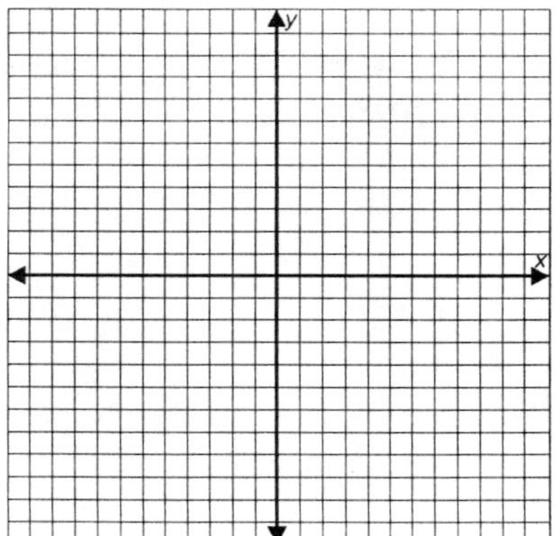

Example 2: How does the calculator plot ordered pairs for the function f(x) = X + 2 when X ≤ 5?

Solution: Enter the function at the Y1 = prompt. To designate the restriction on the domain, X ≤ 5, the function should be entered as displayed at the right. To plot points, the calculator substitutes values for X and evaluates the function to determine the corresponding Y values. When an X value is substituted into the **TEST** expression (X≤5), the calculator **tests** to see if that value is <u>less than</u> or <u>equal</u> to 5. If it is less than or equal to 5, the calculator turns the point *on* and plots it. If the value is <u>greater than</u> 5, the calculator turns the point *off*, preventing it from being displayed as part of the function. The notation for *on* and *off* are **1** and **0**, respectively. With the viewing WINDOW in ZInteger, press [GRAPH] to display the screen at the right. This display should match the hand drawn graph on the previous page.

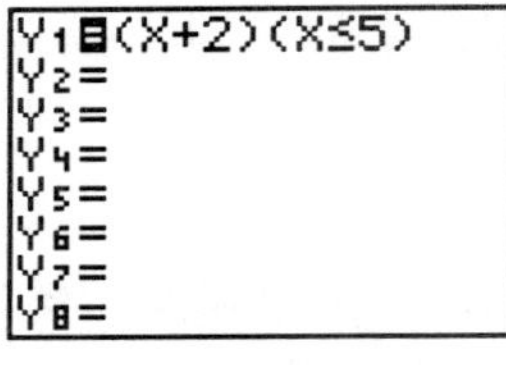

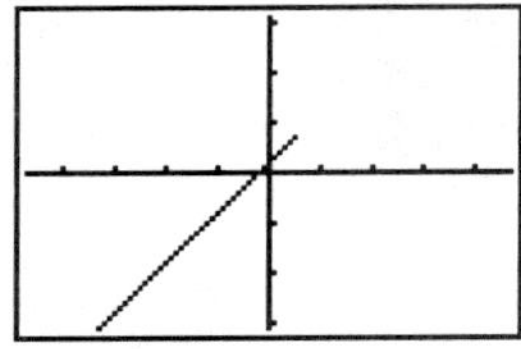

The table below demonstrates the calculator's response to various values of X:

X	(X + 2)(X ≤ 5)	Point on/off?	f(X)	ordered pair
3	(3 + 2)(1)	ON	5	(3,5)
4	(4 + 2)(1)	ON	6	(4,6)
5	(5 + 2)(1)	ON	7	(5,7)
6	(6 + 2)(0)	**OFF**	0	(6,0)
7	(7 + 2)(0)	**OFF**	0	(7,0)
8	(8 + 2)(0)	**OFF**	0	(8,0)

Observe that the last three X-values (6,7,8) returned a zero (and were thus turned off) since they were not part of the domain, X ≤ 5. When the collective group of points is graphed, with those values of X ≤ 5 being plotted, the desired graph is displayed. ◆

Points that are not part of the designated function have an f(X) value of 0 and are plotted on the X-axis where their display is not apparent. NOTE: TI-83/83plus and TI-85/86 users will find the Axes On/OFF command on the **FORMAT** screen. Highlight **AxesOff** on this screen and press [ENTER] to activate the command. (TI-82 users turn the Axes Off on the **WINDOW FORMAT** screen, by pressing [WINDOW] [▶] and cursoring down and over to highlight AxesOff. Press [ENTER] to activate the command and view the result.)

TURN ON the axes TURN OFF the Y1 function before proceeding. DO NOT delete it.

Example 3: Graph the function f(X) = X + 2 when X > 5.
Solution: At the Y2 = prompt enter (X + 2)(X > 5) and press [GRAPH]. The display you see is the section of f(X) = X + 2 when X ≤ 5 that was not graphed in the previous example. Now TURN ON the Y1 = function and press [GRAPH]. By combining the graphs of both f(X) = X + 2 when X ≤ 5 and f(X) = X + 2 when X > 5 you have displayed the graph of f(X) = X + 2 (where there are no restrictions on the domain). By placing restrictions on the domain only a piece of the whole function is graphed. ◆

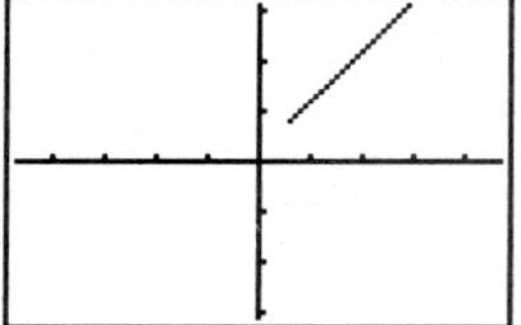

TURN OFF both Y1 and Y2 before proceeding.

Example 4:   Graph the function f(X) = X + 2 when -5 < X < 5.

Solution: Enter (X + 2) at the Y3 = prompt. Now examine
- 5 < X < 5. This would be read: X is greater than negative five
(X>-5) **AND** X is less than five (X<5). Thus at the Y3 = prompt you
should enter (X + 2)(X > -5)(X < 5). Press **[GRAPH]**. Your display
should correspond to the one at the right.

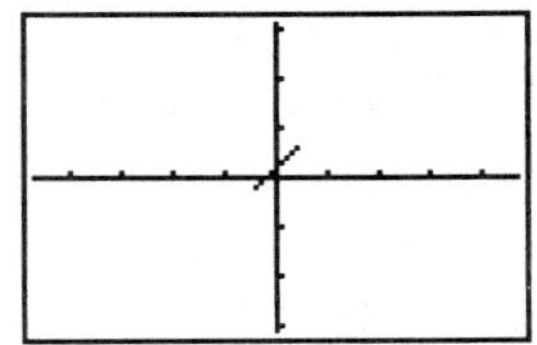

EXERCISE SET

Directions: WITHOUT GRAPHING, match each of the piecewise functions defined below
with their graph. All pictured graphs are in the ZStandard viewing window.

_____1.   f(X) = X² + 1 when X ≥ 0

_____2. $f(X) = \begin{cases} \sqrt{X} + 2 & \text{when } X \geq 0 \\ 2X & \text{when } X < 0 \end{cases}$

_____3. $f(X) = \begin{cases} 3 & \text{when } X \geq 0 \\ -3 & \text{when } X < 0 \end{cases}$

_____4. $f(X) = \begin{cases} X + 4 & \text{when } X \geq 2 \\ -2X - 5 & \text{when } X \leq 0 \end{cases}$

a.   b.   c.   d.

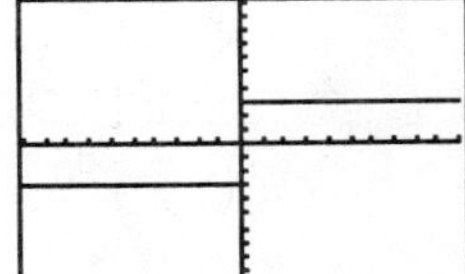 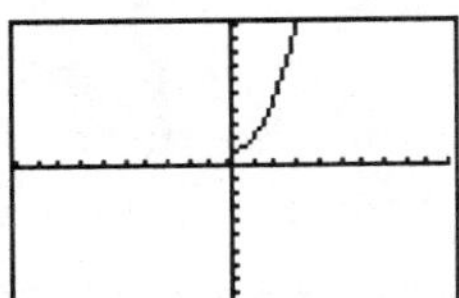 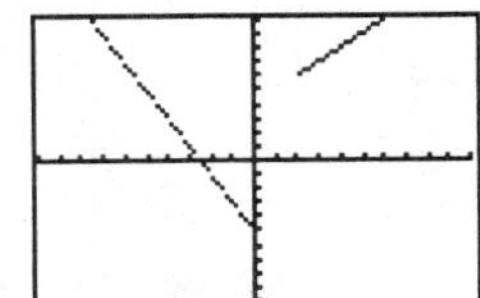 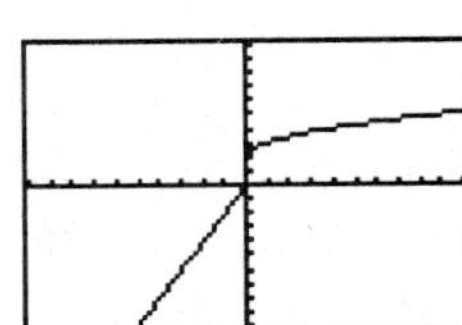

Directions: Graph each piecewise function in the ZStandard viewing WINDOW.

5.   f(X) = -2X² + 5 when X < 1

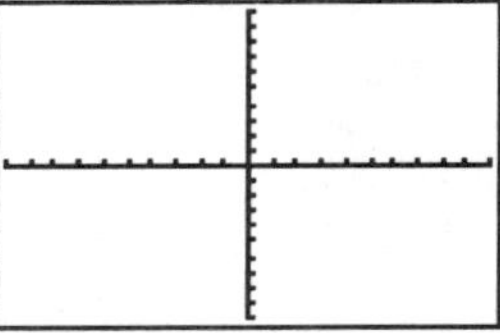

6.   f(X) = |X + 3| when -6 < X < 0

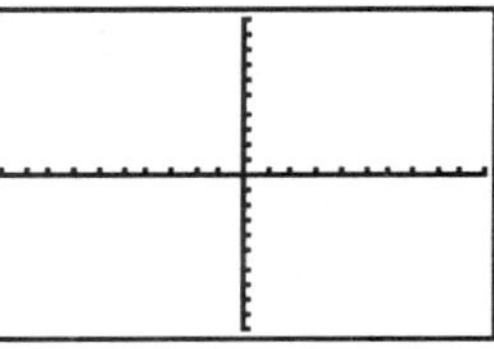

SETTING WINDOWS

When setting WINDOW values for piecewise functions, the Ymin and Ymax values should be determined by the methods discussed in the unit titled "Where Did the Graph Go?". The X-values will, however, be determined by the restrictions placed on the domain in the problem. In the first example, f(X) = X + 2 when X ≤ 5. Thus the Xmax would be at least 5 and the minimum would need to be a value which will yield a "complete" or "satisfactory" graph (as explained in "Where Did the Graph Go?"). Since f(X) = X + 2 is a linear function, it would be acceptable to set the Xmin = -5. The next two examples consider other possibilities.

Example 5:   Determine the viewing WINDOW for f(X) = (X + 6)² + 4 when X < -3.

Solution: Graph the function in the ZStandard viewing WINDOW on the screen at the right. At the **Y=** prompt enter **((X + 6)² + 4)(X < -3).**
Compare your graph to the one pictured. How can you be confident that you are seeing a satisfactory graph of the **piecewise** function?
To assure a satisfactory graph of piecewise functions you must consider the effect of the domain on the function, as well as those points which display the graph's interesting features.

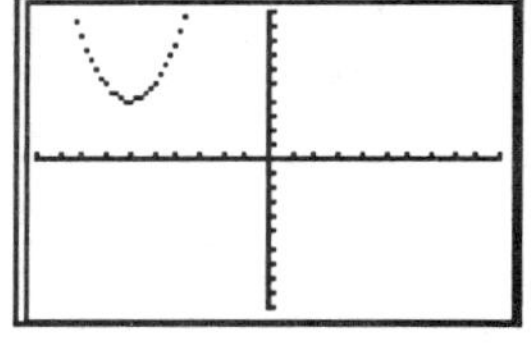

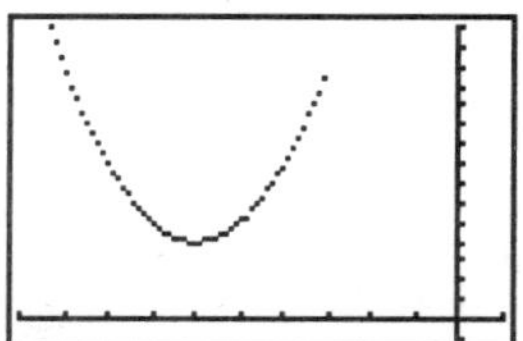

The Xmax should be at least -3 since the domain is restricted to all X < -3. Since this a parabolic function, with vertex at (-6,4), the Xmin needs to be less than -6. Remember, a satisfactory graph displays all the interesting features of the graph. The two interesting features of this graph are its vertex and the point at which the curve terminates. Thus, be sure to include the Y values that correspond to the maximum/minimum values in the domain. Since X < -3 the graph would need to include the point (-3,13). Thus it can be concluded that:
Xmin < -6, Xmax > -3, Ymin < 4 and   Ymax > 13.   The WINDOW would need to be at least [-6,-3] by [4,13]. A slightly larger WINDOW: [-10,1] by [-1,15] was used. See the graph displayed above.   ◆

Example 6:   Consider the graph of the piecewise function:

$$f(X) = \begin{cases} X^2 + 4X & \text{when} \quad -4 \le X < 0 \\ 5 - 2X & \text{when} \quad 0 \le X < 3 \end{cases}$$

Solution: The function should be read as f(X) = X^2 + 4X when -4 ≤ X < 0 **and** f(X) = 5 - 2X when 0 ≤ X < 3. This will be entered on the calculator at the **Y1 =** prompt as (X^2 + 4X)(X≥-4)(X<0) + (5 - 2X)(X≥0)(X<3). Press **[GRAPH]** to display the function. The screen at the right displays the function when graphed in the smallest applicable viewing WINDOW, [-4,3] by [-4,5]. A satisfactory graph is displayed if the selected WINDOW displays the two pieces of the function shown. See the solutions key at the end of the unit for specific directions on selecting an appropriate viewing WINDOW.

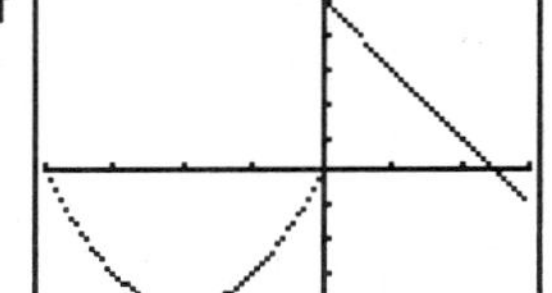

◆

EXERCISE SET CONTINUED

Directions: For each exercise below, complete the following

   a. Determine the Xmin, Xmax, Ymin and Ymax values of the function.
   b. Graph the function using the TEST menu.
   c. State the final viewing WINDOW in the format [Xmin,Xmax] by [Ymin,Ymax].

7. $f(X) = \begin{cases} -2X^2 + 15 & \text{when} \ X \ge 0 \\ |X + 3| & \text{when} \ -12 < X < 0 \end{cases}$

Xmin = _______   Xmax = _______

Ymin = _______   Ymax = _______

WINDOW: [___, ___] by [___, ___]

8. $f(X) = \begin{cases} X^2 + 4X & \text{when} \ -4 \le X < 0 \\ 5 - 2X & \text{when} \ 0 \le X < 3 \\ X - 2 & \text{when} \ 3 \le X < 6 \end{cases}$

Xmin = _______   Xmax = _______

Ymin = _______   Ymax = _______

WINDOW: [___, ___] by [___, ___]

157

9.	The definition of absolute value specifies that $|x| = \begin{cases} x & \text{when } x \geq 0 \\ -x & \text{when } x < 0 \end{cases}$.

Using this definition, write the function $f(x) = |x + 4|$ as a piecewise function.

$$f(X) = \begin{cases} \underline{\hspace{4cm}} \\ \underline{\hspace{4cm}} \end{cases}$$

Establish an appropriate viewing WINDOW and sketch the graph display.

WINDOW: [___, ___] by [___, ___]

Your piecewise function should be entered after the **Y1 =** prompt. To check the accuracy of your piecewise functon, enter $|X + 4|$ after the **Y2 =** prompt. Display both Y1 and Y2. You can verify they are identical by comparing the Y-values in the TABLE feature.

10.	Explore graphing greatest integer functions using the graphing calculator.

Solutions: **1.** 2   **2.** 4   **3.** 1   **4.** 3   **5.** At the Y1 = prompt, enter $(-2X^2 + 5)(X<1)$.

6. At the Y1 = prompt, enter $(abs(X + 3))(X>-6)(X<0)$.

7. At the Y1 = prompt, enter $(-2X^2 + 15)(X\geq0) + (abs(X + 3))(X>-12)(X<0)$, Xmin = -12 and Xmax = ∞, since X$\geq$0, Ymin = -∞ and Ymax = 15. Since (0,15) is the vertex of the quadratic, the WINDOW should be at least [-12,5] by [-10,16].

8. At the Y1 = prompt, enter $(X^2 + 4X)(X\geq-4)(X<0) + (5 - 2X)(X\geq0)(X<3) + (X - 2)(X\geq3)(X<6)$, Xmin = -4, Xmax = 6, Ymin = -∞, Ymax = 5; the WINDOW should be at least [-4,6] by [-4,5].

9. $f(X) = X+4$ when X$\geq$- 4 and $f(X) = -X-4$ when X$<$- 4.   **10.** Answers may vary.

Example 6: Begin by determining the Xmax and Xmin for your viewing WINDOW as discussed in the previous example. Compare the domains: $-4 \leq X < 0$ and $0 \leq X < 3$. Taking the union of these two domains gives $-4 \leq X < 3$ and thus the Xmin and Xmax values are determined. (NOTE: Xmin can be less than -4 and Xmax can be greater than 3 if you wish.) We need to take into consideration the type of functions we are graphing as we consider the Ymax and Ymin . The function $f(X) = 5-2X$ is linear and its only "interesting features" are its starting and stopping points. The starting point would be f(0), since $0 \leq X < 3$ is the domain. Thus our WINDOW needs to include (0,5). The stopping point would be f(3) which indicates our WINDOW should include (3,-1). Based on $f(X) = 5 - 2X$: Ymax $\geq$ 5, Ymin $\leq$ -1. DO NOT forget to take into consideration the graph of $f(X) = X^2 + 4X$. This is a parabolic function. Its "interesting features" will be its starting and stopping points, as well as its vertex. Since the domain is $-4 \leq X < 0$, f(-4) indicates the starting point is (-4,0) and f(0) indicates the stopping point is (0,0). The vertex is (-2,-4). Based on $f(X) = X^2 + 4X$: Ymax $>$ 0, Ymin $<$ -4   CONCLUSION: Ymax $\geq$5 and Ymin $<$ -4. A suggested viewing WINDOW would be [-4,3] by [-4,5]. (Your viewing WINDOW may be larger than the suggested WINDOW but not smaller.)

UNIT 24
RATIONAL FUNCTIONS

*Unit 17 is a prerequisite for this unit. Answers appear at the end of the unit.

In this unit we will examine the graphs of **reduced** rational functions.

The graph of the rational function $f(X) = \dfrac{3X^2 + 6}{X^2 + 1}$ in the **ZStandard** 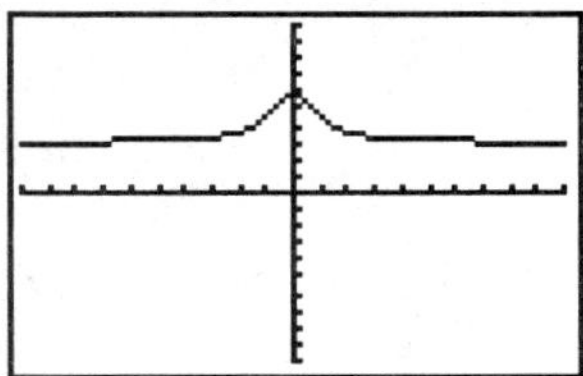

viewing **WINDOW** is displayed at the right.

When tracing along the graph, notice that on either side of the Y-axis
the following occurs: the curve appears to approach $Y = 3$ as the X-values approach
$\pm$ infinity and y appears to have a maximum value of 6.

When examining rational functions, a mere visual display often proves inadequate. Care
must be taken when simply using the visual display. For this reason, we will refer to the
TABLE feature, frequently changing the increments. Examine the coordinates of the graph
as the X-values grow both larger and smaller without bound (i.e. as X approaches positive
infinity and negative infinity). As the X-values approach both positive and negative infinity
the Y-values approach the value 3.

TI-85	TI-85 USERS CONSTRUCT A TABLE USING THE **EVALF** OPTION OF THE CALCULATOR.

This was determined by examining the Y-values in the table as the X-values grow closer to
50. First, set **ΔTbl** to 1, with the table start at 1. Before drawing any conclusions as to
what is happening to the Y-values, reset **ΔTbl** to 0.1. To scroll through the X-values until
X = 50 would be a time consuming task. Instead, change **Indpnt** from **AUTO** to **ASK**. We
will now be able to input specific X-values in the table, pressing **[ENTER]** to return the
corresponding Y-values. Go to the TABLE and enter the numbers 44 through 50 in the
X-column, pressing **[ENTER]** after each entry. It should become apparent that as X grows
large, without bound, Y *approaches* the value of 3.

To examine the Y-values of the graph as X grows smaller without bound (X→ -∞), return to
the top of the TABLE and enter -44 through -50 in the X column. Again, pressing **[ENTER]**
after each entry returns the corresponding Y-values. These, too, approach 3 as X
approaches negative infinity (X→ -∞).

With your pencil, sketch the graph of Y = 3 on the graph of the
function at the right. This line is called a horizontal asymptote. The
graph of this particular function approaches this line but never crosses
it. This means that the distance between the graph and the line
approaches zero as you move **farther and farther out along the line**.
The graph approaches the asymptote from above. 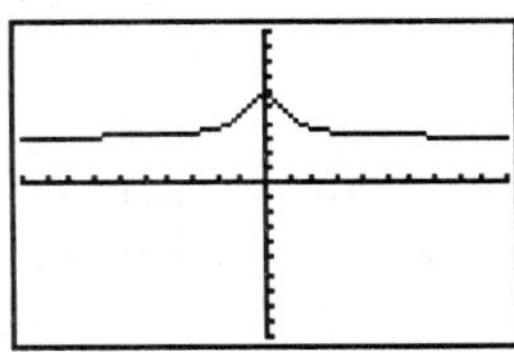

The graph of $f(X) = \dfrac{3X^2 + 6X}{X^2 + 1}$ is displayed in the standard viewing

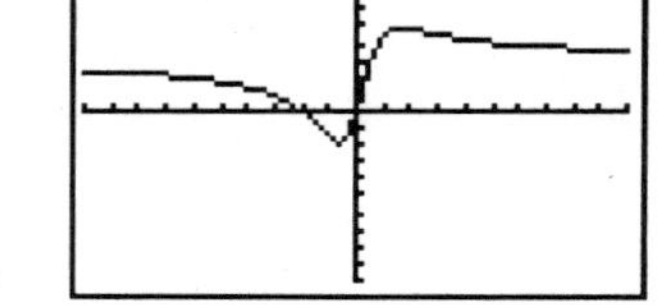

window at the right.

To ascertain if the graph has the same horizontal asymptote of $Y = 3$, either use the TABLE feature or the TRACE feature.

As X grows large without bound ($X \to +\infty$), the graph of f(X) approaches the horizontal asymptote $Y = 3$ from above; as X grows small without bound ($X \to -\infty$) the graph of f(X) approaches the horizontal asymptote $Y = 3$ from below. Note that the graph crosses $Y = 3$. This is acceptable. When discussing asymptotes, we are concerned with the end behavior of the graph.

NOTE: Reset the TABLE to AUTO, TblStart to 0, and change the increment back to 1 unit.

<u>VERTICAL ASYMPTOTES</u>

Before graphing the next function, $f(X) = \dfrac{1}{X - 3}$, examine the TABLE values that it

produces. Enter $\dfrac{1}{X - 3}$ at the Y1 = prompt. Display the TABLE.

When $X = 3$, the Y value reads *ERROR*. This is because 3 is not an element of the domain of the function.   When $X = 3$, the function is not defined. It is at this point that a vertical asymptote occurs. Vertical asymptotes divide the graph into sections.

The graph is displayed in the standard viewing window below left. Since the calculator is in connected MODE, the calculator has connected the two sections of the function across the vertical asymptote. For this reason, the calculator will provide a better display of rational functions when the mode is changed to dot. The following screens display the comparative information. If your calculator utilizes a graph style icon, you should remember that the icon must be reset to dot each time the Y= is cleared and a new expression is entered.

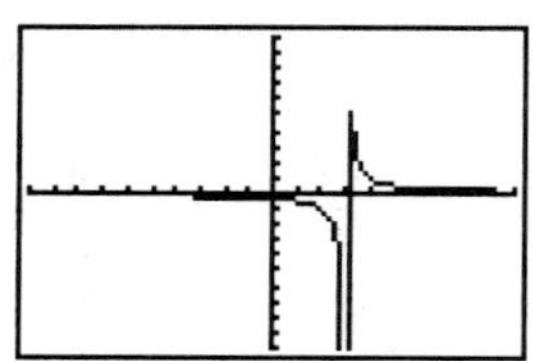

Connected Mode

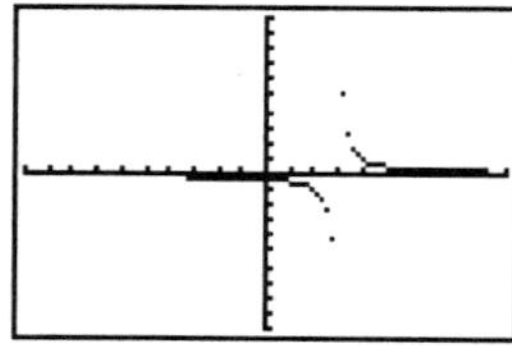

Dot Mode

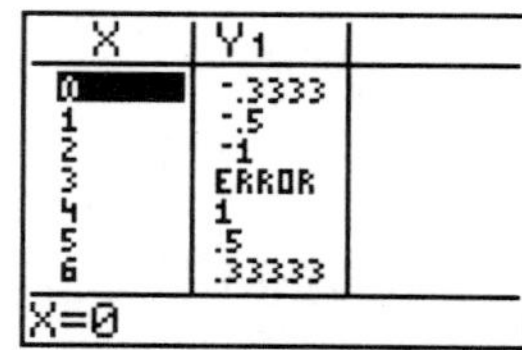

Table

You may TRACE from left to right across the graph and form a conjecture as to what seems to happen to the Y-values as the cursor crosses the vertical asymptote.

To more closely examine the function's behavior around the vertical asymptote of $X = 3$, four TABLES have been displayed. The difference between the tables is the increments (ΔTbl) of the X-values. The independent variable (X) has been incremented by 0.1, 0.01, 0.001 and 0.0001, respectively.

X	Y1
2.7	-3.333
2.8	-5
2.9	-10
3	ERROR
3.1	10
3.2	5
3.3	3.3333

X=3.3

X	Y1
2.97	-33.33
2.98	-50
2.99	-100
3	ERROR
3.01	100
3.02	50
3.03	33.333

X=3.03

X	Y1
2.997	-333.3
2.998	-500
2.999	-1000
3	ERROR
3.001	1000
3.002	500
3.003	333.33

X=3.003

X	Y1
2.9997	-3333
2.9998	-5000
2.9999	-10000
3	ERROR
3.0001	10000
3.0002	5000
3.0003	3333.3

X=3.0003

You should observe the following:

    a. as the graph approaches the asymptote from the left, the Y values approach negative infinity and

    b. as the graph approaches the asymptote from the right the Y values approach positive infinity.

EXERCISE SET

Directions: <u>Without</u> graphing, determine which of the following equations will have one or more vertical asymptotes and write the equation(s) of the asymptote(s) in the blank. If there are no vertical asymptotes, write NONE.

1. $f(X) = \dfrac{11 + X}{12 + 2X}$ __________

2. $f(X) = \dfrac{X - 8}{X^2 - 5X + 6}$ __________

3. $f(X) = \dfrac{5(3 - 2X)}{3}$ __________

4.   <u>Without</u> graphing, answer the following questions that reference the function

$$f(X) = \frac{X - 1}{2} - \frac{3X - 4}{2}:$$

    a. Does the function have a vertical asymptote?________

    b. What requirement is necessary for a vertical asymptote?

161

Slant asymptotes occur in rational functions when the degree of the polynomial in the numerator is one more than the degree of the polynomial in the denominator.

The graph of $f(X) = \dfrac{X^2 - X - 5}{X - 3}$ has a vertical asymptote of $X = 3$ because 3 is a restricted value in the domain.

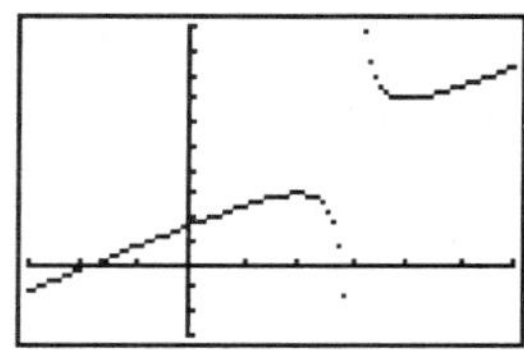

ZBox from the ZOOM menu has been used to adjust the WINDOW to approximately [-3,6] by [-3,10] in the given graphical display.

When the indicated polynomial division is performed,

$f(X) = \dfrac{X^2 - X - 5}{X - 3}$ becomes $f(X) = X + 2 + \dfrac{1}{X - 3}$. As X grows

large without bound, $\dfrac{1}{X - 3}$ decreases in size, i.e. it gets arbitrarily close to zero. This would suggest that for <u>very</u> <u>large</u> values of X,

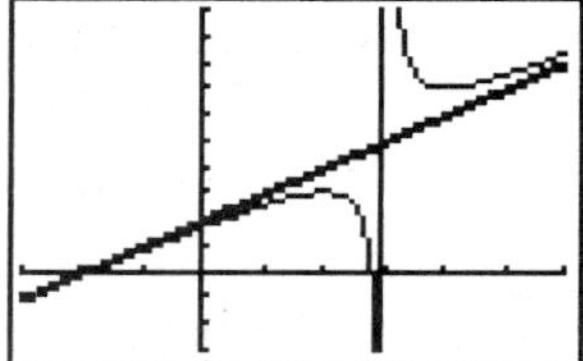

$f(X) = \dfrac{X^2 - X - 5}{X - 3} = X + 2 + \dfrac{1}{X - 3}$ would look much like

the graph of $f(X) = X + 2$. Graph the line $f(X) = X + 2$ at the Y2 = prompt and your graph should look like the one displayed at the right. (Y2 = X + 2 has been displayed as a thick line by using the graph style icon.)

Access the TABLE (make sure the TblStart = 0 and ΔTbl = 1). Scroll up and down the table while observing the values of the functions Y1 and Y2. You should observe that the values of Y1 come arbitrarily close to but never **equal** the values of Y2.

The function $f(X) = X + 2$ is the slant asymptote for $f(X) = \dfrac{X^2 - X - 5}{X - 3}$.

In the previous EXERCISE SET the function $f(X) = \dfrac{X - 1}{2} - \dfrac{3X - 4}{2}$ was examined.

Without graphing you can determine if the function has a slant asymptote. The denominator must contain a variable for the function to have a slant asymptote. This function does not satisfy this requirement.

EXERCISE SET CONTINUED

Directions: Sketch the graph of each function in exercises 5 - 12. Begin in the standard viewing window. If the WINDOW values need to be changed for clarity, record the window coordinates below the screen. Determine the equations of the vertical, horizontal and/or slant asymptotes (if there are none then so state), as well as the domain and range.

5. $f(X) = \dfrac{4X - X^2}{X - 2}$

vertical asymptote(s)_______________________________

horizontal asymptote(s)_______________________________

slant asymptote(s)_______________________________

Domain_______________________________ Range_______________________________

6. $f(X) = -\dfrac{3X - 12}{X - 3}$

vertical asymptote(s)_______________________________

horizontal asymptote(s)_______________________________

slant asymptote(s)_______________________________

Domain_______________________________ Range_______________________________

7. $f(X) = \dfrac{X - 7}{X + 2}$

vertical asymptote(s)_______________________________

horizontal asymptote(s)_______________________________

slant asymptote(s)_______________________________

Domain_______________________________ Range_______________________________

8. $f(X) = \dfrac{X^2 - X + 3}{X - 2}$

vertical asymptote(s)_______________________________

horizontal asymptote(s)_______________________________

slant asymptote(s)_______________________________

Domain_______________________________ Range_______________________________

9. $f(X) = \dfrac{(X - 3)(X - 2)}{X + 3}$

    vertical asymptote(s)__________________________

horizontal asymptote(s)__________________________

slant asymptote(s)__________________________

Domain__________________________ Range__________________________

10. $f(X) = -\dfrac{(2X^2 + 7X)}{X + 4}$

vertical asymptote(s)__________________________

horizontal asymptote(s)__________________________

slant asymptote(s)__________________________

Domain__________________________ Range__________________________

11. $f(X) = \dfrac{3}{X^2 - 4} + \dfrac{2}{5X + 10}$

vertical asymptote(s)__________________________

horizontal asymptote(s)__________________________

slant asymptote(s)__________________________

Domain__________________________ Range__________________________

12. $f(X) = \dfrac{X + 3}{X - 5} + \dfrac{6 + 2X^2}{X^2 - 7X + 10}$

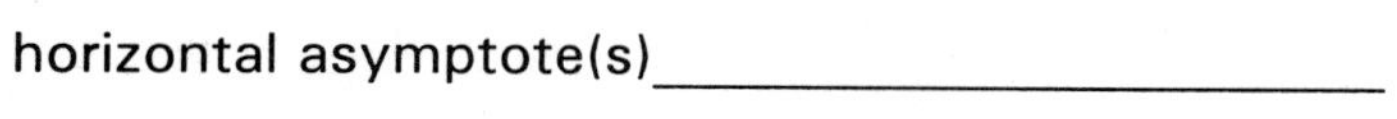

vertical asymptote(s)__________________________

horizontal asymptote(s)__________________________

slant asymptote(s)__________________________

Domain__________________________ Range__________________________

<u>**Solutions:**</u> **1.** X = -6   **2.** X = 2, X = 3   **3.** none   **4.** a. no b. The variable must appear in the denominator.

	Vertical Asymptote	Horizontal Asymptote	Slant Asymptote	Domain	Range
5.	X = 2	none	Y = -X + 2	$\{X \mid X \neq 2\}$	$\Re$
6.	X = 3	Y = -3	none	$\{X \mid X \neq 3\}$	$\{Y \mid Y \neq -3\}$
7.	X = -2	Y = 1	none	$\{X \mid X \neq -2\}$	$\{Y \mid Y \neq 1\}$
8.	X = 2	none	Y = X + 1	$\{X \mid X \neq 2\}$	$\Re$
9.	X = -3	none	Y = X - 8	$\{X \mid X \neq -3\}$	$\Re$
10.	X = -4	none	Y = -2X + 1	$\{X \mid X \neq -4\}$	$\Re$
11.	X = 2, X = -2	Y = 0	none	$\{X \mid X \neq -2 \text{ or } 2\}$	$\{Y \mid Y \neq 0\}$
12.	X = 5, X = -2	Y = 3	none	$\{X \mid X \neq -2 \text{ or } 5\}$	$\{Y \mid Y \neq 3\}$

UNIT 25
THE ALGEBRA OF FUNCTIONS

*Unit 19 is a prerequisite for this unit. Answers appear at the end of the unit.

All graphs are displayed in the standard viewing window.

The four arithmetic operations of addition, subtraction, multiplication, and division can be illustrated easily with functions on the calculator. Consider the two functions, f(X) = 2X - 3 and g(X) = 2X + 4. Enter f(X) at the Y1 = prompt and g(X) at the Y2 = prompt. Press [GRAPH] to display the screen at the right.

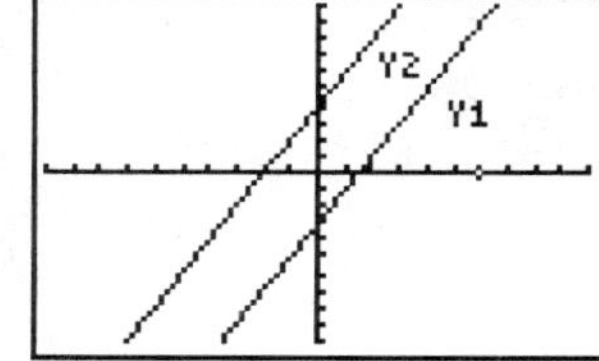

Enter the sum of the two functions, (f + g)(X) as Y1 + Y2 at the Y3 = prompt by using the **VARS** capability of the calculator. Access **VARS** by pressing [VARS] [▸] [1:Function...][1:Y1] [+] [VARS] [▸] [1:Function...] [2:Y2].. The calculator has been instructed to graph the sum of the functions entered at the Y1 = and Y2 = prompts respectively. Upon pressing [GRAPH], the display should correspond to the one at the right.

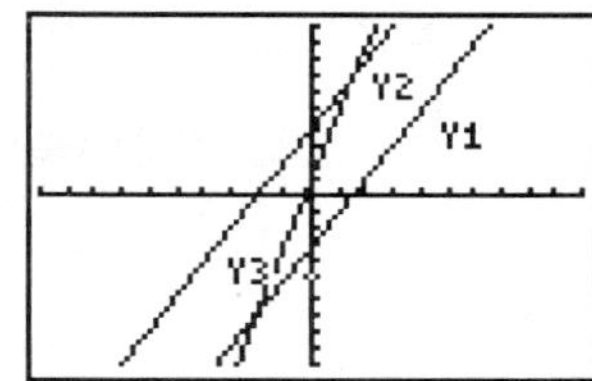

(TI-82) Press [Y =], cursor down to the Y3 = prompt and press [2nd] <Y-VARS> [1:Function...] [1:Y1] [+] [2nd] <Y-VARS> [1:Function...] [2:Y2].

TI-85/86 WHEN y(x) = IS PRESSED, THE SECOND MENU LINE DISPLAYS A y ABOVE THE F2 KEY. INSTEAD OF ACCESSING Y-VARS, PRESS [F2](y) [1] TO ENTER y1.

If you think about the algebraic sum (4x + 1) of the two functions you can easily reconcile the critical parts of the graph of a linear equation, i.e. the slope and the y-intercepts.

To see the difference of the two functions, (f - g)(X), enter the appropriate expression at the Y4 = prompt (using the VARS feature of the calculator). Turn **OFF** the graph of Y3 = and press [GRAPH] to see the difference of the two functions. Your screen should match the one to the right. Find (f - g)(X) algebraically and compare it to the graph. The difference is the constant function f(x) = -7.

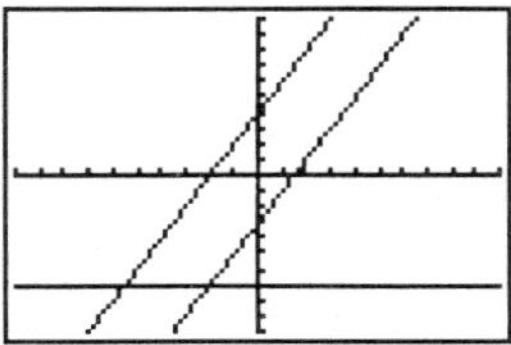

To see the product of the two functions, enter the appropriate expression at the Y5 = prompt. Turn the graph of Y4 = **OFF** and press [GRAPH] to see the graphs of Y1 = , Y2 = , and the function that is the product of the two original functions. Find (f · g)(X) algebraically. Observe the relationship between critical points of all three graphs.

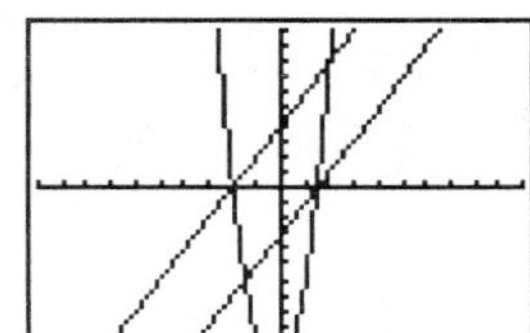

Enter the appropriate expression at the Y6 = prompt for the quotient (f/g)(X). Again, use the VARS capability of the calculator, and turn all other graphs OFF except for Y1 = and Y2 = . Your screen should match the one at the right. The critical features of the graph may be better understood if the calculator is placed in dot mode. Put the

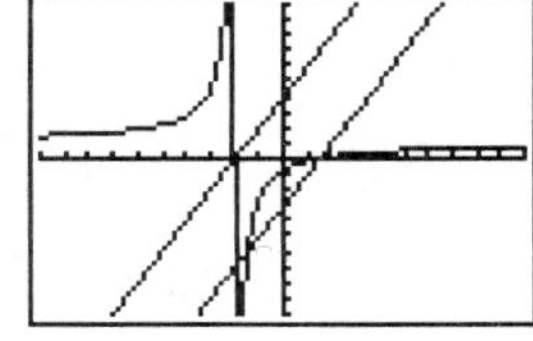

calculator in dot mode and compare the display to the one pictured.
Because this is a rational function (recall the previous unit) care must
be taken in finding the domain. The domain was not addressed in the
previous functions because they were either linear or quadratic and the
domains were all real numbers.

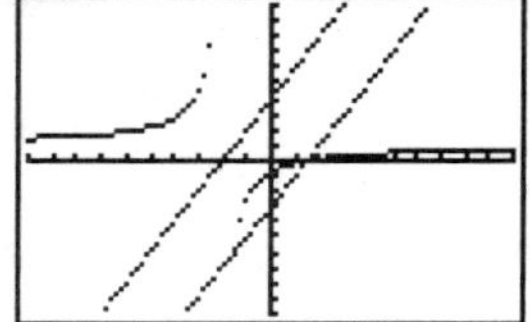

The composition of f(X) and g(X) can be shown algebraically as
$f[g(X)] = 2(2X + 4) - 3 = 4X + 5$. Enter this at the Y7 = prompt as
2*Y2 - 3 or as Y1(Y2), taking advantage of the function notation
available on the TI-82/83/83plus/86 calculators. Again, ensure that
all but the graphs of Y1 = , Y2 = and Y7 = are turned off as you
compare your display to the one pictured.

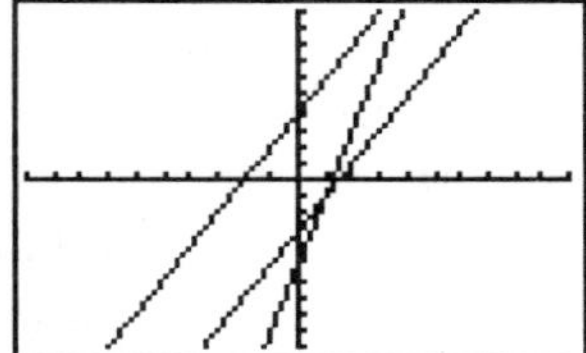

EXERCISE SET

1. The graphs of $f(X) = \sqrt{X + 3}$ and $g(X) = X^2$ will be used as reference graphs for
    the remaining exercises. Sketch the graphs of these functions on the grids provided
    and state the domains.

DOMAIN of f(X):________________   DOMAIN of g(X):________________

2. (f + g)(X)

    Domain: ________________

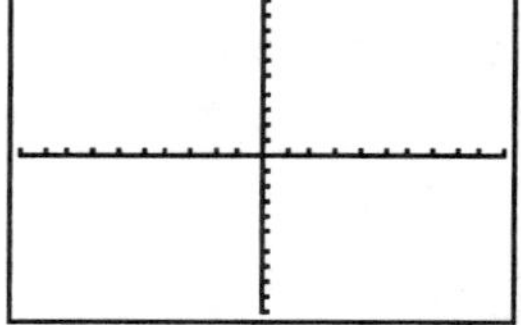

3. (f - g)(X)

    Domain: ________________

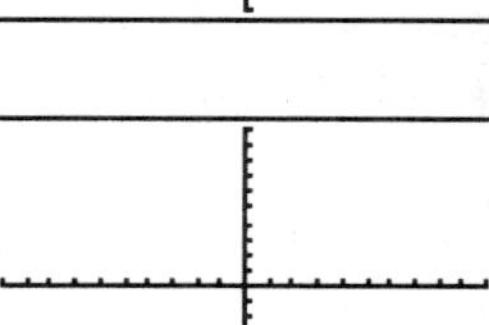

4. (f · g)(X)

    Domain: ________________

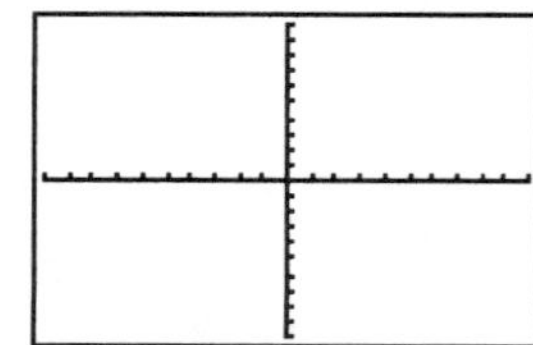

5.   (g - f)(X)

    Domain: ___________________

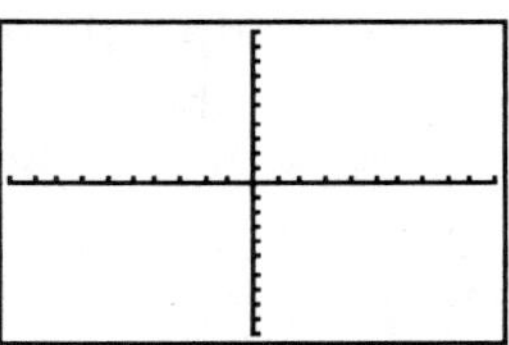

6.   (f ∘ g)(X)

    Domain: ___________________

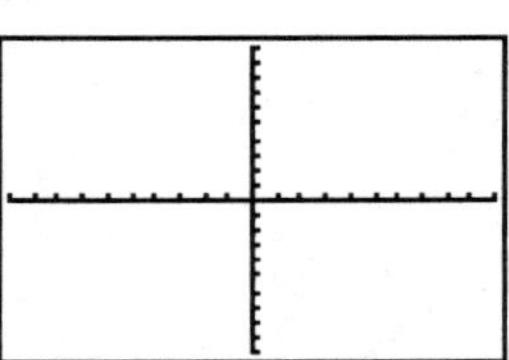

7.   If (f ∘ g)(X) has domain $\Re$, does that guarantee (g ∘ f)(X) also has domain $\Re$?

8.   Graph (g ∘ f)(X) .

    a. Domain: ___________________

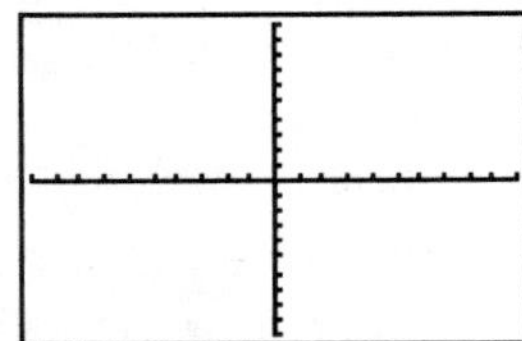

    ✍ b. Is the composition of functions commutative? Justify your answer.

    ✍ c. What accounts for the difference in domain between   (f ∘ g)(X) and

    (g ∘ f)(X) ?

✍ 9. Find f(2)   Explain how to use the TABLE or Value/EVAL/evalF features of the
                     calculator.

✍10. Find (g ∘ f)(3)   Explain the use of the TABLE or Value/EVAL features of the
                     calculator to accomplish this.

11. **Application:** The profit or loss for a publishing company on a textbook supplement is indicated by the function $y = 10x - 15000$. If the revenue generated is $14 per book, then the revenue function would be $R(X) = 14X$. In producing the text, the company has an initial expenditure of $15000 as well as $4 per book to print. Thus the cost function would be $C(X) = 4X + 15000$. Let Y1 be the revenue function, Y2 be the cost function and Y3 be R(X) - C(X). Sketch the graph of all three functions, specifying the size of the viewing window.

[_____ , _____] by [_____ , _____]

a. Find the break even point.

b. At what point in production does the revenue exceed the cost?

✍ c. Why does the profit function never exceed the revenue function?

12.   In general, should you be able to predict the domain of $(f + g)(X)$ by examining the domains of f(X) and g(X)?

13.   In general, should you be able to predict the domain of $(f \cdot g)(X)$ by examining the domains of f(X) and g(X)?

<u>Solutions:</u> **1.** domain of f(X): $\{X \mid X \geq -3\}$, domain of g(X): $\Re$ **2.** $\{X \mid X \geq -3\}$

3. $\{X \mid X \geq -3\}$ **4.** $\{X \mid X \geq -3\}$ **5.** $\{X \mid X \geq -3\}$ **6.** $\Re$

7. No   **8. a.** $\{X \mid X \geq -3\}$ **b.** Composition is not commutative, if it were the graphs would be identical. **c.** In f(g(x)), the range of the g functions is the domain of the f function. In g(f(x)), the range of the f function becomes the domain of the g function.

9. $f(2) \approx 2.2361$ **10.** 6

11. [0,10000] by [0,100000] break even at 1500 texts, revenue > cost at 1500 texts because profit is revenue - cost.

12. Yes, the domain should be the union of the two domains from the original functions.

13. Yes, the domain should be the union of the two domains from the original functions.

170

UNIT 26
EXPONENTIAL AND LOGARITHMIC FUNCTIONS

*Unit 19 is a prerequisite for this unit. Answers appear at the end of the unit.

This unit will examine the graphs of exponential functions (to include base e) and their inverses, logarithmic functions. All graphs should be displayed on the ZDecimal x 2 viewing window: i.e. [-9.4, 9.4] by [-6.2, 6.2], with both scales set to 1 for TI-82/83/83plus calculators. (This window will be [-12.6,12.6] by [-6.2,6.2] on the TI-85/86.)

1.　　An exponential function is a function of the form $Y = f(X) = b^X$ where b is a positive real number not equal to 1.

2.　　Examine the function when b (the base) is larger than 1, $b > 1$. GRAPH and TRACE each function below. Sketch the display carefully.

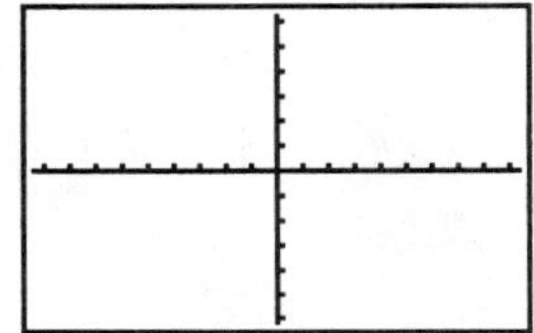 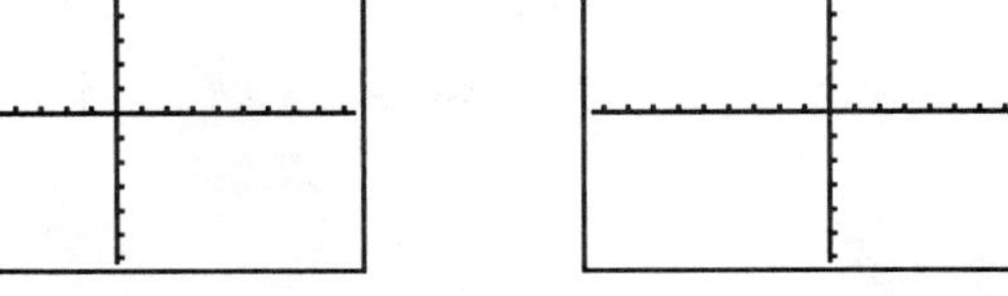 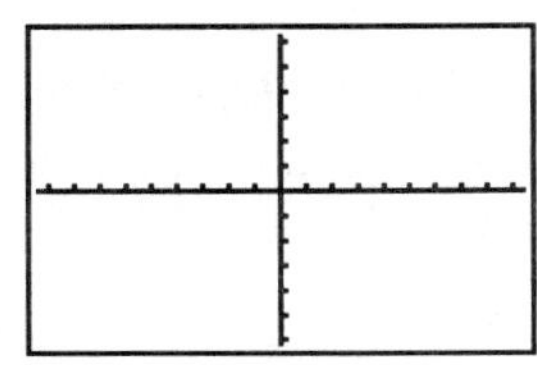

$Y = 2^X$　　　　　$Y = 3^X$　　　　　$Y = 8^X$

3.　　Access the TABLE (increment by 1). Scroll up and down while examining the values for Y. This can also be accomplished by carefully examining X and Y values while tracing.

    a. As X increases, what happens to Y?

    b. As X decreases, what happens to Y?

    c. Will the value of Y ever be equal to 0? Why or why not?

    d. State the domain and the range of the functions.

    Domain:___________　　　　Range:___________

    e. How are the above 3 graphs

    similar:

    different:

4. Examine the function when b (the base) is larger than 0 but less than 1, $0 < b < 1$.
GRAPH and TRACE each function below. Sketch the display carefully.

$$Y = (1/10)^X \qquad Y = (1/2)^X \qquad Y = (4/5)^X$$

5. Access the TABLE (ensure that it is set to increment by 1) or use TRACE. Scroll up
and down while examining the values for Y.

a. As X increases, what happens to Y?

b. As X decreases, what happens to Y?

c. Will the value of Y ever be equal to 0? Why or why not?

d. State the domain and the range of the functions.

Domain:___________   Range:__________

e. How are the above 3 graphs

similar:

different:

6. The exponential functions graphed thus far are continuous. Moreover, they are either
continuously increasing or continuously decreasing. Explain what the word
"continuous" means in the above contexts, i.e. define continuous, continuously
increasing function, and continuously decreasing function.

7. All of the graphs above have the same Y-intercept. What is it? __________

8. Why is the point with coordinates (0,1) on each of the above graphs?

✍9.   If an exponential function is multiplied by a constant, what do you THINK will happen to the graph?

10.   Graph $Y1 = 2^X$ and $Y2 = 6(2^X)$ on the calculator and sketch the display. Compare Y2 to Y1.

   a. Was the domain or range affected?_______

   b. Did the Y-intercept change?________

   c. What is the Y-intercept of Y2?_______

11.   Predict the Y-intercept for $Y = 8(2^X)$._______

12.   What if a constant were added to the function $Y = 2^X$? Describe the manner in which the constant will shift the graph.

   positive constant:

   negative constant:

13.   Graph the exponential functions $Y = 2^X + 3$ and $Y = 2^X - 3$ on the calculator and sketch the display. Were your predictions from #12 correct?

14.   Graph the exponential function $Y = 4^X - 18$ on the calculator and sketch the display.

   a.   State the domain:___________________________

   b.   State the range:___________________________

   c.   State the Y-intercept:___________

   d.   State the X-intercept:___________

15.   Consider the **equation** $4^X = 18$. Graphically solve this equation using the ROOT/ZERO feature of the CALC menu. Circle the root on the displayed graph and record the solution:
   X = _______

16. To solve this same equation algebraically will require the use of the Change of Base
Formula: $\log_b n = \dfrac{\log n}{\log b}$.

$$4^X = 18$$
$$X = \log_4 18$$
$$= \frac{\log 18}{\log 4}$$
$$\approx 2.084962501$$

Compare the approximation above with the root to the equation in 15 and the
X-intercept of the graph of the equation in 14. They should all have the same
approximation.

EXPONENTIAL INVERSES

The inverse of an exponential function is a logarithmic function. To find the inverse of the
exponential function $(Y = b^X)$ interchange the X and Y variables. This yields the equation
$X = b^Y$ $(b > 0, b \neq 1)$ which is defined as the logarithmic function $Y = \log_b X$. Until your
text addresses Properties of Logarithms, you will not have an algebraic method for solving
this equation for Y. Until that time, the **DrawInv** option from the DRAW menu will be used
to draw the inverses of these functions.

17. Graph $Y = 2^X$. Following the instructions in the INVERSE
section of the unit entitled "Functions", draw the inverse of Y1.
Sketch the final display of both functions and label them as Y1
and INV Y1 on the graph.

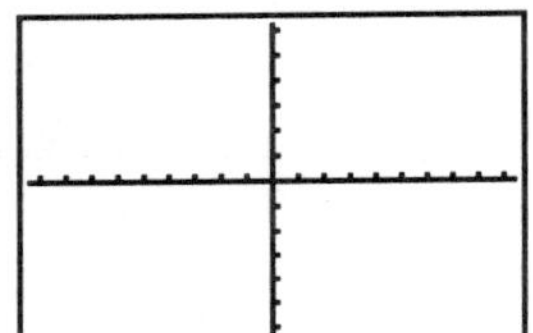

    Y1: Domain:____________ Range:____________

    INV Y1: Domain:____________ Range:____________

18. Why are the domain and range in the graph of INV Y1 the reverse of the domain and
range of the Y1 graph?

    Recall that the Y1 graph had a Y-intercept of 1 and no X-intercept. Although you
    cannot TRACE on INV Y1, what appear to be the X and Y intercepts for the **inverse
    of Y1**?

19. The defined function $Y = 2^X$ has as its inverse function $X = 2^Y$. This equation must
be solved for Y in order to actually graph the inverse function as opposed to merely
drawing the function. Logarithmic notation and properties allow us to do that.
Solving $X = 2^Y$ using logarithmic properties yields:

$$X = 2^Y$$
$$\log X = \log 2^Y \qquad \textit{(take the logarithm of both sides)}$$
$$\log X = Y(\log 2) \qquad \textit{(Logarithmic Property: } \log_b P^n = n\log_b P\textit{)}$$
$$\frac{\log X}{\log 2} = Y$$

20. Graph $Y = \dfrac{\log X}{\log 2}$ and verify the X and Y intercepts you stated in #14.

21. Use the TABLE feature to quickly complete the following tables for $Y = 2^X$ and $X = 2^Y$ and thus verify that the X and Y values of the ordered pairs are reversed in functions that are inverses of one another.

$Y = 2^X$

X	Y
1	
2	
3	
4	

$X = 2^Y$

X	Y
2	
4	
8	
16	

22. Graph $Y = 0.5^X$. Use either the DRAW menu to display the inverse of $Y = 0.5^X$ or graph the inverse of $Y = 0.5^X$ using logarithmic properties to solve for Y. Sketch the final display of both functions and label them as Y1 and INV Y1 on the graph.

Y1: Domain:___________   Range:___________

    Y-intercept:_________

INV Y1: Domain:___________   Range:___________

    X-intercept:_________

BASE- e EXPONENTIAL FUNCTIONS

The irrational number **e** occurs in the mathematical modeling of natural events. Its value is approximately equal to 2.71828182846, and it is often used as the base of an exponential function.

23. Graph $Y = 2^X$, $Y = e^X$ and $Y = 3^X$ on the screen at the right. Why does the graph of $Y = e^X$ rise faster than the graph of $Y = 2^X$ but not as fast as the graph of $Y = 3^X$?

24. A common occurrence of the use of **e** in mathematical modeling is in the computation of compound interest, specifically continuous compounding. When an amount of money invested grows exponentially, the formula for computing the value of the investment is $A = Pe^{rt}$ where interest is compounded continuously.

25. If an initial investment of $2000 is placed in an account earning 4.5% interest
compounded continuously, write an equation that will model the value of the
investment at the end of X years.

____

26. Graph the equation from #25 on the screen at the right.
(Viewing WINDOW HINT: Since X represents the number of
years, set the Xmax and Xmin to display non-negative values of
X. To make the interpretation of the graphical display
"friendlier", you may want to use the space width formulas
discussed in the unit entitled "Preparing to Graph". Remember,
the initial investment is $2000; this will affect the Ymax.)

27. TRACE along the graph and determine the value of the investment (to the nearest
dollar) after

1 year________   2 years __________   5 years __________   10 years __________

28. Explain how to use the TABLE to determine value of the investment after 3 years 3
months.
Record this value: ________

29. Summarize what you have learned in this unit. Your summary should:
   a. state the definition of an exponential function,
   b. discuss the graphs of exponential functions when $b > 0$ (your discussion should
      address domain, range and Y-intercept)
   c. discuss the graphs of exponential functions when $0 < b < 1$ (your discussion
      should address domain, range and Y-intercept)
   d. discuss why the case of $b = 1$ is excluded in the definition of an exponential
      function
   e. state the definition of a logarithmic function (i.e. the inverse of an exponential
      function)
   f. discuss the relationship between the domain, range and intercepts of an
      exponential function and its inverse.

2. 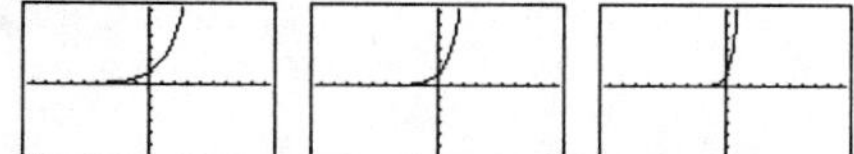

3a. As X increases, Y increases. **3b.** As X decreases, Y decreases.

3c. No: a non-zero base raised to any power is a non-zero number. **3d.** D:$\mathbb{R}$, R: $\{Y \mid Y > 0\}$
3e. Answers may vary.

4. 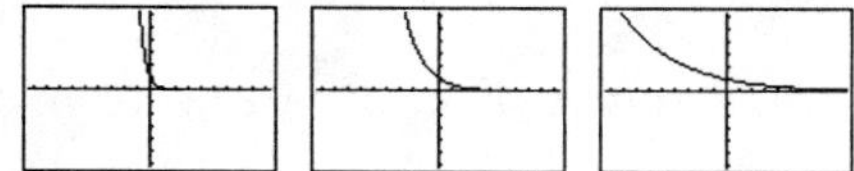

5a. As X increases, Y decreases. **5b.** As X decreases, Y increases.

5c. No: a non-zero base raised to any power is a non-zero number. **5d.** D:$\mathbb{R}$, R: $\{Y \mid Y > 0\}$
5e. Answers may vary.
6. A continuous function is one in which there are no gaps (or holes). Continuously increasing and/or decreasing functions have no turns (i.e. no relative maximums or minimums).
7. 1 **8.** When the exponent is zero the value of the exponential expression is one.
9. Answers may vary. **10a.** No **10b.** Yes **10c.** 6 **11.** 8 **12.** positive constant: shifts graph up; negative constant: shifts graph down
13.

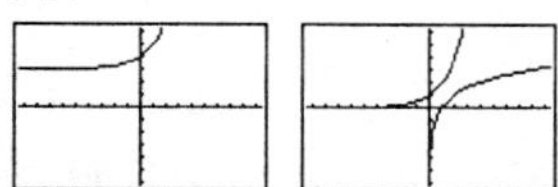

14.   D:$\mathbb{R}$, R: $\{Y \mid Y > -18\}$, Y-intercept: -17,
X-intercept: 2.084962501

15. ROOT: 2.084962501   **17.** Y1:see 3d.; INV Y1: domain:$\{X \mid X > 0\}$, range:$\mathbb{R}$

18. Because the X and Y variables were interchanged. The X-intercept = 1 and there is no
Y- intercept.

22. Y1: same as 5d., Y-intercept = 1; INV Y1: domain:$\{X \mid X > 0\}$, range: $\mathbb{R}$, X-intercept = 1

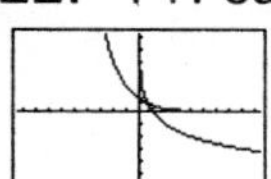

25. Y = 2000e$^{.045x}$

26. WINDOW: Xmin = 0, Xmax = 94, Xscl = 0, Ymin = 0, Ymax = 5000, Yscl = 0

27. 1yr = 2092, 2yr = 2188, 5yr = 2505, 10yr = 3137

28. The TABLE will need to be incremented by .01 since 3 yrs. 3 mos. is 3.25 years. The value after 3.25 yrs. is $2315.
29. Answers may vary.

*Unit 20 is a prerequisite for this unit. Answers appear at the end of the unit.

The unit entitled "Discovering Parabolas" explored the graphs of equations of the form $y = ax^2 + bx + c$, where a, b, and c are real numbers. Graphs of these equations were parabolas that opened either up or down, and were functions. We will now examine the graphs of parabolas that are not functions, which open left and right.

In the unit "Discovering Parabolas" the following conclusions were drawn:

Graphs of equations of the form $y = ax^2 + bx + c$ are called parabolas. Completing the square in x yields an equation of the form $y = a(x - h)^2 + k$.

If $a > 0$, the parabola opens up; the vertex of the parabola has coordinates (h,k); this vertex is the minimum point of the graph.

If $a < 0$, the parabola opens down; the vertex of the parabola has coordinates (h,k); this vertex is the maximum point of the graph.

Changing the size of $|a|$ affects the width of the parabola; i.e. the larger $|a|$, the slimmer the parabola; the smaller $|a|$ the flatter the parabola.

Changes in the value of h result in horizontal shifts; changes in the value of k result in vertical shifts.

EXERCISE SET

Directions:

a. Set the viewing window to ZDecimal X 2, i.e. [-9.4, 9.4] by [-6.2, 6.2], with both scales set to 1. (This window will be [-12.6,12.6] by [-6.2,6.2] on the TI-85/86.)

b. Graph each equation below; copy the graph screen in the space provided.

c. State the coordinates of the vertex, the parabola's orientation (**up** or **down**), and the effect of the coefficient **a** on the width of the parabola (compare it to the reference parabola below) as **wider**, **narrower**, or **same**.

Reference parabola:

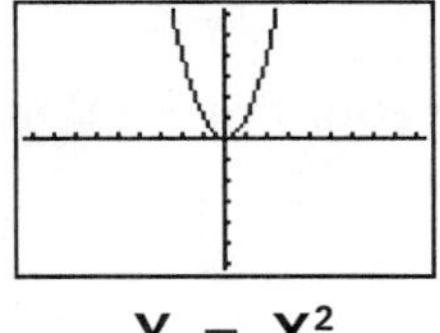

$$Y = X^2$$

1. $Y = X^2 + 4$

Vertex: _______
Orientation: _______
Width: _______

$Y = (X + 1)^2 - 4$

Vertex: _______
Orientation: _______
Width: _______

$Y = (-\tfrac{1}{3})X^2 - 3$

Vertex: _______
Orientation: _______
Width: _______

2. Each of the graphs in #1 are functions. Each graph passes the vertical line test: any vertical line drawn in the plane of the graph will intersect the graph in AT MOST one point. It also means that every X-value has one and only one corresponding Y-value. Enter the polynomial $5(X - 3)^2 + 2$ at the Y1 = prompt on the calculator. Examine the symmetry about the vertical line X = 3, both graphically and numerically through the TABLE feature.

Set the TABLE to begin at 0 and to be incremented by 1.

When X = 2, Y = _____. When X = 4, Y = _______.

✍ Is this a one-to-one function? Why or why not?

3. We will now investigate equations of the form $x = a(y - k)^2 + h$. Consider the equation $X = Y^2$. Because the calculator is set up to graph equations that are solved for the variable Y, solve this equation for Y :

Graph each equation at the Y1 and Y2 prompts, respectively, and then sketch the individual graphs of Y1, Y2 and the combined graphs of Y1 and Y2 below.

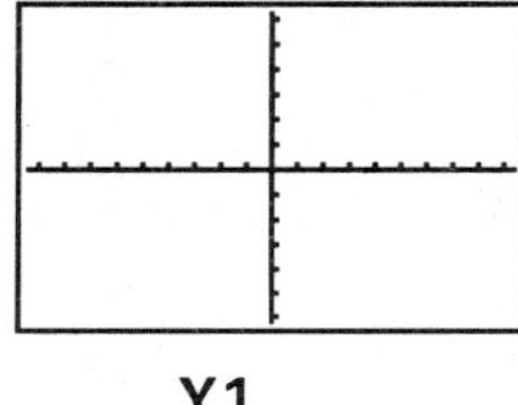

Y1

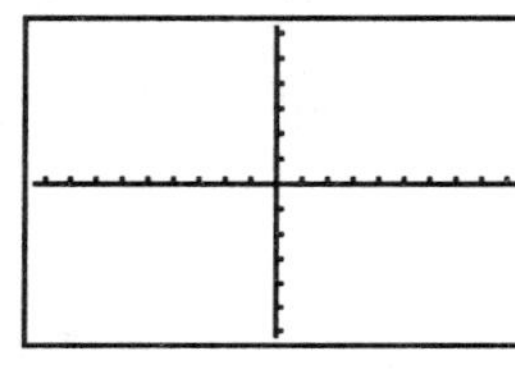

Y2

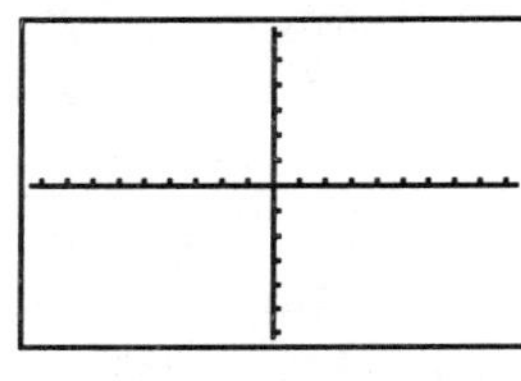

Y1 and Y2

NOTE: The above combined graphs of Y1 and Y2, $X = Y^2$, will be the reference graph in future examples and exercises.

✍ 4. Each of the individual graphs is a function. Is the curve produced by the two graphs a function? Why or why not?

5. Consider the equation $X = 2(Y - 1)^2 + 4$. In order to graph it on the calculator, the equation must be solved for Y. This will be done by using the Square Root Property:

$$x = 2(y - 1)^2 - 4$$
$$x + 4 = 2(y - 1)^2$$
$$\frac{x + 4}{2} = (y - 1)^2$$
$$\pm\sqrt{\frac{x + 4}{2}} = y - 1$$
$$1 \pm\sqrt{\frac{x + 4}{2}} = y$$

Graph $1 + \sqrt{\dfrac{X + 4}{2}}$ at the Y1 = prompt and $1 - \sqrt{\dfrac{X + 4}{2}}$ at the Y2 = prompt.

Sketch the graphs in the space below. Answer the following questions about the **CURVE** represented by these combined functions.

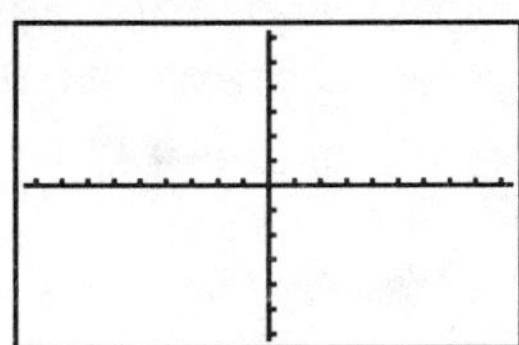

a. Does the parabola open <u>right</u> or <u>left</u>? _________

Is the graph <u>wider</u> or <u>narrower</u> than the reference graph (see #3) of $X = Y^2$? ______

b. What are the coordinates of the vertex? ________

What are the coordinates of the vertex of the reference graph? ________

c. Compare the <u>original</u> equations, $X = Y^2$ and $X = 2(Y - 1)^2 + 4$. Compare the constants and the coefficients and **EXPLAIN** what caused the horizontal and vertical shifts.

Directions: For each equation below,
    a. Solve for the variable Y (you should obtain <u>two</u> equations).
    b. Enter these equations at the Y1 = and Y2 = prompts respectively.
    c. Using the ZDecimal x 2 viewing window, sketch the graphs of the equations in the space provided.
    d. Specify "right" or "left" for the orientation of the curves **AND** state the vertex of the parabola represented by the curves.

6. $X = Y^2 - 1$

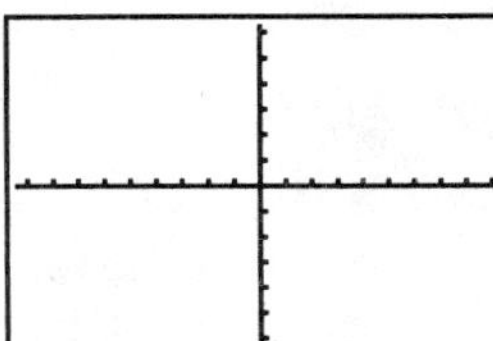

    vertex: ________

    orientation: ________

7. $X = -Y^2 + 2$

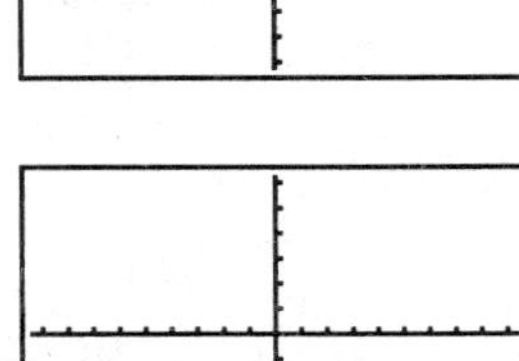

    vertex: ________

    orientation: ________

8. Consider the equation $X = (Y - 3)^2 - 4$.

PREDICT the vertex: ________   and the orientation: ________

✍ What **two** equations need to be graphed on the calculator to verify your predictions? Is the graph of the combined equations a function? Why or why not?

9. We have been investigating equations of the form $x = ay^2 + by + c$, that have been written in the form $x = a(y - k)^2 + h$. Using the conclusions drawn at the beginning of this unit from "Discovering Parabolas" as your model, make the following generalizations:

    the coordinates of the vertex:___________

    the orientation of the graph: (opens left/right under what conditions)

    the size of the graph (as compared to the reference graph $X = Y^2$):
    Hint: THINK about what happens as $|a|$ gets larger or smaller than 1.

Example: Consider the square root function $Y = 2 + \sqrt{3 + X}$. Use the calculator to graph the function whose display is at the right. Observe that the domain is $\{x \mid x \geq -3\}$ and the range is $\{y \mid y \geq 2\}$.

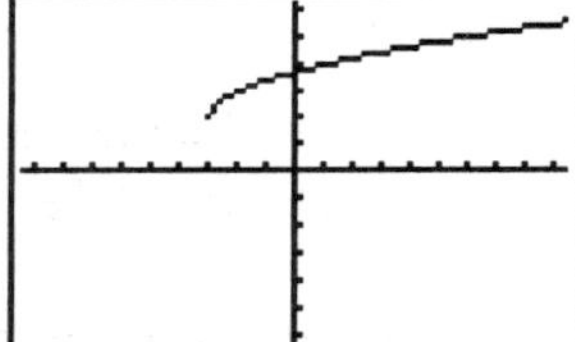

Turn off the graph by placing the cursor over the equal sign and pressing **[ENTER]**.

TI-85/86   USE **[F5](SELCT)** TO TURN GRAPHS ON AND OFF.

Now, enter the expression $Y = 2 - \sqrt{3 + X}$ at the Y2 = prompt to display the graph at the right. Observe that the domain is $\{x \mid x \geq -3\}$ and the range is $\{y \mid y \leq 2\}$.

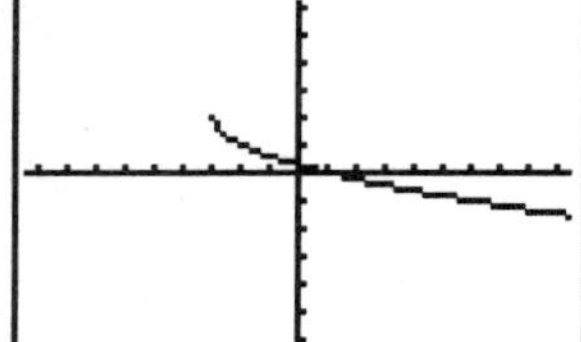

Turn the graph of Y1 on and look at the curve produced by the two square root functions. It is a parabola that opens to the right with vertex at (-3,2).

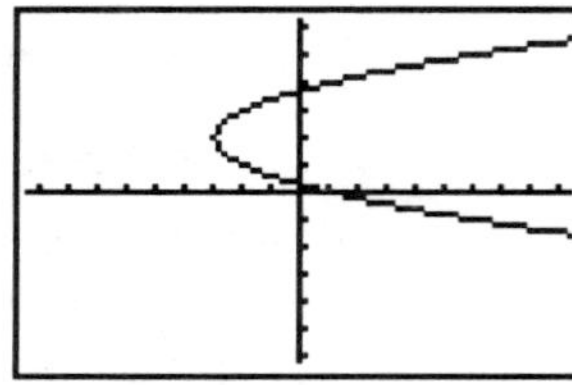

Directions: Match each graph in 10-13 with its corresponding equation(s). Some graphs may correspond to more than one equation.

10.______ 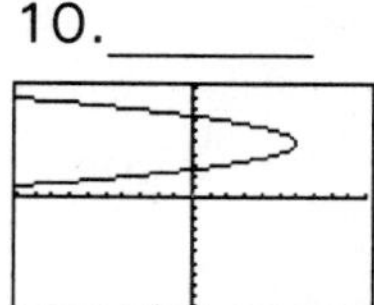 11.______ 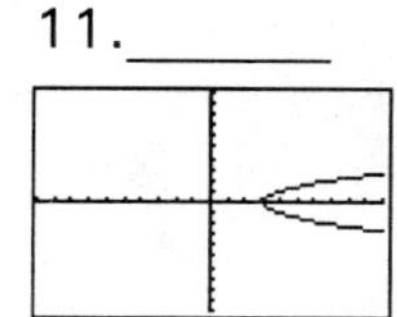 12.______ 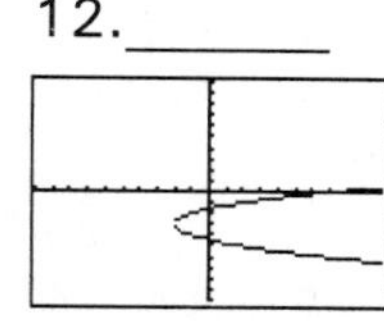 13.______ 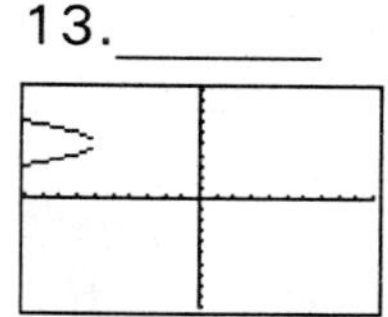

a. $5 \pm \sqrt{-X - 6} = Y$

b. $X = Y^2 + 3$

c. $-3 \pm \sqrt{X + 2} = Y$

d. $5 \pm \sqrt{6 - X} = Y$

e. $X = -(Y + 5)^2 - 6$

f. $X = -(Y - 5)^2 + 6$

14. Write the equation for the parabola that has vertex (-4, -3) and opens right.

Hint: Begin with $Y = ? + \sqrt{? + x}$ and $Y = ? - \sqrt{? + x}$. Combine the equations and write the parabolic equation in the form $x = a(y - k)^2 + h$.

15. Write what you believe would be the equation for a parabola with vertex (-4, -3) that opens left. You should write this equation in standard form.

Solve the equation you wrote for the variable y, graph it on the calculator. Sketch your graph in the space below.

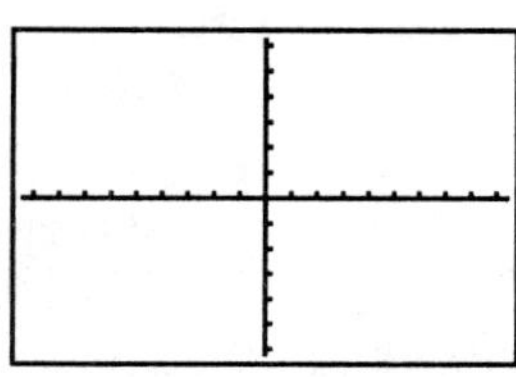

16. a. Write the equation of the square root function with domain $X \geq -4$ and range $Y \geq -3$.

b. Write the equation of the square root function with domain $X \geq -4$ and range $Y \leq -3$.

c. Combine these two equations and write the parabolic equation associated with the curve produced when the two functions are graphed.

d. The above equation is that of a parabola that opens ____ and has vertex (,   ).

✍17. Reconcile the graphing of parabolic equations that are functions and those that are not functions. Your discussion should include the points that are considered critical to the graphing of the conic (i.e. x/y intercepts, max/min points, etc.) What part do the focus and directrix play in the graphing?

✍18. Summarize the relationship between the square root function(s) and parabolic curves.

SOLUTIONS: **1.** (0,4), up, same   (-1,-4), up, same   (0,-3), down, wider

2. The function is NOT one-to-one because it does not pass a horizontal line test, i.e. there are 2 different X-values for the Y-value of 7.

3. 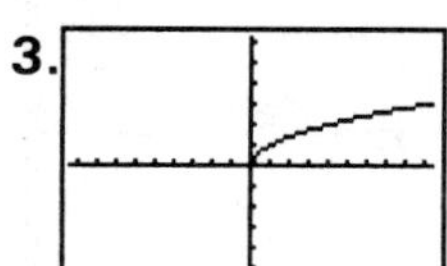 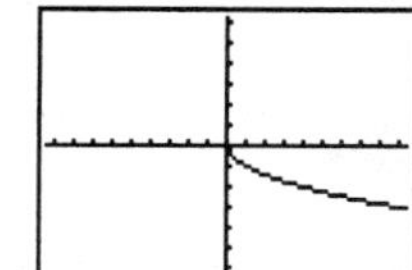 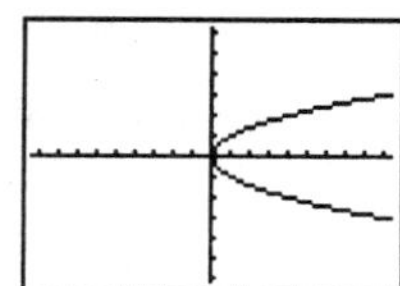

4. No, because for each x value in the domain there are two y values in the range: it fails vertical line test. Notice the curve (made up of graphs of **two** square root functions) is a parabola. However, two expressions had to be entered on the calculator to obtain the appropriate curve. It no longer opens up or down, but rather to the right, with the vertex at the origin.

5a. right, narrower   **5b.** (-4,1) (0,0)   **5c.** The constant added inside the quantity before squaring results in a vertical shift, whereas, the constant term added outside the quantity squared results in a horizontal shift.

6. (-1,0), right   **7.** (2,0), left   **8.** (-4,3), opens righ $Y = 3 + \sqrt{4 + X}$ *and* $Y = 3 - \sqrt{4 + X}$.
It is not a function

9. (h,k), If a > 0 the parabola opens right. If a < 0 the parabola opens left. As the absolute value of "a" increases, the graph becomes more narrow. As the absolute value of "a" decreases, the graph becomes wider.

10. d, f   **11.** b   **12.** c   **13.** a, e

14. $y = -3 \pm \sqrt{-4 + x}$ $x = (y + 3)^2 - 4$ **15.** $y = -3 \pm \sqrt{4 - x}$

16. $y = -3 + \sqrt{4 + x}$ $y = -3 - \sqrt{4 + x}$ $x = (y + 3)^2 - 4$ right, vertex:(-4,-3)

17. Answers may vary.   **18.** Answers may vary.

CORRELATION CHART

UNIT * Instructors are encouraged to use Units marked by an * to introduce concepts.	PRE-REQ. UNIT	CORRELATING CONCEPT
#28 Trigonometric Functions: Amplitude, Phase Shifts and Translations, p.187	#19	Examines the effects of constants on the graphs of trigonometric functions.
#29 Predict-A-Graph, p.197	#28	Provides practice in identifying different types of trigonometric graphs.
#30 Graphical Explorations: Trigonometric Identities, p.201	#28	Graphs of equations are examined to determine identities.
#31 Graphical Solutions: Trigonometric Equations, p.205	#30	Graphical solutions using the root/zero and/or intersect features.
#32 Polar Graphing, p.209	#28	Several polar graphs are examined, as well as the graphs of conics in polar form.
#33 Parametric Graphs of Conics, p.215	#32	Circles, ellipses, and hyperbola will be graphed using parametric equations.
#34 Roots of Complex Numbers, p.221	#33	Use of the TABLE feature in finding the n nth roots of a complex number.

Unit Title	TI-82 Keys	TI-83/83plus Keys	TI-85/86 Keys
#28:Trigonometric Functions: Amplitudes, Phase Shifts and Translations, p.187	ZOOM 7:ZTrig	ZOOM 7:ZTrig	GRAPH/ZOOM ZTRIG
#29:Predict-A-Graph, p.197	No new keys.	No new keys.	No new keys.
#30:Graphical Explorations: Trigonometric Identities, p.201	No new keys.	No new keys.	No new keys.
#31:Graphical Solutions: Trigonometric Equations, p.205	No new keys.	No new keys.	No new keys.
#32:Polar Graphing, p.209	MODE Pol	MODE Pol	MODE Pol PolarC
#33:Parametric Graphs of Conics, p. 215	MODE Par	MODE Par	MODE Param
#33:Roots of Complex Numbers, p.221	No new keys.	No new keys.	No new keys.

UNIT 28
TRIGONOMETRIC FUNCTIONS:
AMPLITUDE, PHASE SHIFTS, AND TRANSLATIONS

*Unit 19 is a prerequisite for this unit. Answers appear at the end of the unit.

REVIEW

When translations and stretches of graphs were previously examined, the following conclusions were drawn:

a. $y = f(x) + c$:   translates vertically "c" units from the graph of $y = f(x)$, possibly affecting the range

b. $y = f(x + b)$:   translates horizontally "b" units from the graph of $y = f(x)$, possibly affecting the domain

c. $y = a \cdot f(x)$:   the size of the graph is affected by "a" but the domain and range are not affected

GRAPHS INVOLVING SINE AND COSINE

The sine and cosine functions are periodic functions. As the value of x increases, the values of the two functions continuously repeat. When graphing the functions, the period and amplitude of the functions are the focal point. The period of the function is the value p where $f(x + p) = f(x)$, and the amplitude is the maximum distance of the curve from the axis (determined by (maxY - minY)/2). The calculator MODE can either be set to **Radian** or **Degree.** In radian mode, the X values will be rounded values of a constant times π. Thus when determining the period of a function by examining the graph display, having the calculator set to **Degree** MODE facilitates the process. Before proceeding, set **MODE** to **Degree** and the viewing WINDOW to **ZTrig.** In **ZTrig,** the Xscl = 90 and the min and max values are such that each move of the cursor is equal to 7.5 degrees. It is important to remember that **ZTrig** must be reset each time the calculator is changed from **Degree** to **Radian.**

✍1.   Use TRACE to establish the period and amplitude of $Y = \sin X$ and $Y = \cos X$.

$Y = \sin X$

$Y = \cos X$

amplitude:________________   amplitude:________________

period:________________   period:________________

symmetric with respect to:________   symmetric with respect to:________

What is the basic difference in the graphs?

2.   Each of the groups of functions below is in the form $y = a \cdot f(x)$. Sketch each graph
      in a different color on the display below. The reference graph, listed first in the
      series, is graphed for you.

   Y = sin X, Y = 3 sin X,   Y = cos X, Y = 3 cos X,
   Y = 0.5 sin X   Y = 0.5 cos X

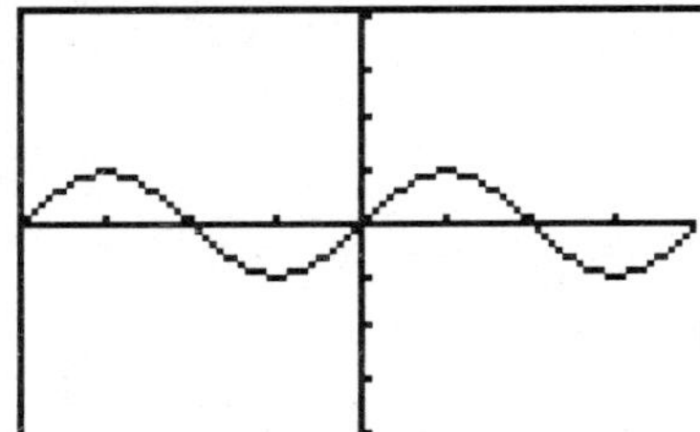   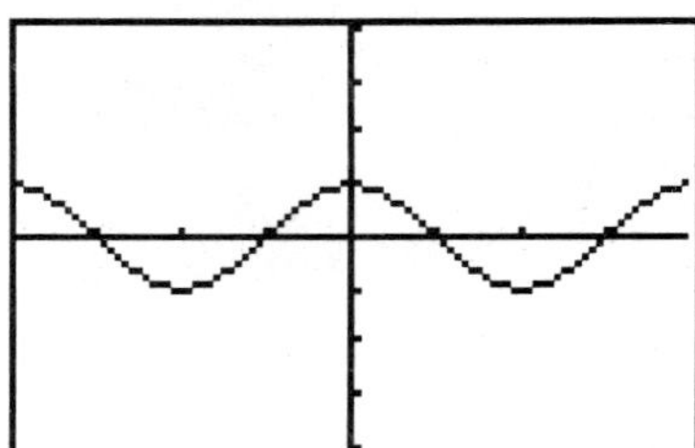

✍ 3.   In general, the amplitude of $y = a \cdot f(x)$ is determined by ________ .

        If $y = \sin x$ and $y = \cos x$ have extreme values of y such that $-1 \leq \sin x \leq 1$ and
        $-1 < \cos x \leq 1$, respectively, then $y = a \cdot \sin x$ and $y = a \cdot \cos x$ have extreme
        values of y such that ________________ and ________________ , respectively.

        Does a > 0 affect the period of either curve?

        Discuss the relationship between the amplitude of the function and the range of the
        function.

✍ 4.   Compare the graphs of Y = 3 sin X and
        Y = -3 sin X. Are either the period or
        amplitude affected when a < 0?   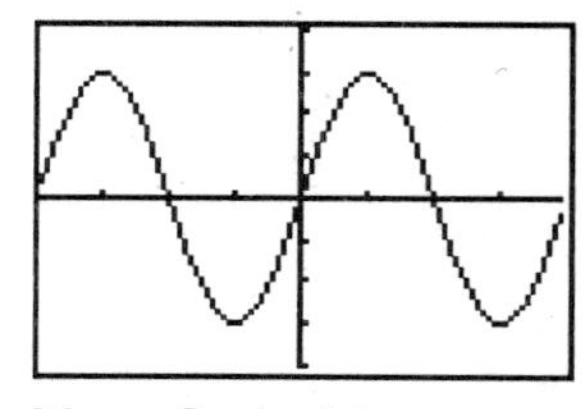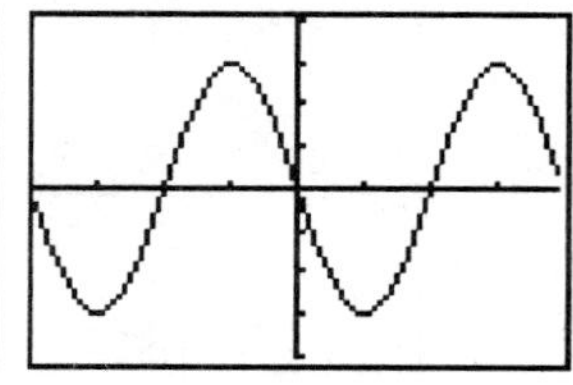

                                                      Y = 3 sin X   Y = -3 sin X

        In general, the graph is reflected across the X-axis when a < 0.

5. Individually graph each of the functions that is in the form $y = f(bx)$. Use TRACE to examine the effect of $b > 0$ on the period of the function. Record your findings both in terms of degrees and radians (π format), in the following table.

Function	Period in Degrees	Period in Radians	Function	Period in Degrees	Period in Radians
Y = sin X	360	2π	Y = cos X	360	2π
Y = sin .75X			Y = cos .75X		
Y = sin .5X			Y = cos .5X		
Y = sin .25X			Y = cos .25X		
Y = sin 2X			Y = cos 2X		
Y = sin 3X			Y = cos 3X		
Y = sin 4X			Y = cos 4X		

State the algebraic relationship between radians and degrees that is used in converting from one method of measurement to the other.

6. Compare the graphs of Y = sin 2X, Y = sin -2X and Y = cos 2X, Y = cos -2X. Sketch each graph in a different color on the display below. The first graph in each pair is already displayed.

Y = sin 2X, Y = sin -2X

Y = cos 2X, Y = cos -2X

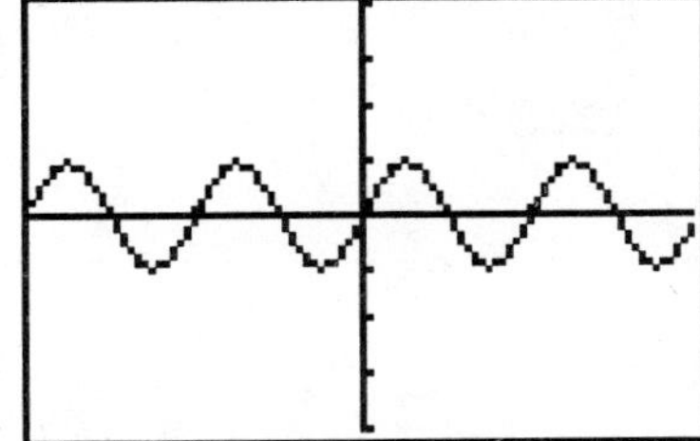

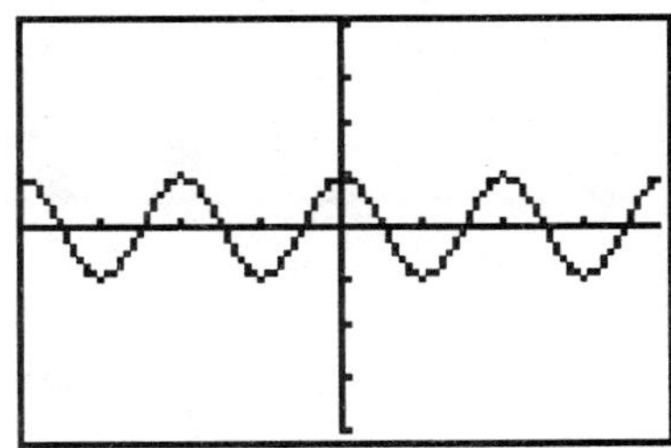

Explain the effect (or lack of effect) of $b < 0$ on each pair of graphs.

Classify the sine and cosine functions as either even or odd. Will the size of "b" affect the classification of even or odd?

✍ 7. Each group of functions below is in the form y = f(x + c). Sketch each graph in a
different color on the display below. The reference graph is listed first in the series.

Y = sin X, Y = sin (X + 45),   Y = cos X, Y = cos (X + 45),
Y = sin (X - 45)   Y = cos (X - 45)

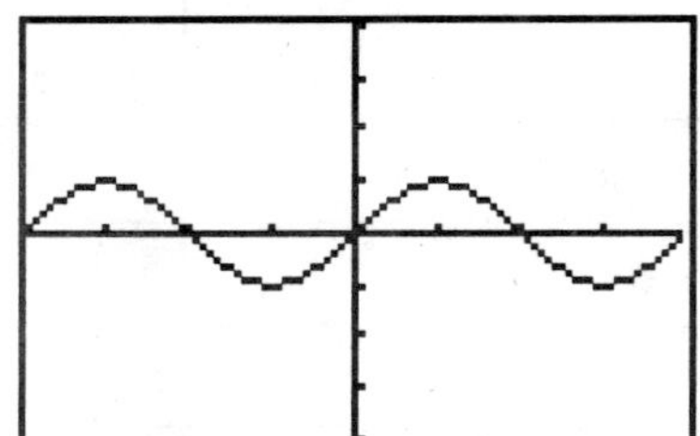   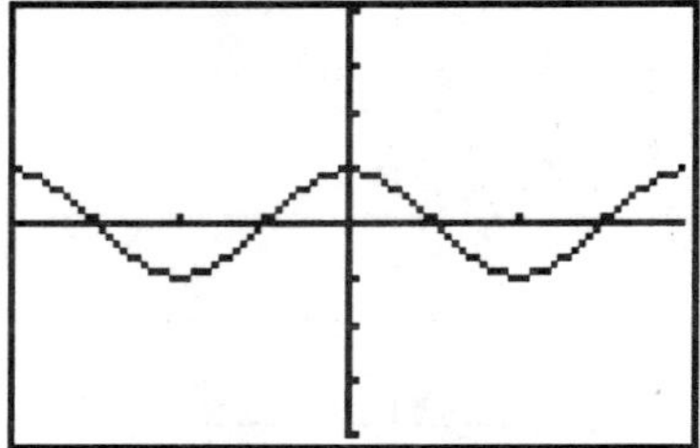

TRACE and describe the effect of "c" on the graph.

8. Change the **MODE** to **Radian**, reset the WINDOW to **ZTrig**, and graph the following:
Y = sin X, Y = sin (X + π/6), Y = sin (X + π/4), Y = sin (X + π/2),
Y = sin (X + π)
(Suggestion: sketch each graph in a different color.)

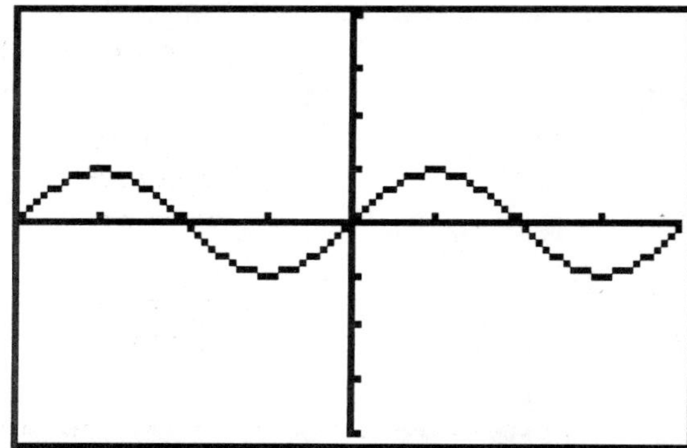

Observe the effect of the constant on the graph of the sine curve.

The value "c" is the phase shift of the graph. Shifting occurs based on the sign of "c". When c > 0, the graph shifts "c" units to the ____________(left/right) and when "c" < 0, the graph shifts "c" units to the ____________(left/right).

✍ 9. Describe the effect if the constant "c" is a multiple of 2π.

NOTE: Change the **MODE** back to **Degree** and reset the WINDOW to **ZTrig**.

✍ 10. Each group of functions below is in the form y = f(x) + d. Sketch each graph in a different color on the display below. The reference graph is listed first in the series.

NOTE: Be sure to enter these equations a Y = sin(X) + 2, etc. or the graph displayed will actually be Y = sin(X + 2).

Y = sin X, Y = sin X + 2,
Y = sin X - 2

Y = cos X, Y = cos X + 3,
Y = cos X - 3

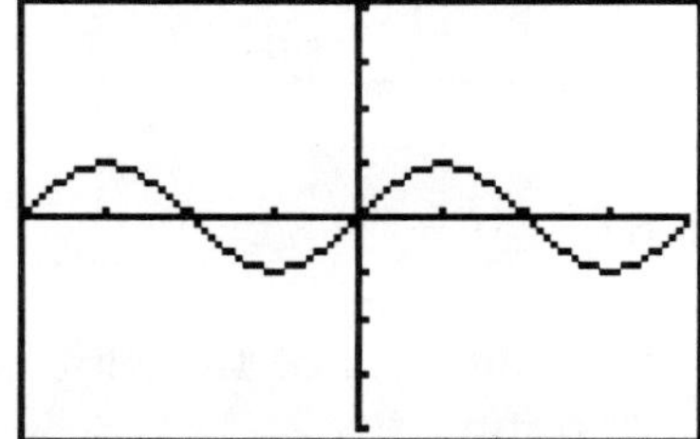

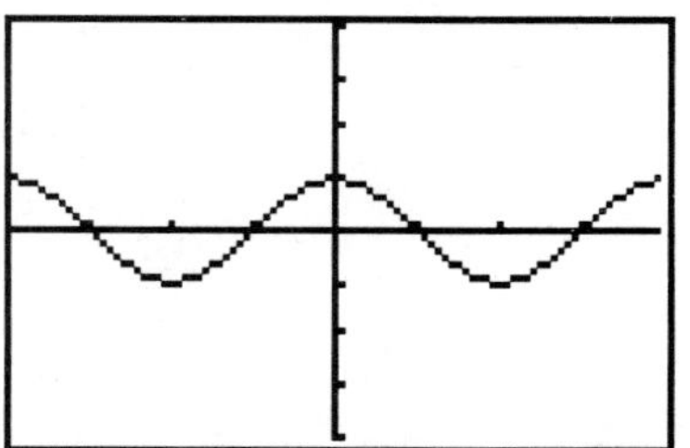

Does the constant "d" affect the period or the amplitude?

GRAPHS INVOLVING TANGENT AND COTANGENT

The sine and cosine functions are the basic trigonometric functions. The other four functions can be defined in terms of these two functions. Consider the definition of the six trigonometric functions for the acute angle X:

$$\sin X = \frac{opposite}{hypotenuse} \qquad \tan X = \frac{opposite}{adjacent} \qquad \csc X = \frac{hypotenuse}{opposite}$$

$$\cos X = \frac{adjacent}{hypotenuse} \qquad \cot X = \frac{adjacent}{opposite} \qquad \sec X = \frac{hypotenuse}{adjacent}$$

From these six functions of the acute angle X, the following are determined by the sine and cosine functions:

$$\tan X = \frac{sinX}{cosX} \qquad\qquad \cot X = \frac{cosX}{sinX} = \frac{1}{tanX}$$

$$\sec X = \frac{1}{cosX} \qquad\qquad \csc X = \frac{1}{sinX}$$

The graphing calculator has sine, cosine and tangent function keys, however all other functions will need to be entered in terms of the above definitions. Thus Y = cot X will be entered as Y = 1/tanX.

✎ 11.   Establish the period of Y = tan X and Y = cot X. The graph of each is displayed below. Use TRACE to determine the period.

Y = tan X

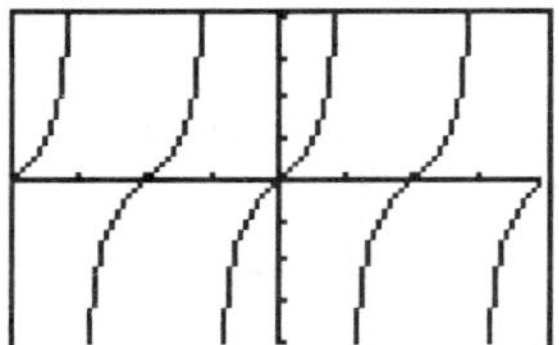

Y = cot X

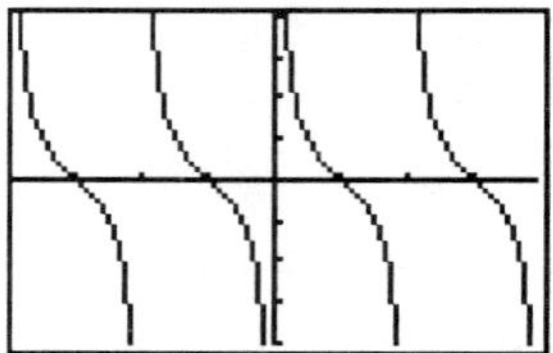

period:___________________

period:___________________

Describe the basic difference(s) in the graphs. Use the definitions of tangent and cotangent from the previous page to support your description.

Where do the vertical asymptotes occur? (Be aware that it may sometimes be necessary to display these graphs in dot mode.)

12.   Each of the groups of functions below is in the form y = a · f(x). Sketch each graph in a different color on the display below. The reference graph, listed first in the series, is graphed for you.

Y = tan X, Y = 2 tan X, Y = .2 tan X

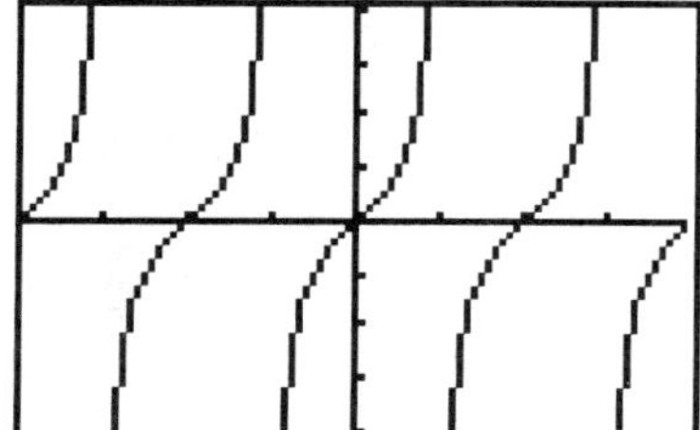

Y = cot X, Y = 2 cot X, Y = .2cot X

✍ 13. In general, if a > 0, what effect does "a" have on the graph of y = a · f(x)?

✍ 14. Compare the graphs of Y = tan X and Y = - tan X.

Y = tan X   Y = - tan X

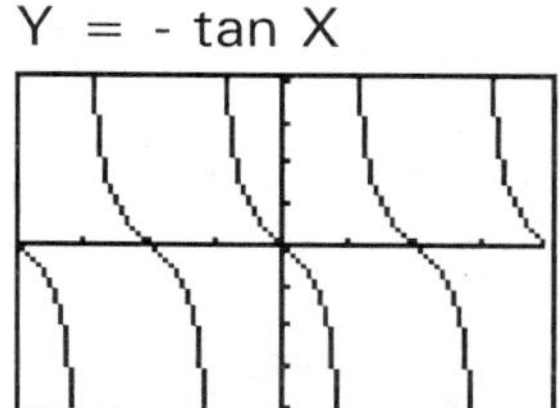

Describe the affect of a < 0.

15.   Individually graph each of the functions that is in the form y = f(bx). Use TRACE to
       examine the effect of b > 0 on the period of the function. Record your findings
       both in terms of degrees and radians (π format), in the following table.

Function	Period in Degrees	Period in Radians
Y = tan X	180	π
Y = tan .75X		
Y = tan .5X		
Y = tan .25X		
Y = tan 2X		
Y = tan 3X		
Y = tan 4X		

Function	Period in Degrees	Period in Radians
Y = cot X	180	π
Y = cot .75X		
Y = cot .5X		
Y = cot .25X		
Y = cot 2X		
Y = cot 3X		
Y = cot 4X		

✍ 16. Compare the graphs of Y = tan 2X, Y = tan -2X and Y = cot 2X, Y = cot -2X.
       Sketch each graph in a different color on the display. The first graph in each pair is
       already displayed.

Y = tan 2X, Y = tan -2X   Y = cot 2X, Y = cot -2X

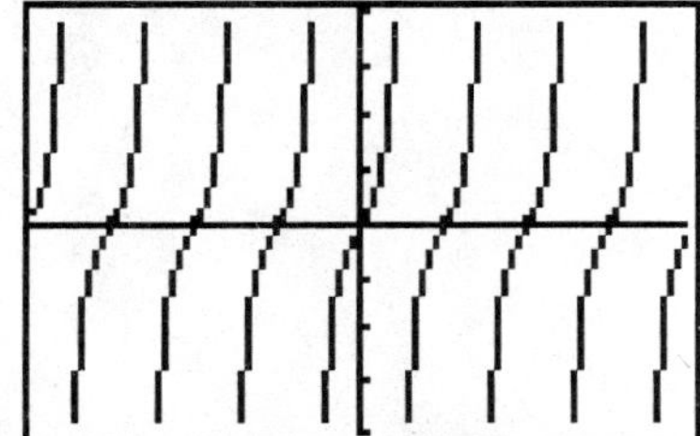

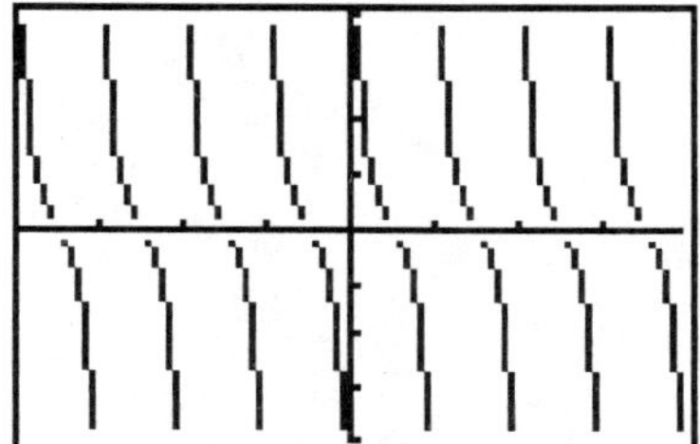

Explain the effect of b < 0 on each pair of graphs.

Classify the tangent and cotangent functions as even or odd. Will the size of "b" affect the classification of even or odd?

✍️ 17. Graph Y = - tan X and Y = tan -X on the screen at the right. Justify the display.

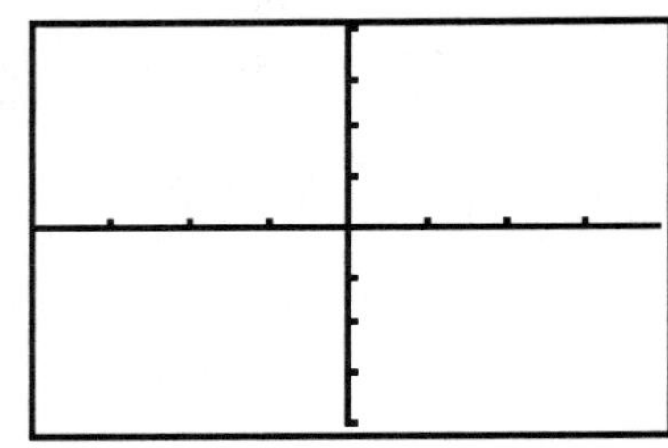

✍️ 18. Each group of functions below is in the form y = f(x + c). Sketch each graph in a different color on the display below. The reference graph is listed first in the series. (dot mode will be necessary.)

Y = tan X, Y = tan (X + 45)
Y = tan (X - 45)

Y = cot X, Y = cot (X + 45),
Y = cot (X - 45)

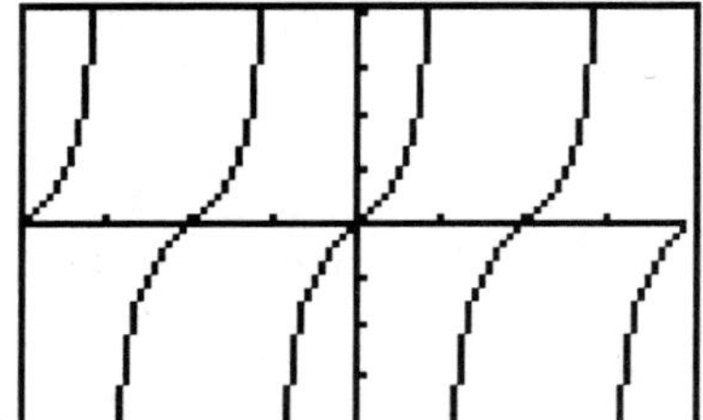

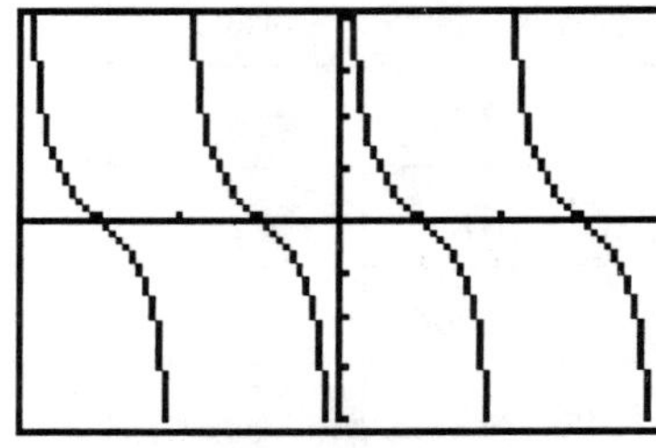

TRACE and describe the effect of "c" on the graph.

✍️ 19. Explore the relationship between the graph of Y = sin X and Y = csc X by graphing both functions on the same viewing window. As X approaches the zeroes of Y = sin X, explore what happens to the graph of Y = csc X.

✍20. Algebraic sums, differences, products and quotients of periodic functions produce
periodic functions. Determine the period of each component function and the period
of the combined function. Based on this information determine an algorithm for
finding the period of combined functions.

a. $Y = \sin 2X - \sin .75X$

b. $Y = \dfrac{\cos 3X}{\cos .25X}$

c. $Y = 2 \sin 4X + 3 \sin X$

Solutions:

1. amp. = 1 and period = 2π on both curves, sine curve is symmetric with respect to origin while cosine curve is symmetric with respect to the y-axis.

3. a, $-a < \sin X \leq a$, and $-a \leq \cos X \leq a$, "a" does not affect the period; the amplitude of the function and its range are the same

4. $a < 0$ does not effect period or amplitude

5.

Function	Period in Degrees	Period in Radians
$Y = \sin X$	360	2π
$Y = \sin .75X$	480	$8\pi/3$
$Y = \sin .5X$	720	4π
$Y = \sin .25X$	1440	8π
$Y = \sin 2X$	180	π
$Y = \sin 3X$	120	$2\pi/3$
$Y = \sin 4X$	90	$\pi/2$

Function	Period in Degrees	Period in Radians
$Y = \cos X$	360	2π
$Y = \cos .75X$	480	$8\pi/3$
$Y = \cos .5X$	720	4π
$Y = \cos .25X$	1440	8π
$Y = \cos 2X$	180	π
$Y = \cos 3X$	120	$2\pi/3$
$Y = \cos 4X$	90	$\pi/2$

1 radian = $180°/\pi$, 1 degree = $\pi/180°$ radians

6. b<0 reflects the graph across the x-axis, both functions are odd, "b" does not affect classification of even or odd

7. "c" shifts the graph left/right, c>0 shifts left whereas c<0 shifts right

8. "c" shifts the graph left/right, c>0 shifts left whereas c<0 shifts right

9. 2π shifts the graph one complete period and thus is the same graph.

10. affects the amplitude

11. both functions have period π, Vertical asymptotes appear at multiples of π/2 for the tangent function and at multiplies of π for the cotangent function.

13. The larger the value of a, the more the graph of the function straightens out and thus approaches the appearance of a vertical line. This occurs within each period.

14. a<0 reflects the graph across the y-axis: the magnitude of $|a|$ affects the graph the same way as in 13 above.

15.

Function	Period in Degrees	Period in Radians
Y = tan X	180	π
Y = tan .75X	240	4π/3
Y = tan .5X	360	2π
Y = tan .25X	720	4π
Y = tan 2X	90	π/2
Y = tan 3X	60	π/3
Y = tan 4X	45	π/4

Function	Period in Degrees	Period in Radians
Y = cot X	180	π
Y = cot .75X	240	4π/3
Y = cot .5X	360	2π
Y = cot .25X	720	4π
Y = cot 2X	90	π/2
Y = cot 3X	60	π/3
Y = cot 4X	45	π/4

16. b<0 reflects the graph across the y-axis, The functions are both odd and the size of b does not affect this classification.

17. Answers may vary.

18. "c" shifts the graph left when added and right when subtracted

19. Answers may vary.

20. Answers may vary.

*Unit 28 is a prerequisite for this unit. Answers appear at the end of the unit.

The ability to link a graphical display with its equation is a useful tool. It is suggested that this unit be completed without the aid of your calculator.

I.   MATCHING: Match each graph with its trigonometric function. All graphs are displayed in the **ZTrig** viewing window.

A. 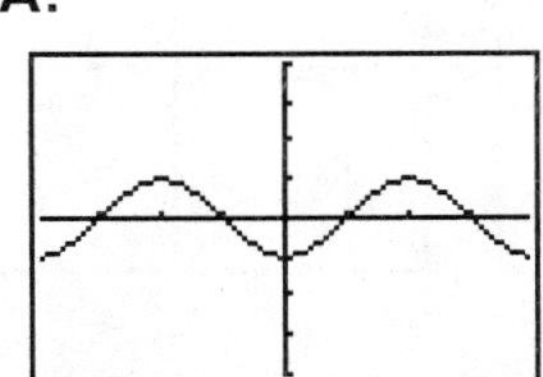 **B.** 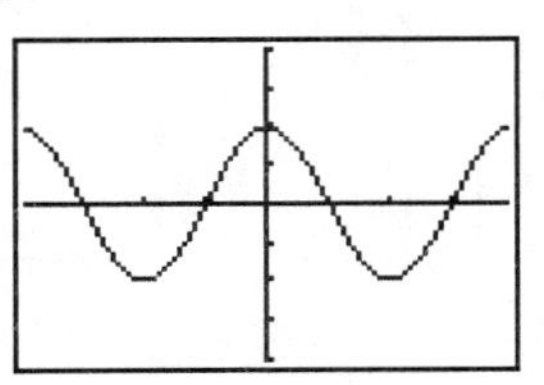 **C.** 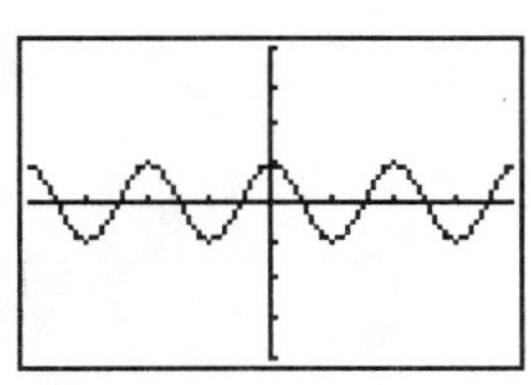 **D.** 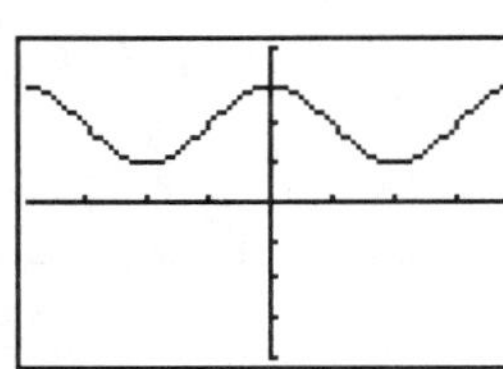

E. 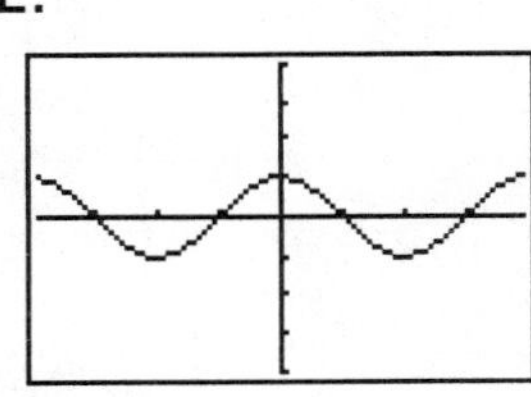 **F.** 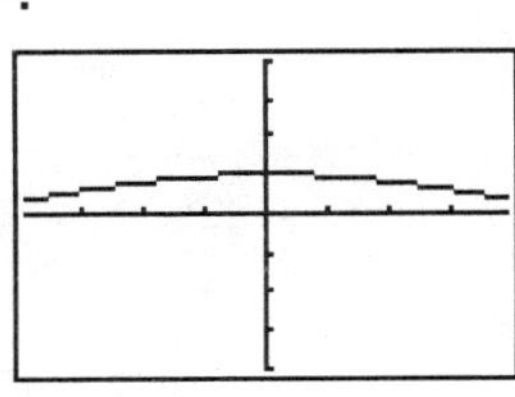 **G.** 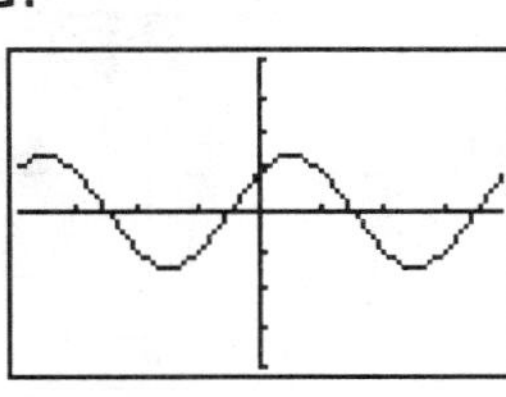 **H.** 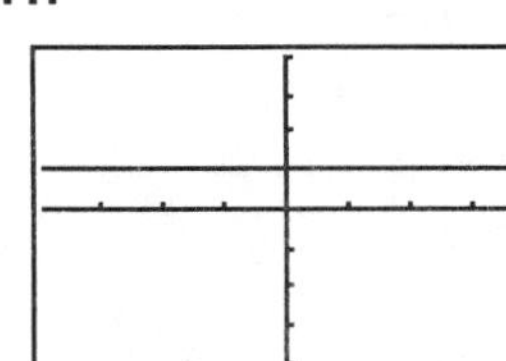

I. 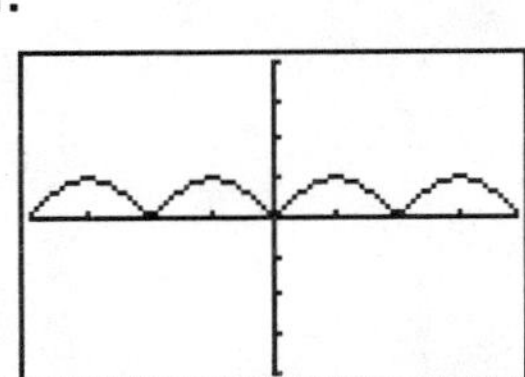 **J.** 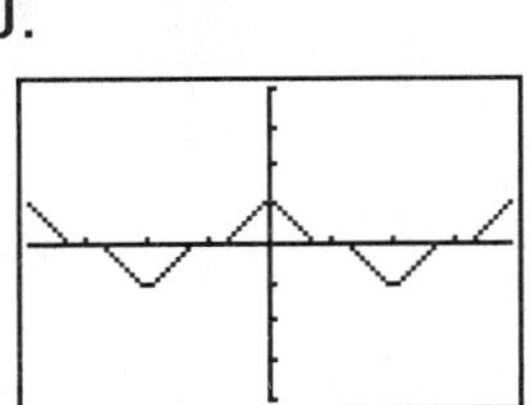 **K.** 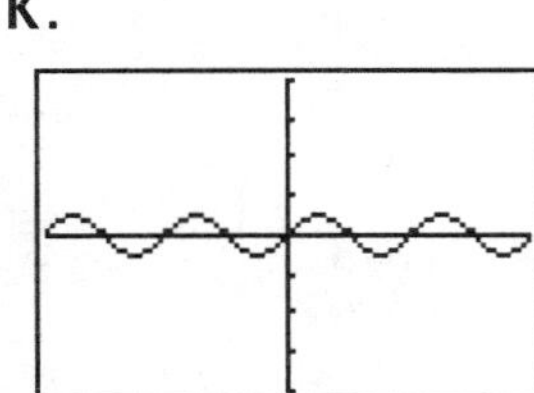 **L.** 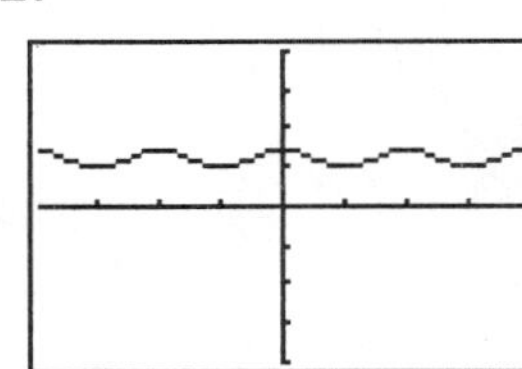

_____ 1. $-\cos X$

_____ 2. $\cos (X + 2)$

_____ 3. $\cos X + 2$

_____ 4. $\sin^2 X + \cos^2 X$

_____ 5. $\cos^3 X$

_____ 6. $\sin X + \cos X$

_____ 7. $\cos 0.2X$

_____ 8. $\cos 2X$

_____ 9. $2 \cos X$

_____ 10. $\sqrt{1 + \cos^2 X}$

_____ 11. $\sqrt{1 - \cos^2 X}$

_____ 12. $\sin X \cos X$

II.   MATCHING: Match each graph with its trigonometric function. All graphs are displayed in the **ZTrig** viewing window.

A.
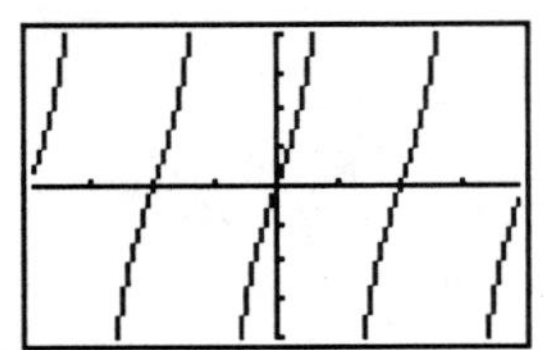

B.
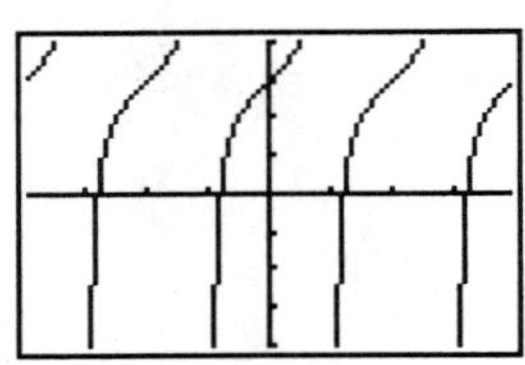

C.
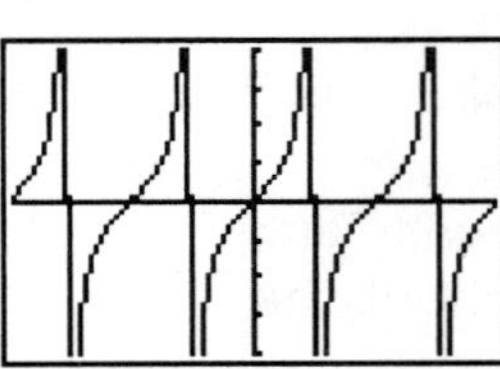

D.
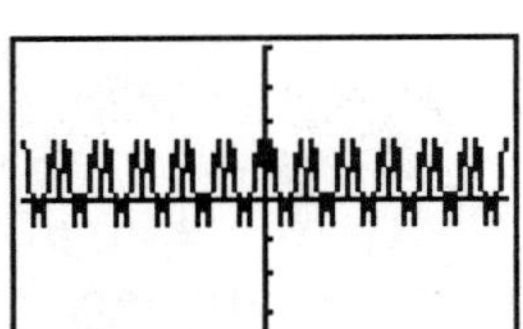

E.
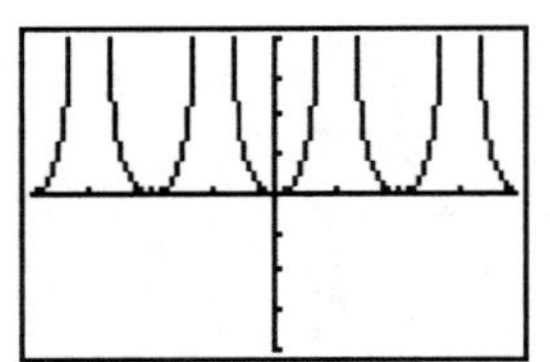

F.
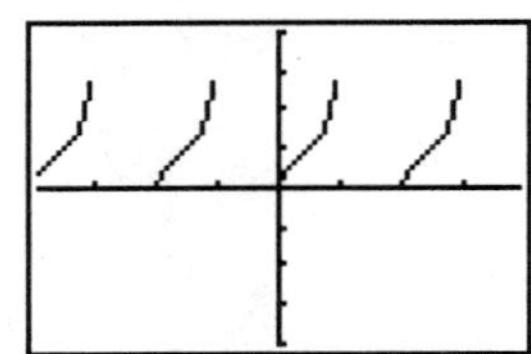

G.
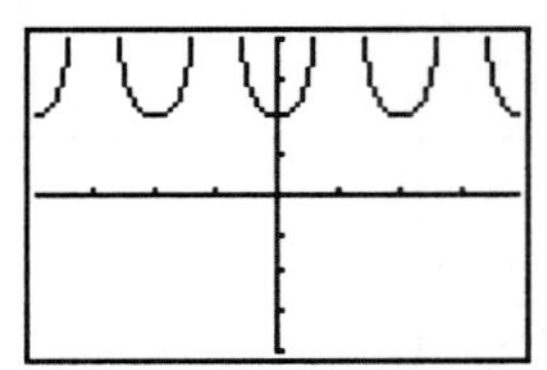

H.
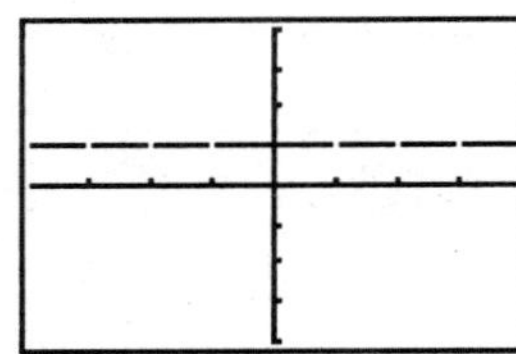

I.
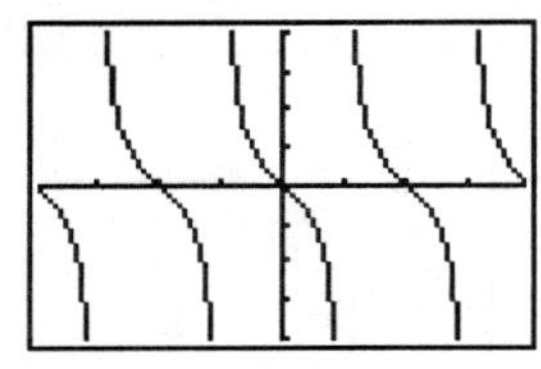

J.
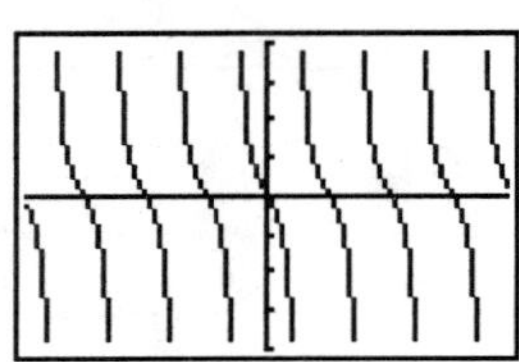

_____ 1. $\tan X^2$

_____ 2. $\tan X + 3$

_____ 3. $\tan (X + 3)$

_____ 4. $\tan^2 X \cot^2 X$

_____ 5. $\tan^2 X + 2$

_____ 6. $\tan^2 X$

_____ 7. $3 \tan X$

_____ 8. $- \tan X$

_____ 9. $- \tan 2X$

_____ 10. $\sqrt{\tan X}$

III. **DIRECTIONS:** The reference function, indicated below the screen, is graphed in bold. Write the equation for the remaining function that is graphed. The calculator display is set for **Degree MODE** in the **ZTrig** viewing window.

1.

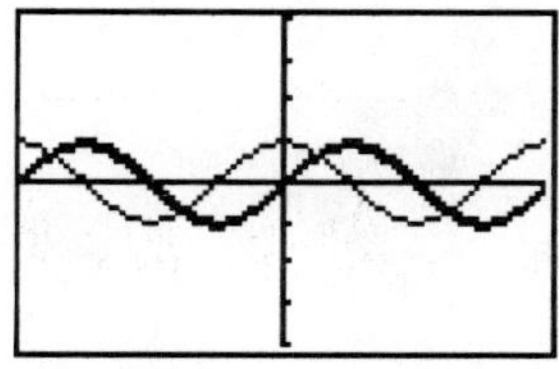

Y = sin X

____

2.

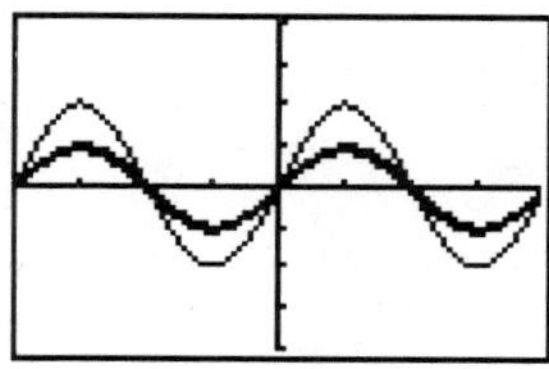

Y = sin X

____

3.

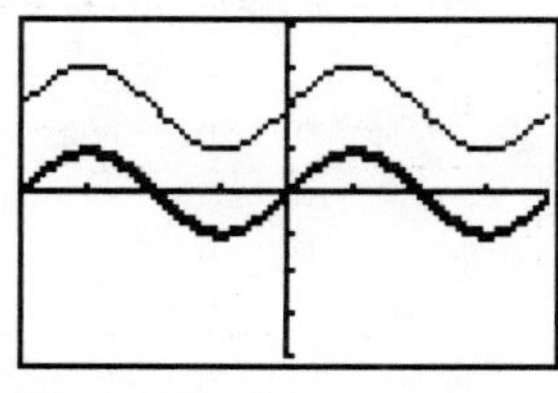

Y = sin X

____

4.

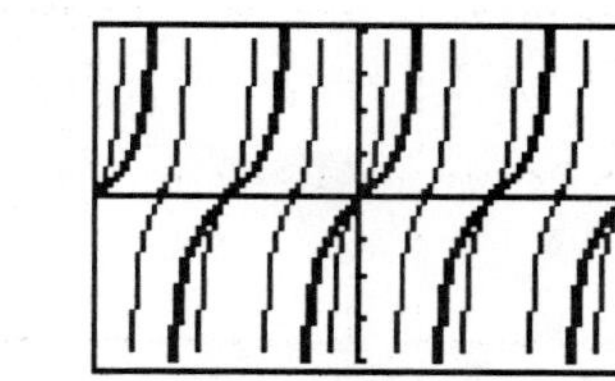

Y = tan X

____

5.

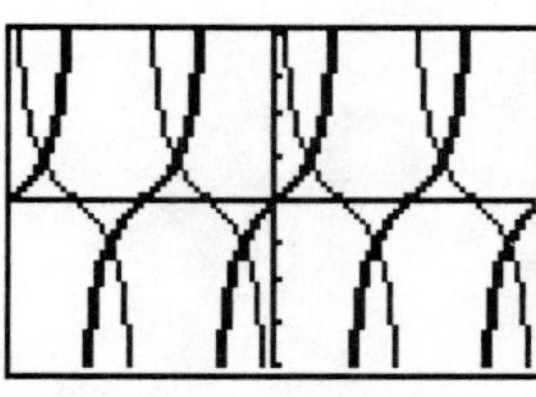

Y = tan X

____

6.

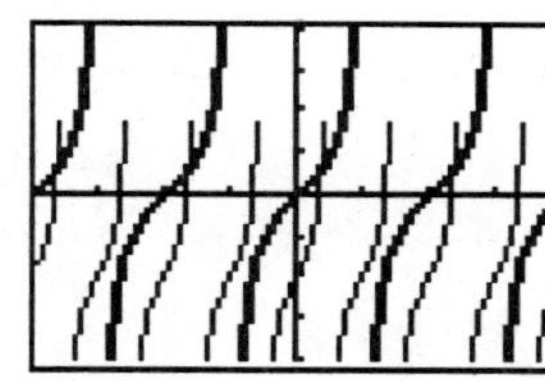

Y = tan X

____

UNIT 30
GRAPHICAL EXPLORATIONS:
TRIGONOMETRIC IDENTITIES

*Unit 28 is a prerequisite for this unit. Answers appear at the end of the unit.

Identities are equations that are true for <u>all</u> values of the variable (for which the expressions within the equation are defined). The graphs of the expression on each side of the equation can be used to confirm or deny the fact that a given equation is an identity.

One of the fundamental identities states that $\sin^2 X + \cos^2 X = 1$. To verify this using a graphing calculator, first ensure all options at the far left are highlighted when **MODE** is pressed.

Enter $\sin^2 X + \cos^2 X$ at the Y1= prompt and the number 1 at the Y2= prompt. Press **[ZOOM] [7:ZTrig]** to display the graphs of the two expressions.

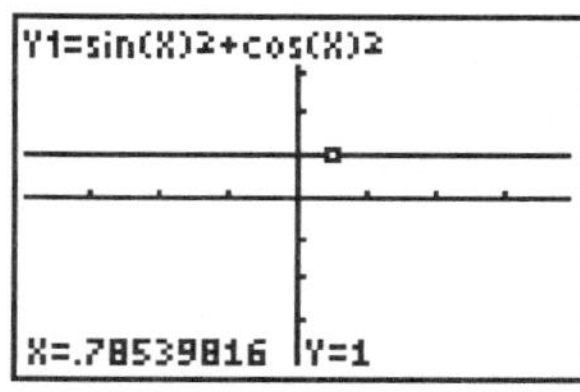

Your screen display should match the one pictured at the right. To verify that this is the graph of two functions rather than one, press **[TRACE]** to locate the cursor with the coordinates displayed. Notice the equation displayed in the upper left hand corner of the screen (a number 1 is displayed in the upper right corner on the TI-82). This indicates that the cursor is on the graph of the expression entered at the Y1= prompt. Press the down arrow key and again observe the equation in the upper left hand corner of the display screen. It is the equation entered at the Y2= prompt, however, the X and Y coordinates did not change even though the cursor was moved from one graph to the other. This verifies the fact that the graph of $\sin^2 X + \cos^2 X$ is the same as the graph of the line $y = 1$ and thus the two expressions are equal. If your calculator has a table feature, you could also check to verify that for each x-value all the values in the Y1 and Y2 columns are equal.

EXERCISE SET

✎1.   Does $\sin X = \sin X + 6$?   Sketch the graph on the axes provided.
Explain how you obtained your answer.

Directions: Match each of the following expressions to the appropriate graph.

a. sin X

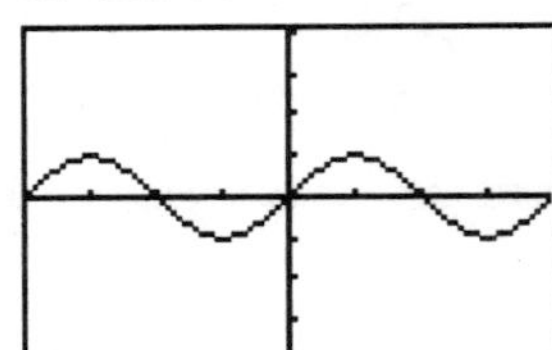

b. cot X

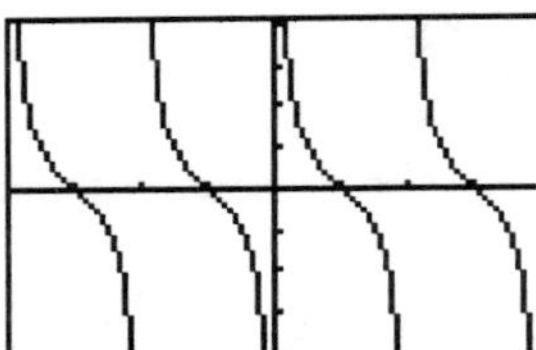

_____2.   cot X sec X

_____3.   tan X csc X

_____4. $\dfrac{1 - \cos^2 X}{\sin X}$

c. cos X

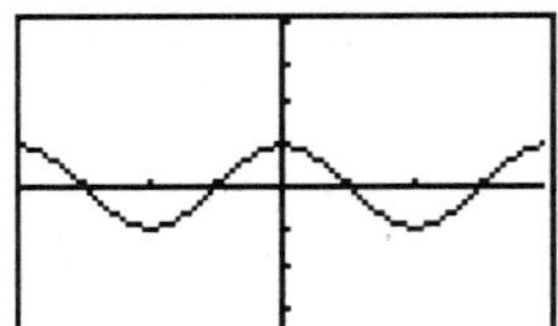

d. -sec X

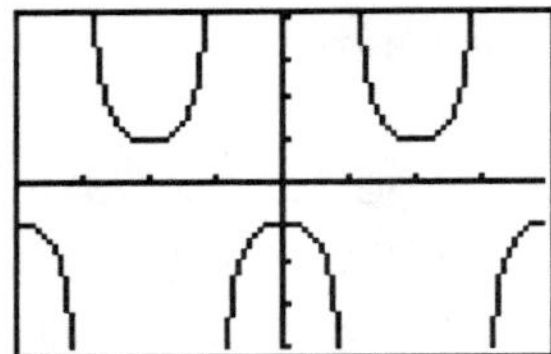

_____5.   sinX secX

_____6. $\dfrac{\sin X}{\cos^2 X - 1}$

_____7.   cosX cscX

_____8. $\dfrac{\cos X}{\sin^2 X - 1}$

e. sec X

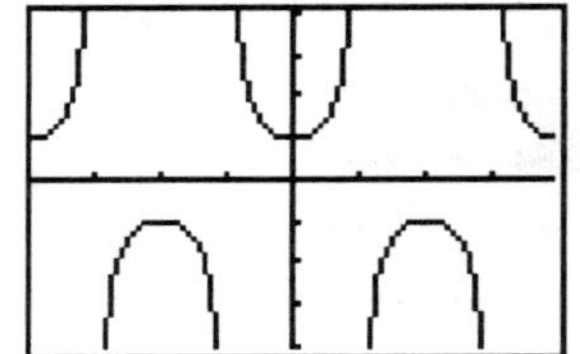

f. tan X

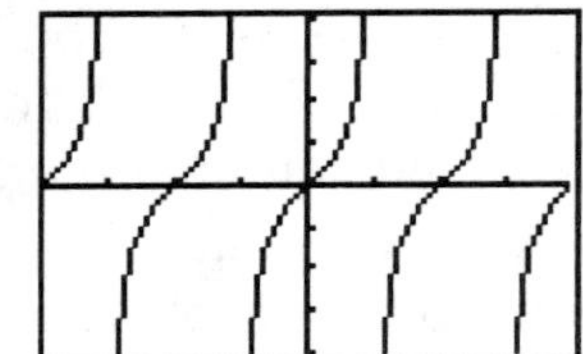

g. -csc X

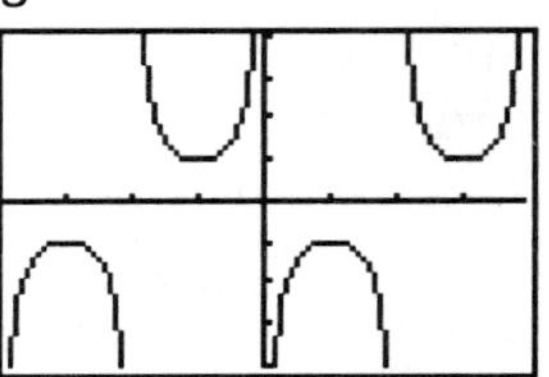

h. csc X

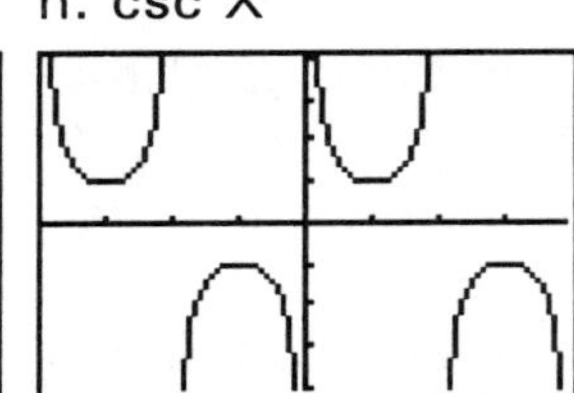

9.   Factor each expression below and verify that the original problem and your factorization are graphically equivalent. Sketch the graph displayed.

a. $\sin^2 X + \sin X \cos X =$

____

b. $\sin^2 X - 9 =$

____

c. $\sin^2 X + \sin X - 6 =$

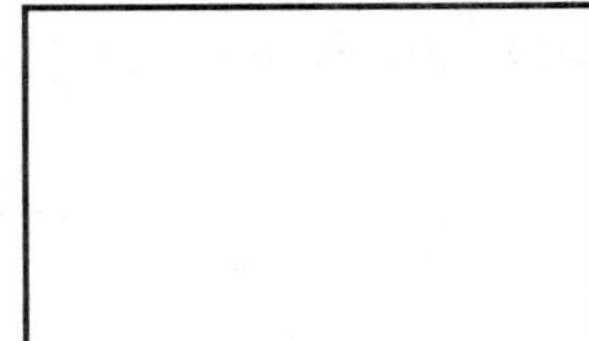

____

d. $\cos^4 X - \sin^4 X =$

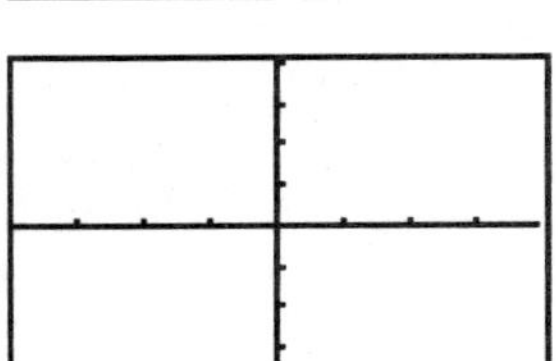

____

10. State the values of X that will make the equation $\cos X = \sqrt{1 - \sin^2 X}$ an identity.
     $(0 \leq X \leq 2\pi)$

____

✍11. Verify with the calculator that $\cos X = \sin(\pi/2 - X)$. Record your method(s) below.

✍ 12.   Is $\dfrac{1}{3}\tan^4 X + \dfrac{1}{2}\tan^2 X = \dfrac{1}{3}\sec^4 X - \dfrac{1}{6}\sec^2 X$ an
          identity?

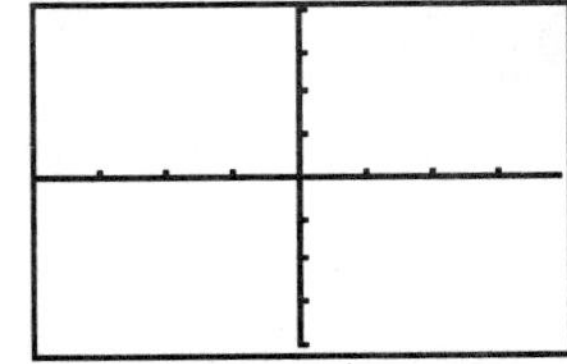

After sketching the display on the screen at the right, discuss
the various methods available on the calculator for disproving
the proposed identity. What constant C, expressed as a fraction, should be added to
the left side of the equation to complete the identity?

<u>**Solutions:**</u> **1.** Answers may vary **2.** h **3.** e **4.** a **5.** f **6.** g **7.** b **8.** d

9. $\sin X(\sin X + \cos X)$, $(\sin X - 3)(\sin X + 3)$ WINDOW: [-4,4] by [10,1],

$(\sin X + 3)(\sin X - 2)$ WINDOW: [-6, 6] by [-10, 1], $(\cos^2 X - \sin^2 X)(\cos^2 X + \sin^2 X)$

10. $[-\pi/2, \pi/2]$ **11.** Answers may vary **12.** $C = 1/6$

UNIT 31
GRAPHICAL SOLUTIONS: TRIGONOMETRIC EQUATIONS

*Unit 30 is a prerequisite to this unit. Answers appear at the end of the unit.

The focus of this unit is solving trigonometric equations. Unlike identities, which are true for all values of the variable, trigonometric equations are true for <u>some</u> values of the variable.

Example 1: Solve the equation 4 cos X - 2 = 0.

Solution: Algebraically:
$$4 \cos X = 2$$
$$\cos X = 1/2$$

On the calculator, one solution of cos X = 1/2 is obtained by entering $\cos^{-1}$ (1/2) on the home screen. In degree mode, X = 60° or X = 60(Π/180) = Π/3 when expressed in radians. Because $\cos^{-1}$(1/2) gives only a first quadrant solution and since cosine is also positive in quadrant IV, we also have X = 300° or 11Π/6 radians as an additional solution in the interval [0,2Π). This angle has 60° as its reference angle and is in quadrant IV.

By graphing both sides of cos X = 1/2, it is clear that there are infinitely many other solutions. Enter cos X at the Y1 = prompt and 1/2 at the Y2 = prompt. Display the graph in the **ZTrig** viewing WINDOW with the MODE set to **Degree**. Using TRACE, the initial graph indicates that X has values of ± 300° and ± 60° (or ± 11Π/6 and ± Π/3 radians). Thus, solutions will be displayed in the general form as X = Π/3 + 2nΠ and X = 11Π/6 + 2nΠ where n is an integer and X is expressed in radians.

Graphically: To graphically solve 4 cos X - 2 = 0, the non-zero side is entered at the Y1 = prompt followed by the use of ROOT/ZERO to calculate the solutions. Tracing confirms the previously determined values with the root of 60° (Π/3) displayed on the graph at the right.

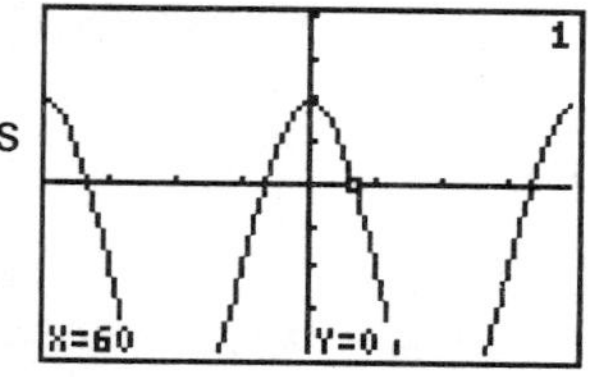

◆

Example 2:. Graphically solve sin X = 1 - cos 2X on the interval [0,2Π) using the INTERSECT option of the calculator.

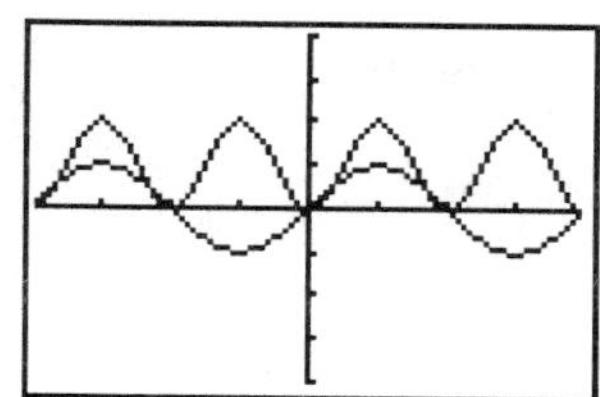

Solution: The initial graph is pictured in the **ZTrig** viewing WINDOW. Because the points of intersection are somewhat difficult to see, we need a better window within the interval $[0, 2\pi)$. **ZBOX** can be used to determine a better viewing window.

The solutions are 0, 30, 150 and 180 degrees. ◆

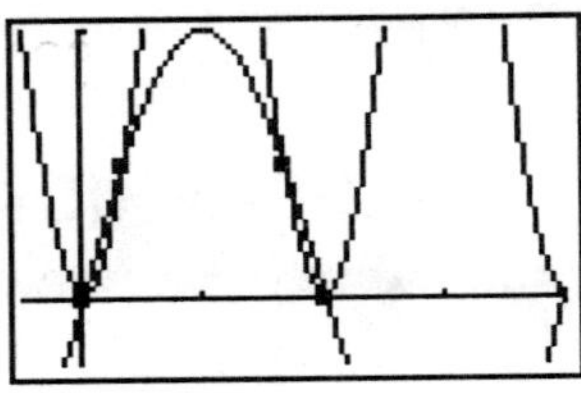

EXERCISE SET

1.   a. Algebraically solve $6 \sec^2 X + 5 \sec X + 1 = 0$ and confirm your algebraic solution by solving graphically.

algebraic solution: graphical solution:

✍ b. Why are there no values of X such that $\sec X = -1/2$ or $\sec X = -1/3$?

Directions: Solve each equation graphically by the method of your choice and record the general solution in radian form. Sketch the graph displayed in the **ZTrig** viewing WINDOW (or an appropriate window that displays the solutions).

2. $4 \sin^2 X = 1$

Solutions:_______________________________________

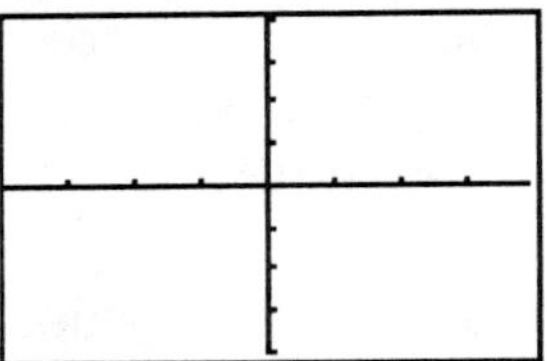

3. $2 \sin X \cos X + 2 \sin X + \cos X + 1 = 0$

Solutions:_______________________________________

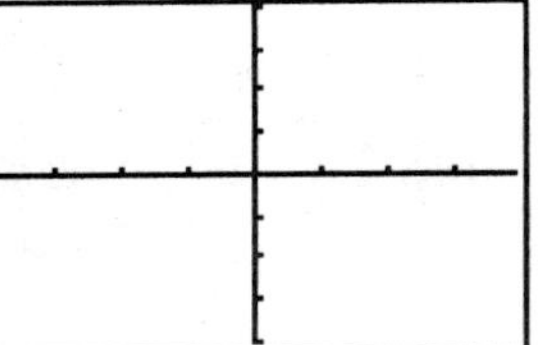

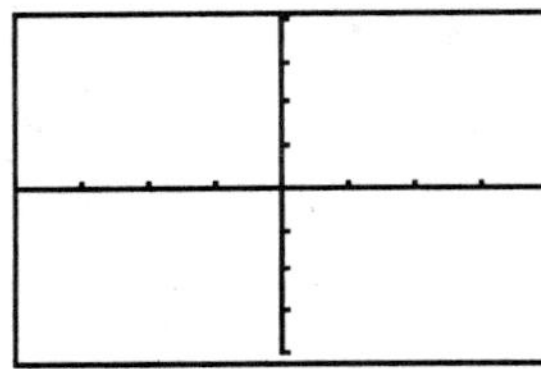

4. $7 \cos 2X = \cos 4X - 8$

Solutions:_________________________________

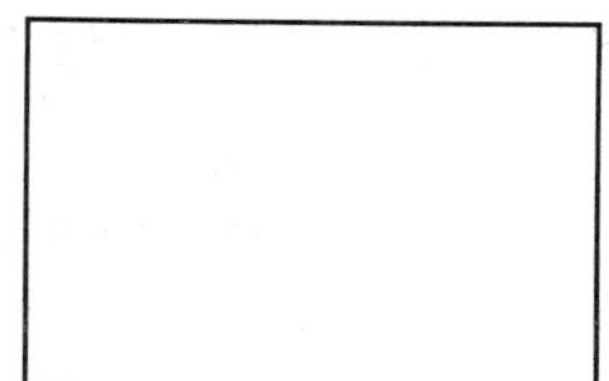

[___ , ___] by [___ , ___]

5. $2 \tan X = \tan (X/2)$

Solutions:_________________________________

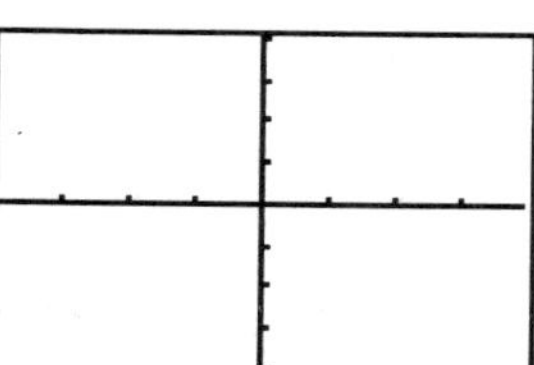

6. $\sin X + \csc X = \cos X$

Solutions:_________________________________

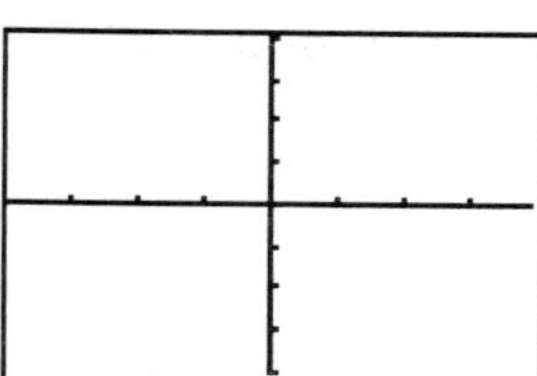

7.   Find the solutions of $\sqrt{3} \sin \dfrac{X}{3} + \sqrt{1 - \cos 2X} - \sqrt{3} = 0$, in radian measurement to the nearest hundredth in the interval $[0, 4\pi]$.

Solutions:_________________________________

Sketch the graph in the viewing window $[-\pi/2, 4\pi]$ by $[-4, 2]$.

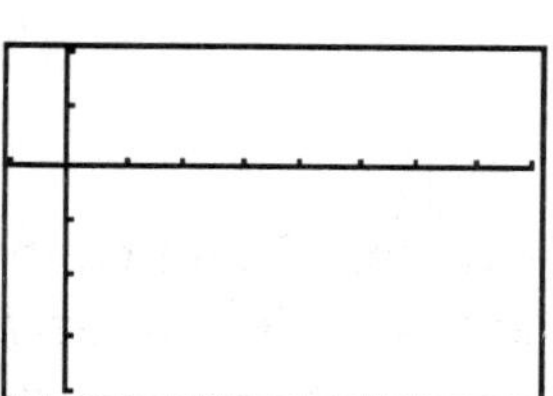

8.   Examine each group of equations graphically in the **ZTrig** viewing window. Determine which ones are conditional equations and which ones represent identities. (Label "C" or "I".)

$\sin(X + 2) = \sin X + \sin 2$ _____ $\sin(X + 2) = \sin X \cos 2 + \cos X \sin 2$ _____

$\sin(X - 2) = \sin X \cos 2 - \cos X \sin 2$ _____

$\cos(X + 2) = \cos X + \cos 2$ _____ $\cos(X + 2) = \cos X \cos 2 - \sin X \sin 2$ _____

$\cos(X - 2) = \cos X \cos 2 + \sin X \sin 2$ _____

$$\tan(X + 2) = \tan X + \tan 2 \underline{\quad\quad} \qquad \tan(X + 2) = \frac{\tan X + \tan 2}{1 - \tan X \tan 2} \underline{\quad\quad}$$

$$\tan(X - 2) = \frac{\tan X - \tan 2}{1 + \tan X \tan 2} \underline{\quad\quad}$$

The equations that represent identities are called the **Sum and Difference Formulas**. These identities will be useful in scientific applications.

<u>**Solutions:**</u>   **1a.** ϕ **1b.** Answers may vary.   **2.** $\pi/6 + 2n\pi$, $5\pi/6 + 2n\pi$, $7\pi/6 + 2n\pi$, $11\pi/6 + 2n\pi$

3. $\pi + 2n\pi$, $11\pi/6 + 2n\pi$ **4.** $\pi/2 + 2n\pi$, $3\pi/2 + 2n\pi$, WINDOW: [-6, 6] by [-10, 4]

5. $0 + 2n\pi$ **6.** ϕ

7. 0.98, 2.93, 3.28, 6.15, 6.49, 8.44

8. The conditional equations are $\sin(X + 2) = \sin X + \sin 2$, $\cos(X + 2) = \cos X + \cos 2$, and $\tan(X + 2) = \tan X + \tan 2$.   **9.** Answers may vary.

UNIT 32
POLAR GRAPHING

*Unit 28 is a prerequisite for this unit. Answers appear at the end of the unit.

In the rectangular coordinate system, each point P in the coordinate plane is associated with an ordered pair of real numbers (x,y). The directed distance from the y-axis is represented by x, and y represents the directed distance from the x-axis. In polar graphing, this point P is associated with an ordered pair (r,θ) where r is the directed distance from the origin and the measure of the angle between the positive x-axis and the ray from the origin through point P is designated as θ. If θ includes all angles, then (r,θ), a coordinate of point P, is also represented by $(r,\theta + k\cdot360°)$ where k is a positive integer.

1.　　Begin this unit by setting MODE to **Degree** and **Pol** for polar graphing. If the MODE is set for polar graphing, then pressing **[ZOOM] [6:ZStandard]** will automatically set the following WINDOW values: θmin = 0°, θmax = 360°, θstep = 7.5, Xmin = -10, Xmax = 10, Xscl = 1, Ymin = -10, Ymax = 10, and Yscl = 1.

TI-85/86　　MODE SHOULD BE SET TO HIGHLIGHT DEGREE, **PolarC**, AND **Pol**. THETA IS LISTED ON THE DISPLAYED MENU ON THE MENU BAR WHEN [R(θ) =] IS PRESSED.

✐2.　　Graph $r = \theta$. Sketch the graph on the display at the right. <u>Without</u> adjusting the viewing window, make a conjecture about the type of graph displayed. (i.e. Is the appearance linear, parabolic, exponential, logarithmic, etc.?)

✐3.　　Set Xscl = 0 and Yscl = 0 and with your cursor centered at the origin, ZOOM OUT once and copy the display. What window values were affected by zooming out?

4.　　Again, center the cursor at the origin and ZOOM OUT. Sketch the graph display.

5.　　ZOOM OUT one final time and sketch the display. TRACE around the entire Spiral of Archimedes and observe the values of θ.

209

6.   In the same viewing window as #5 ([-640,640] by [-640,640]), graph each of the
        following spirals.
        $r = 2\theta$ $r = -2\theta$ $r = \tfrac{1}{2}\theta$

✍7.   In general $r = a\theta$ will be a Spiral of Archimedes if $a > 0$.
        The effect of $a < 0$ is to _________________________.

        As $|a|$ increases the spiral _________________________.

        As $|a|$ decreases the spiral _________________________.

        NOTE: Return to the ZStandard viewing window for polar graphing.

8.   Graph $r = \sin 3\theta$. Set Xscl = 0 and Yscl = 0 and ZOOM IN
        once. Sketch the graph on the display at the right. Equations
        of the form $r = a \sin 3\theta$ form a three-leaved rose.

✍9.   In this same viewing window, graph each of the following and sketch their display.
        $r = 2 \sin 3\theta$ $r = -2 \sin 3\theta$ $r = \tfrac{1}{2} \sin 3\theta$

        The effect of $a < 0$ is to _________________________.

        As $|a|$ increases the rose _________________________.

        As $|a|$ decreases the rose _________________________.

✍10.   Explore $r = \sin a\theta$ for $1 \leq a \leq 8$. What can you conclude about the effect of a on
        the graph of $r = \sin a\theta$?

11. Compare r = sin 2θ and r = cos 2θ. Graph each on the given screen with the
viewing window set to [-2.5,2.5] by [-2.5,2.5] with Xscl and Yscl = 1.

r = sin 2θ

r = cos 2θ

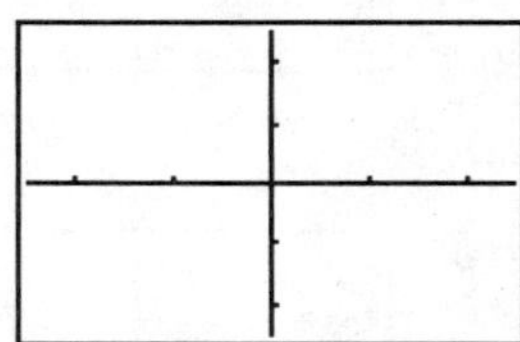

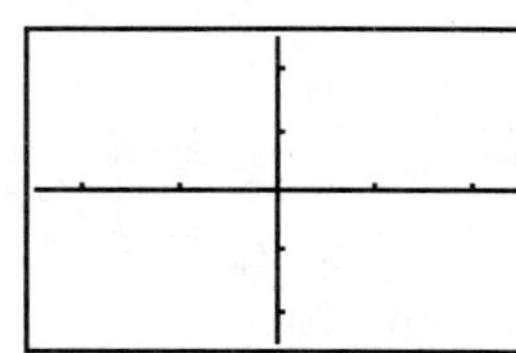

✍ Do the graphs of the polar sine and cosine have the same relationship to the
y-axis as the sine and cosine graphs on a rectangular coordinate system have?

12. Graph r = 1 + 2 cos θ in a [-3,3] by [-3,3] polar viewing
window and sketch the graph. This is the graph of a Limacon
whose general form is r = a + b cos θ, a < b.

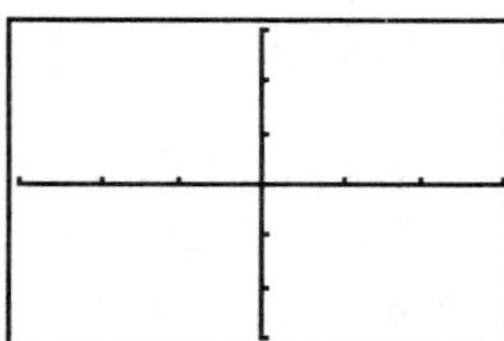

13. Graph r = -1 + 2 cos θ. Is this the same graph as r = 1 + 2 cos θ? (HINT: use
the TABLE to verify)

✍14. Graph the following pairs of polar graphs:

r = 2 + 3 cos θ   r = 2 + 2 cos θ   r = 3 + 2 cos θ
r = 3 + 4 cos θ   r = 1 + cos θ   r = 2 + cos θ

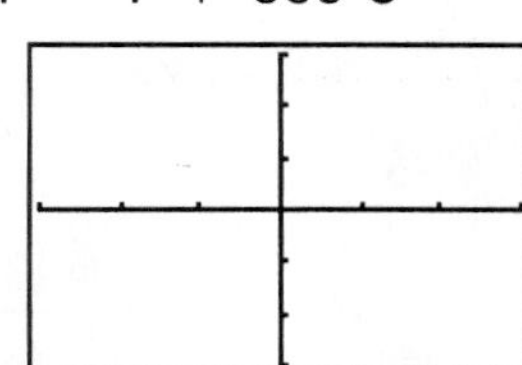

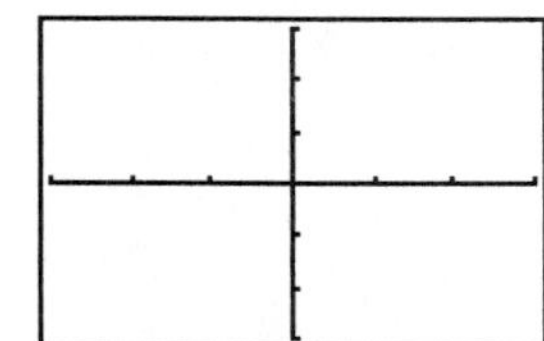

When a < b, as *a* and *b* both increase, the graph _______________________________.

When a = b, as *a* and *b* both increase, the graph _______________________________.

When a > b, as *a* and *b* both increase, the graph _______________________________.

Do <u>all</u> graphs of the form r = a + b cos θ, where a < b, have the same shape
regardless of the size of a and b?

Do <u>all</u> graphs of the form r = a + b cos θ, where a = b, have the same shape
regardless of the size of a and b?

Do <u>all</u> graphs of the form r = a + b cos θ, where a > b, have the same shape
regardless of the size of a and b?

15. Graph the Hyperbolic Spiral $r = \dfrac{2}{\theta}$ using ZOOM IN or ZOOM OUT to find an appropriate viewing window.

[_____ , _____] by [_____ , _____]

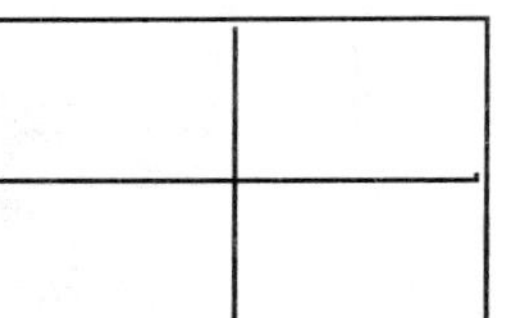

16. Graph the Logarithmic Spiral $r = 2^4\,\theta$ using ZOOM IN or ZOOM OUT to find an appropriate viewing window.

[_____ , _____] by [_____ , _____]

17. Polar equations in the form $r = \dfrac{ep}{1 \pm e \sin \theta}$ and $r = \dfrac{ep}{1 \pm e \cos \theta}$ represent the graphs of conics. The relationship between e and the value 1 (ie. $e < 1$, $e = 1$, or $e > 1$) determines the type of conic (parabola, ellipse, or hyperbola). Match each equation to the appropriate graph.

_____ i. $\quad r = \dfrac{4}{2 - \cos \theta}$ 　　　　 _____ vii. $\quad r = \dfrac{1}{1 - \cos \theta}$

_____ ii. $\quad r = \dfrac{4}{2 + \cos \theta}$ 　　　　 _____ viii. $\quad r = \dfrac{1}{1 + \cos \theta}$

_____ iii. $\quad r = \dfrac{2}{1 + \sin \theta}$ 　　　　 _____ ix. $\quad r = \dfrac{3}{1 - 2 \cos \theta}$

_____ iv. $\quad r = \dfrac{2}{1 - \sin \theta}$ 　　　　 _____ x. $\quad r = \dfrac{5}{1 + 3 \cos \theta}$

_____ v. $\quad r = \dfrac{4}{3 - \sin \theta}$ 　　　　 _____ xi. $\quad r = \dfrac{3}{1 + 2 \cos \theta}$

_____ vi. $\quad r = \dfrac{4}{3 + \sin \theta}$ 　　　　 _____ xii. $\quad r = \dfrac{5}{1 - 3 \sin \theta}$

a. 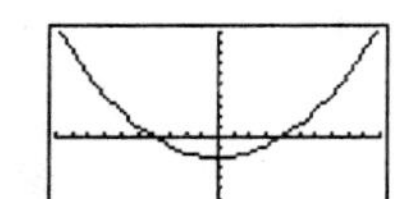 b. 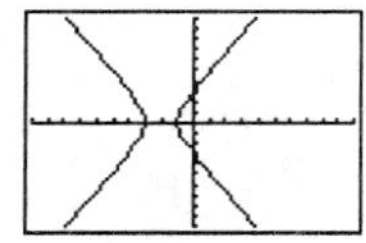 c. 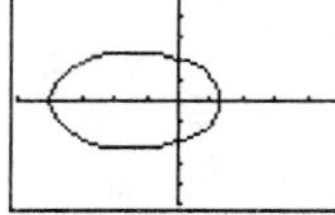 d. 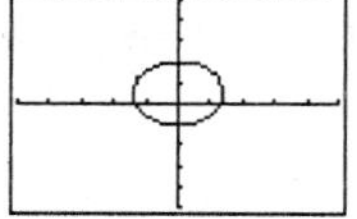 e. 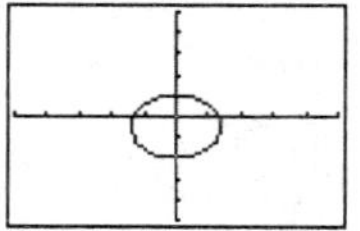 f. 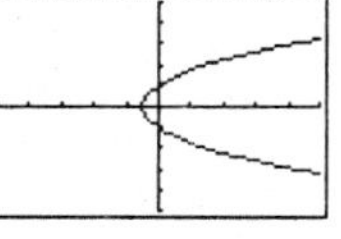

g. 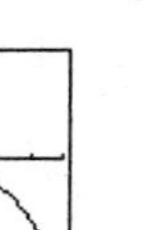 h. 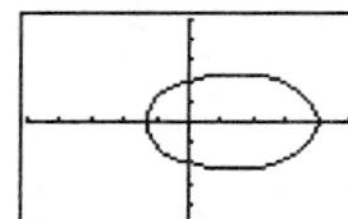 i. 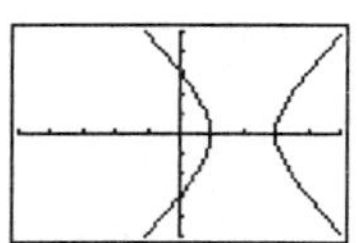 j. 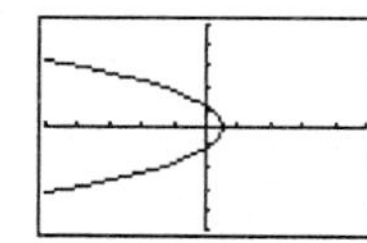 k. 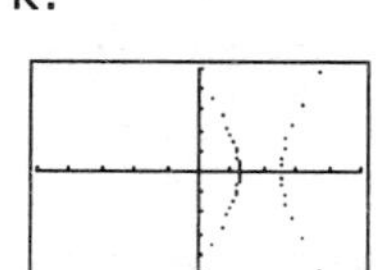 l. 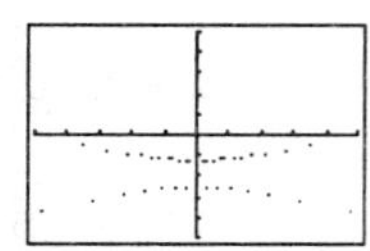

18. When necessary, write each of the equations in #17 in the form $r = \dfrac{ep}{1 \pm e \sin \theta}$ or $r = \dfrac{ep}{1 \pm e \cos \theta}$ and determine e's relationship (e > 1, e = 0, e < 1) to the type of conic.

e < 1:_______________________________

e = 1:_______________________________

e > 1:_______________________________

Solutions: **7.** a < 0 flips the graph across the x-axis <u>and</u> across the y-axis, as |a| increases so does the size of the spiral and as |a| decreases the spiral decreases

9. a < 0 flips the graph across the x-axis, as |a| increases so does the size of the spiral and as |a| decreases the spiral decreases

10. If "a" is even then the number of leaves is 2a, and if "a" is odd the number of leaves is "a".

11. yes   **13.** no, values in table do not correspond

14. If a < b, the graph increases in size as both a and b increase.
   If a = b, the graph increases in size as both a and b increase.
   If a > b, the graph increases in size as both a and b increase.

15. ZOOM IN until the window is approximately [-.04,.04] by [-.04,.04].

16. ZOOM OUT until the window is approximately [-10240, 10240] by [-10240,10240].

17. **i.** h   **ii.** c   **iii.** g   **iv.** a   **v.** d   **vi.** e   **vii.** f   **viii.** j   **ix.** b   **x.** k   **xi.** i   **xii.** l

18. e < 1:ellipse,   e = 1:parabola,   e > 1: hyperbola

*Unit 32 is a prerequisite for this unit. Answers appear at the end of the unit.

Parametric equations define a graph by expressing the variables x and y in terms of a third variable. The third variable is called the parameter. We will designate the third variable as t and express the variables x and y a functions of t: $f(t) = x$ and $g(t) = y$.

Parametric equations allow us to represent as a function curves whose equations are not the graph of a function. To parameterize circles, ellipses and hyperbolas, we can define f(t) and g(t) in terms of trigonometric functions.

Recall that the six trigonometric functions defined by the right triangle with acute angle of measure t and radius of measure r as displayed at the right are as follows:

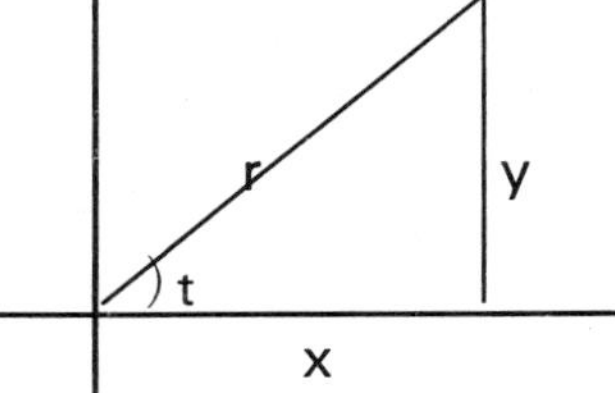

$$\sin t = \frac{opp}{hyp} = \frac{y}{r} \qquad \csc t = \frac{hyp}{opp} = \frac{r}{y}, y \neq 0$$

$$\cos t = \frac{adj}{hyp} = \frac{x}{r} \qquad \sec t = \frac{hyp}{adj} = \frac{r}{x}, x \neq 0$$

$$\tan t = \frac{opp}{adj} = \frac{y}{x}, x \neq 0 \qquad \cot t = \frac{adj}{opp} = \frac{x}{y}, y \neq 0$$

Note: Access the MODE screen of your calculator and ensure that all the options on the left are hightlighted *except* Func. On that line, Par (or Parem, depending on your calculator) should be highlighted.

Circles

The equation of a circle centered at the origin with a radius of r can be parameterized by using the definition of the sine and cosine of the angle t given above:

$$\cos t = \frac{x}{r} \qquad \qquad \sin t = \frac{y}{r}$$
$$r \cdot \cos t = x \qquad \qquad r \cdot \sin t = y$$

and the trigonometric identity $\sin^2 t + \cos^2 t = 1$.

Example 1: Find a parameterization of $x^2 + y^2 = 9$ and display the graph on the calculator.

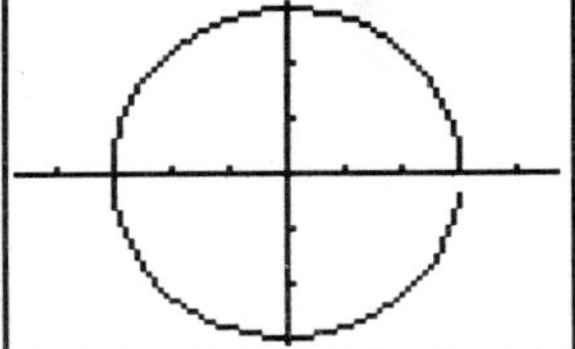

Solution: The circle with equation $x^2 + y^2 = 3^2$ has a radius r equal to 3. The parameterization is $x = 3 \cos t$ and $y = 3 \sin t$ since $(3 \cos t)^2 + (3 \sin t)^2 = 9(\cos^2 t + \sin^2 t) = 9 \bullet 1 = 9 = 3^2$.

With the calculator in **Par**ametric MODE the parameterization is graphed in the ZDecimal viewing window at the right. The Tmin = 0, Tmax $= 2\pi$ and Tstep = 0.15.

◆

Parametric equations graphed with a graphing calculator must have the minimum and maximum values defined for the variable *t* as well as for the variables *x* and *y*. An incomplete graph can result from an inappropriate range for *t*. The increment for *t*, called the t-step, determines how many points are to be plotted when constructing the graph. Thus, a small t-step, between 0.05 and 0.15 in radian MODE, must be selected to produce a smooth graph.

Note: Because you are graphing a circle on a rectangular screen, you must be careful to select viewing windows that produce square results. On the TI-82/83/83plus, the ZDecimal and ZInteger windows produce square screens. Other windows may be squared up using the ZSquare feature under the ZOOM menu. TI-85/86 users will need to apply to ZSQR feature under the ZOOM menu to all screens given.

A circle with radius *r* and center at (h,k), $(x - h)^2 + (y - k)^2 = r^2$, has as its graph the parameterization

$$x = r \cos t + h \qquad y = r \sin t + k$$

Example 2: Find a parameterization of $(x - 2)^2 + (y + 1)^2 = 9$ and graph on the calculator.

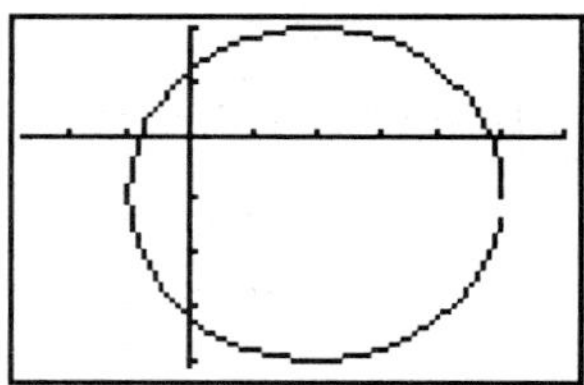

Solution: The equation has a radius of 3, r = 3, and its center is at (2, -1), h = 2 and k = -1. One parameterization would be x = 3 cos t + 2 and y = 3 sin t - 1.

The graph is displayed at the right in the window [-2.7,6] by [-4.1,2]. (The selected window is another version of the ZDecimal window. Note that the differences between the Xmax and Xmin and the Ymax and Ymin are the same as the differences between those same quantities in the ZDecimal window.) ◆

The parameterization of a circle was based on the trigonmetric identity $\sin^2 x + \cos^2 x = 1$. Trigonometric identities can be used to parameterize ellipses and hyperbolas.

Ellipses

The ellipse $\dfrac{x^2}{a^2} + \dfrac{y^2}{b^2} = 1$, with center at (0,0), can be parameterized by the equations

$$x = a \cos t \qquad \text{and} \qquad y = b \sin t$$

since $\quad \dfrac{(a\cos t)^2}{a^2} + \dfrac{(b\sin t)^2}{b^2} = \dfrac{a^2\cos^2 t}{a^2} + \dfrac{b^2\sin^2 t}{b^2} = \cos^2 t + \sin^2 t = 1.$

Example 3: Find a parameterization of the ellipse with the equation $\dfrac{x^2}{4} + \dfrac{y^2}{9} = 1$ and graph on the calculator.

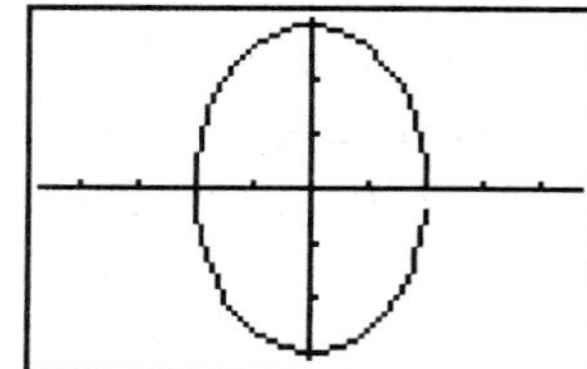

Solution: From the equation, a = 2 and b = 3. Thus, one parameterization would be x = 2 cos t, y = 3 sin t.
A square window is not essential for an elllipse, thus either ZTrig or ZDecimal will produce a satisfactory graph (ZDecimal was used for consistency with the previous examples.) The window has been defined with $0 \le T \le 2\pi$. As suggested before, you should

experiment with Tstep values between .05 and 0.15. Many textbooks will suggest a Tstep of $\dfrac{\pi}{24}$. ◆

An ellipse $\dfrac{(x-h)^2}{a^2} + \dfrac{(y-k)^2}{b^2} = 1$, with center at (h,k), can be parameterized to

$$x = a \cos t + h \quad \text{and} \quad y = b \sin t + k.$$

Example 4: Find a parameterization of $\dfrac{(x-2)^2}{9} + \dfrac{(y-3)^2}{16} = 1$ and graph on the calculator.

Solution: This ellipse has as its center (2,3). Thus, h = 2, k = 3, a = 3, and b = 4, yielding a parameterization of x = 3 cos t + 2 and y = 4 sin t + 3. Neither the ZDecimal or ZTrig window are satisfactory. The window displayed at the right is [-1, 5] by [-1, 7]. ◆

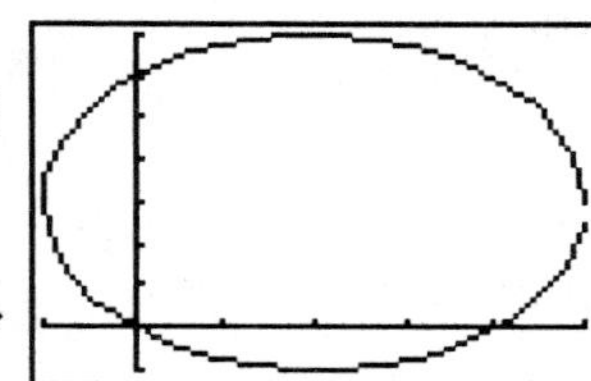

Hyperbola

A hyperbola $\dfrac{x^2}{a^2} - \dfrac{y^2}{b^2} = 1$ with center at (0,0) can be parameterized by

$$x = a \sec t = \frac{a}{\cos t} \quad \text{and} \quad y = b \tan t, \ 0 \le t \le 2\pi$$

using the trigonometric identity $\tan^2 t + 1 = \sec^2 t$. Similarly, a hyperbola with center at (h,k), $\dfrac{(x-h)^2}{a^2} - \dfrac{(y-k)^2}{b^2} = 1$, can be parameterized by

$$x = a \sec t + h = \frac{a}{\cos t} + h \quad \text{and} \quad y = b \tan t + k \ \ (0 \le t \le 2\pi).$$

Example 5: Find a parameterization and calculator graph of $\dfrac{(x+1)^2}{9} - \dfrac{(y-2)^2}{16} = 1$.

Solution: This hyperbola has its center at (-1,2). Thus, h = -1, k = 2, a = 3 and b = 4, yielding $x = \dfrac{3}{\cos t} - 1$ and y = 4 tan t + 2 as one parameterization.

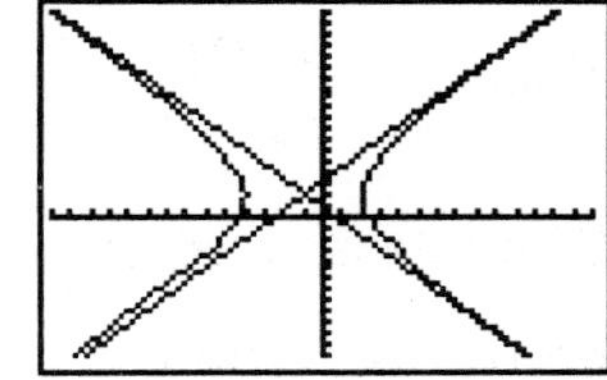

Graphed in the window [-14,14] by [-13, 19] we appear to also have the asymptotes created by the fundamental rectangle that is constructed when we graph by hand. Remember, at $\dfrac{\pi}{2}$ and $\dfrac{3\pi}{2}$ these functions are not defined. Watch carefully as points are plotted on the screen. The calculator is actually connecting two points on either side of the asymptote. If the graphing mode is changed from connected to dot these "lines" , which are indeed not part of the graph, will not appear. ◆

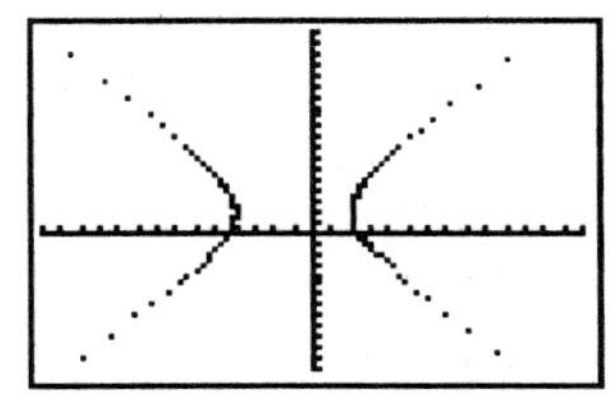

An hyperbola $\dfrac{y^2}{a^2} - \dfrac{x^2}{b^2} = 1$ with center at (0,0) can be parameterized by

$$x = b \tan t \quad \text{and} \quad y = \dfrac{a}{\cos t}, \quad 0 \le t \le 2\pi.$$

Similarly, an hyperbola with center at (h,k), $\dfrac{(y-k)^2}{a^2} - \dfrac{(x-h)^2}{b^2} = 1$, can be parameterized by

$$x = b \tan t + h \quad \text{and} \quad y = \dfrac{a}{\cos t} + k \quad (0 \le t \le 2\pi).$$

Example 6: Find a parameterization and calculator graph of $\dfrac{(y+1)^2}{9} - \dfrac{(x-2)^2}{16} = 1$.

Solution: This hyperbola has its center at (2,-1). Thus, h = 2, k = -1, a = 3 and b = 4,

yielding x = 4 tan t + 2 and $y = \dfrac{3}{\cos t} - 1$ as one parameterization.

The graph is displayed in the window [-10,14] by [-10, 10] in dot mode.

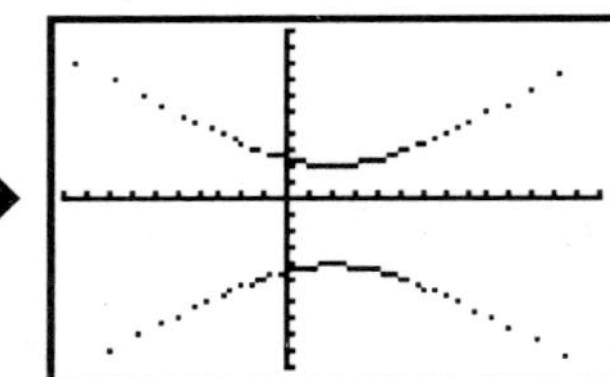

Exercise Set

Directions: Determine a parameterization where $0 \le T \le 2\pi$ for each equation below and sketch the graph. Record window values for each graph.

1.　　$x^2 + y^2 = 36$

　　　x = _______________　　　y = _______________

　　　[_____ , _____] by [_____ , _____]

2.　　$x^2 + (y + 3)^2 = 25$

　　　x = _______________　　　y = _______________

　　　[_____ , _____] by [_____ , _____]

3.　　$(x - 2)^2 + (y - 5)^2 = 4$

　　　x = _______________　　　y = _______________

　　　[_____ , _____] by [_____ , _____]

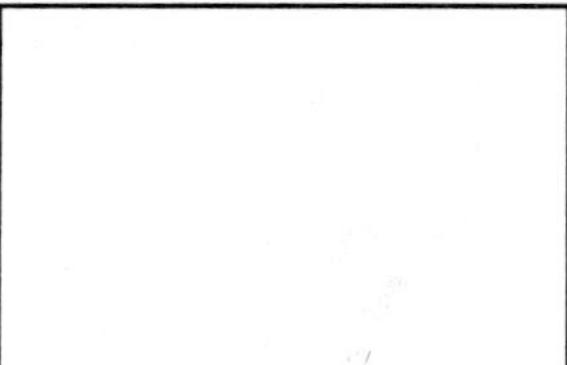

4. $\dfrac{x^2}{4} + \dfrac{y^2}{9} = 1$

x = _______________   y = _______________

[____ , ____] by [____ , ____]

5. $\dfrac{(x-4)^2}{25} + \dfrac{(y-1)^2}{36} = 1$

x = _______________   y = _______________

[____ , ____] by [____ , ____]

6. $\dfrac{(x+3)^2}{16} + \dfrac{(y-2)^2}{25} = 1$

x = _______________   y = _______________

[____ , ____] by [____ , ____]

7. $\dfrac{x^2}{9} - \dfrac{y^2}{4} = 1$

x = _______________   y = _______________

[____ , ____] by [____ , ____]

8. $\dfrac{y^2}{4} - \dfrac{x^2}{9} = 1$

x = _______________   y = _______________

[____ , ____] by [____ , ____]

9. $\dfrac{(x-2)^2}{49} - \dfrac{y^2}{16} = 1$

x = _______________   y = _______________

[____ , ____] by [____ , ____]

10. $\dfrac{(y-3)^2}{4} - \dfrac{(x+1)^2}{9} = 1$

x = ________________   y = ________________

[____ , ____] by [____ , ____]

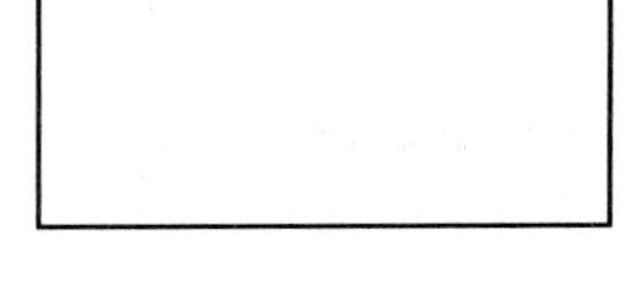

Solutions: **1.** x = 6 cos t, y = 6 sin t;   [-10.6, 10.6] by [-7, 7]

2. x = 5 cos t, y = 5 sin t - 3;   [-7.6, 7.6] by [-8, 2]

3. x = 2 cos t + 2, y = 2 sin t + 5; [-3.3, 7.3] by [0, 7]

4. x = 2 cos t, y = 3 sin t;   [-4.7, 4.7] by [-3.1, 3.1]

5. x = 5 cos t + 4, y = 6 sin t + 1; [-10, 10] by [-10, 10]

6. x = 4 cos t - 3, 5 sin t + 2; [-10, 10] by [-10, 10]

7. $x = \dfrac{3}{\cos t}$, y = 4 tan t; [-4.7, 4.7] by [-3.1, 3.1]

8. x = 2 tan t, $y = \dfrac{3}{\cos t}$; [-10, 10] by [-10, 10]

9. $x = \dfrac{7}{\cos t} + 2$, y = 4 tan t; [-10, 15] by [-10, 10]

10. x = 2 tan t + 3, $y = \dfrac{-3}{\cos t} - 1$; [-10, 15] by [-10, 10]

UNIT 34
ROOTS OF COMPLEX NUMBERS

*Unit 33 is a prerequisite for this unit. Answers appear at the end of the unit.

This unit examines nth roots of complex numbers in trigonometric form. If a + bi is a complex number, then the trigonometric form of that number is r(cos θ + i sin θ) where $r = |a+bi| = \sqrt{a^2 + b^2}$. De Moivre's Theorem, used for computing integral powers of complex numbers, states:

if n is any real number, then $[r(\cos \theta + i \sin \theta)]^n = r^n(\cos n\theta + i \sin n\theta)$.

An extension of De Moivre's Theorem makes it possible to compute the n nth roots of a complex number in trigonometric form:

$$\sqrt[n]{r}\left(\cos \frac{\theta + k \cdot 360^o}{n} + i \sin \frac{\theta + k \cdot 360^o}{n} \right) \text{ where k} = 0,1,2,3,\ldots,n\text{-}1.$$

Example 1:   Find the five fifth roots of $-\sqrt{2} + \sqrt{2}\,i$.

Solution: Express $-\sqrt{2} + \sqrt{2}\,i$ in trigonometric form: 2(cos 135° + i sin 135°), and apply the extension of De Moivre's Theorem.

$$\sqrt[5]{2}\left[\cos \frac{135 + k \cdot 360^o}{5} + i \sin \frac{135 + k \cdot 360^o}{5} \right]$$

Each of the five roots r_1, r_2, r_3, r_4, r_5 is determined by letting k = 0,1,2,3,4. Thus

$$\sqrt[5]{2}\ (\cos 27^o + i \sin 27^o\ ) \approx 1.0235 + .5215i$$
$$\sqrt[5]{2}\ (\cos 99^o + i \sin 99^o\ ) \approx -.1797 + 1.1346i$$
$$\sqrt[5]{2}\ (\cos 171^o + i \sin 171^o\ ) \approx -1.1346 + .1797i$$
$$\sqrt[5]{2}\ (\cos 243^o + i \sin 243^o\ ) \approx -.5215 - 1.0235i$$
$$\sqrt[5]{2}\ (\cos 315^o + i \sin 315^o\ ) \approx .8122 - .8122i$$

Observe that each of the roots r_1 through r_5 has an absolute value of approximately 1.15. Thus, these roots are located on a circle with center at the origin and radius of 1.15.

(Note also that $\sqrt[5]{2} \approx 1.15$.)

◆

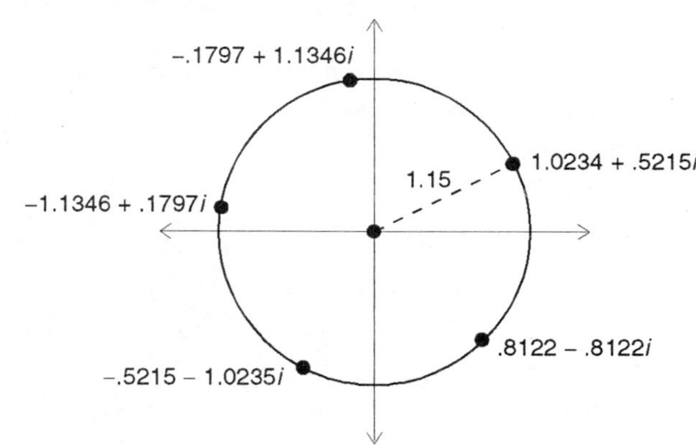

In general, the nth roots of a complex number z are located at equal intervals on a circle of radius $\sqrt[n]{r}$ where z = r(cos θ + i sin θ).

These roots can be displayed both graphically and in table form when the calculator is in parametric MODE.

TABLE

Set the calculator to both **Degree** and **Parametric** MODE.

To display the roots of complex numbers using a circle whose radius is $\sqrt[n]{r}$, recall that the parametric equations $x = r \cos t$ and $y = r \sin t$ define the graph of a circle $x^2 + y^2 = r^2$.

To determine the *n nth* roots of a complex number, let $X1_T = \sqrt[n]{r} \cos T$ and $Y1_T = \sqrt[n]{r} \sin T$.

Press [Y=] and enter $\sqrt[5]{2} \cos T$ after $X1_T =$ and $\sqrt[5]{2} \sin T$ after $Y1_T =$ to determine the five fifth roots of $-\sqrt{2} + \sqrt{2}i$ that were previously computed "by hand."

To view these roots in the TABLE, set **TblStart** = 27 and **ΔTbl** = 360/5. In general, set **TblStart** = θ/n and **ΔTbl** = 360/n.
    NOTE: The angle θ must be the <u>actual</u> angle, not the <u>reference</u> angle.

Display the TABLE as shown at the right. The X1T column provides the "a" values in the complex number a + bi and the Y1T column provides the "b" values.

T	X1т	Y1т
27	1.0235	.5215
99	-.1797	1.1346
171	-1.135	.1797
243	-.5215	-1.023
315	.81225	-.8123
387	1.0235	.5215
459	-.1797	1.1346

T=27

Complex Roots:
1.0235 + .5215 i
-.1797 + 1.1346 i
-1.135 - .1797 i
-.5215 - 1.023 i
.81225 - .8123 i

GRAPH

To display the graph of the circle with radius $\sqrt[5]{2}$ that contains these roots, the WINDOW should be set with Tmin=0, Tmax=360 and Tstep=360/5 (in general, 360/n). Initially set xMin/xMax and yMin/yMax to $\pm\sqrt[5]{2}$ (in general, $\pm\sqrt[n]{r}$) accordingly.

Press [GRAPH] to display the graph at the right. The graph is not displayed as a circle because insufficient points are graphed when the Tstep is too large.

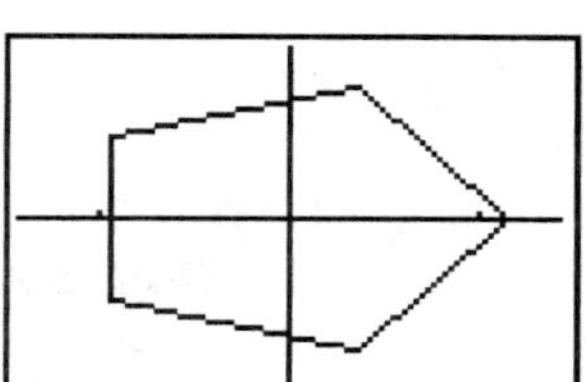

To adjust the WINDOW settings so that a circle of radius $\sqrt[5]{2}$ is displayed, return to Example 1 and observe that the initial angle is 27°, with subsequent angles occurring at 72° intervals. If the Tstep is a divisor of both 27 <u>and</u> 72, then the initial angle of 27° will appear on the circle as well as all angles of the form 27 + 72k. In general, the Tstep must be a divisor of θ and 360/n. Because 9 is the largest divisor of 27 and 72, the Tstep was set to 9 for the display at the right and **ZSquare** was then applied to produce a true circle.
TRACE to verify the remaining four fifth roots.

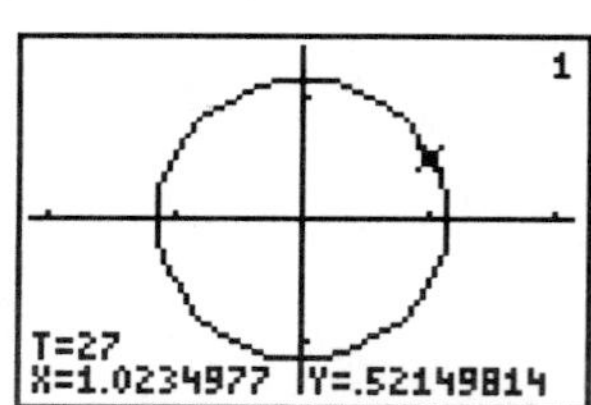

Example 2: Use the TABLE feature to determine the five fifth roots of 1 (1 + 0i), also called the five fifth roots of unity. Fill in the values on the TABLE and record the roots in a + bi form.

Solution: $\sqrt[n]{r} = \sqrt[5]{1}$. Begin by letting X1T = $\sqrt[5]{1}$ cos T and Y1T = $\sqrt[5]{1}$ sin T. Set the table to start at 0 and increment by 72 (360/5).

The roots, as displayed in the table, are:

These roots are 1, 0.3090 + .9511i, -0.8090 + 0.5878i, -0.8090 - 0.5878i, and 0.3090 - 0.9511i.

To graphically display the five fifth roots of 1 on a unit circle, the Tstep will need to be a divisor of 360/5 = 72 and θ = 0. Since 0 is divisible by all real numbers except itself, any divisor of 72 is an acceptable Tstep value. The divisors of 72 are 1,2,3,4,6,8,9,12,18,24,36, and 72. A Tstep of 72 does not produce a circle, however, the roots are clearly apparent. A Tstep value less than or equal to 24 will produce a unit circle, tracing will verify the roots displayed in the table.

T	X1T	Y1T
0	1	0
72	.30902	.95106
144	-.809	.58779
216	-.809	-.5878
288	.30902	-.9511
360	1	0
432	.30902	.95106

T=0

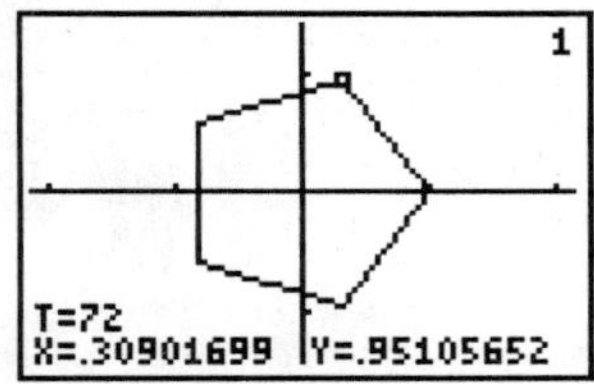

Tstep = 72

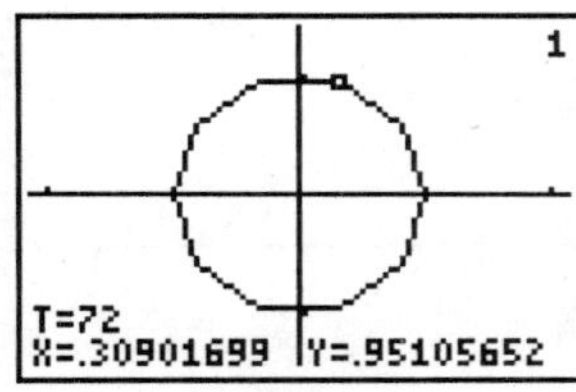

Tstep = 36

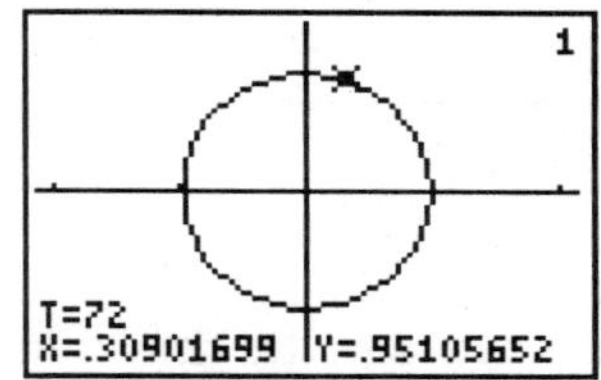

Tstep = 24

NOTE: Another form of the original problem would be stated $x^5 = 1$.

$\blacklozenge$

EXERCISE SET

Directions: Display the roots in TABLE form, completing the displayed table, and then record the roots in a + bi form.

1.   Find the four fourth roots of -8 - 8$\sqrt{3}$i, i.e. solve $x^4 = -8 - 8\sqrt{3}i$.

ROOTS ARE:

T	X1T	Y1T

____

____

____

____

2.   Find the four fourth roots of $-8 + 8\sqrt{3}i$, i.e. solve $x^4 = -8 + 8\sqrt{3}i$.

ROOTS ARE:	T	X1T	Y1T
___________			
___________			
___________			
___________			

3.   Compare the solutions to $x^4 = -8 - 8\sqrt{3}i$ and $x^4 = -8 + 8\sqrt{3}i$. Based on your observations, if the solutions to $x^5 = -\sqrt{2} + \sqrt{2}\,i$ are

$$1.0235 + .5215i \quad -1.1346 + .1797i \quad .8122 - .8122i$$
$$-.1797 + 1.1346i \quad -.5215 - 1.0235i,$$

can you write the solutions to $x^5 = -\sqrt{2} - \sqrt{2}\,i$?

___________________   ___________________   ___________________

___________________   ___________________

Verify these roots with the calculator.

4.   Find the two square roots of $-4i$.

ROOTS ARE:	T	X1T	Y1T
___________			
___________			

5.   Verify the answers in #4 either by performing the multiplication by hand (TI-82 users) or with the calculator (TI-83/83plus/85 users). Show either the hand computation below or copy the screen display of your calculator.

6. Find the four fourth roots of unity.

ROOTS ARE:

T	X1ᴛ	Y1ᴛ

7. Find the eight eighth roots of unity.

ROOTS ARE:

T	X1ᴛ	Y1ᴛ

8. Based on the results of #6 & 7, can a partial list of roots be determined for the twelve twelfth roots of unity or even the 4n 4nth roots of unity? In general, if you know the roots of unity for index "k", is it possible to create an accurate partial list of roots of unity for index kn?

9. Based on the results of #1,2 & 3, is it possible to formulate an algorithm for computing like roots of complex conjugates? Since these three problems only addressed fourth roots, be sure to explore other roots before drawing any conclusions.

<u>**Solutions:**</u>

1. $n = 4$, $r = 16$, $\sqrt[n]{r} = 2$, TblStart $= 60$, ΔTbl $= 90$
   $1 + 1.7321i$, $-1.732 + i$, $-1 - 1.732i$, $1.7321 - i$

2. $n = 4$, $r = 16$, $\sqrt[n]{r} = 2$, TblStart $= 30$, ΔTbl $= 90$
   $1.7321 + i$, $-1 + 1.7321i$, $-1.732 - i$, $1 - 1.732i$

3. Solutions are complex conjugates of those listed.

4. $n = 2$, $r = 4$, $\sqrt[n]{r} = 2$, TblStart $= 135$, ΔTbl $= 180$
   $-1.4142 + 1.4142i$, $1.414 - 1.414i$

5. Remember as you verify: you are computing with approximations and thus your result will be "close" but not exact.

6. $n = 4$, $r = 1$, $\sqrt[n]{r} = 1$, TblStart $= 0$, ΔTbl $= 90$
   $1 + 0i$, $0 + i$, $-1 + 0i$, $0 - i$

7. $n = 8$, $r = 1$, $\sqrt[n]{r} = 1$, TblStart $= 0$, ΔTbl $= 45$
   $1 + 0i$, $.7071 + .7071i$, $0 + i$, $-.7071 + .7071i$, $-1 + 0i$, $-.7071 - 7071i$, $0 - i$, $.7071 - .7071i$

8. Answers may vary.

9. Answers may vary.

CORRELATION CHART

UNIT	PRE-REQ. UNIT	CORRELATING CONCEPT
#35 Matrices, p. 229	#3	Matrix operations and applications to complex numbers.
#36 Combinatorics and Probability, p. 241	#18	A calculator/graphics approach to combinatorics and probability
#37 Statistics: Plotting Paired Data, p.247	#17	Paired data and the types of statistical plots available for display are explored.
#38 Frequency Distributions, p.257	#37	Frequency distributions and the use of the calculator to construct histograms and box-and-whisker plots.
#39 Line of Best Fit, p.267	#38	A calculator approach to graphing linear regression equations and interpreting displayed data.

Unit Title	TI-82 Keys	TI-83/83plus Keys	TI-85/86 Keys

#35: Matrices, p. 229	MATRX
	  EDIT
	  MATH
	   1:det
	
	   5:identity
	   8:rowSwap(
	   9:row + (
	   0:*row(
	   A:*row + (
	x^{-1}
		MATRX
		  EDIT
		  MATH
		   1:det(
		
		   5:identity(
		
		  C:rowSwap(
		   D:row + (
		   E:*row(
		   F:*row + (
		x^{-1}
			MATRX
			  EDIT
			  MATH
			   det
			  OPS
			   ident
			   rSwap
			   rAdd
			   multR
			   mRAdd
			x^{-1}

#36: Combinatorics and Probability, p. 241	MATH/PRB
	  2:nPr
	  3:nCr
		MATH/PRB
		  2:nPr
		  3:nCr
			MATH/PROB
			  nPr
			  nCr

#37: Statistics: Plotting Paired Data, p.247	STAT
	  1:Edit
	  4:ClrList
	STAT PLOT
	  1:Plot 1
	MEM
	  2:Delete
	  3:List
	ZOOM
	  9:ZoomStat
		STAT
		  1:Edit
		  4:ClrList
		STAT PLOT
		  1:Plot 1
		MEM
		  4:ClrAllLists
		
		ZOOM
		  9:ZoomStat
			STAT
			  EDIT
			   CLRxy
			
			  DRAW
			   SCAT
			   xyLINE
			ZOOM
			  ZDATA

#38: Frequency Distributions, p.257	STAT/CALC
	  1:1-Var Stats
	  3:SetUp
		STAT/CALC
		  1:1-Var Stats
			STAT/CALC
			  OneVa(TI-86)
			STAT/DRAW
			  HIST

#39: Line of Best Fit, p.267	STAT/ CALC
	  3:SetUp
	  5:LinReg(ax + b)
	VARS
	  5:Statistics/EQ
	   7:RegEQ
		STAT/CALC
		  2:2-Var Stats
		
		  4:LinReg(ax + b)
		VARS
		
		  5:Statistics/EQ
		   1:RegEQ
			STAT/CALC
			  TwoVa(TI-86)
			  LINR
			VARS
			  STAT
			   RegEq

$$\boxed{\begin{array}{c} \textbf{UNIT 35} \\ \textbf{MATRICES} \end{array}}$$

*Unit 3 is a prerequisite for this unit. Answers appear at the end of the unit.

A matrix is a rectangular array of numbers. This unit will explore operations with matrices and their application to systems of linear equations.

To enter the matrix $A = \begin{bmatrix} 2 & 3 & 4 \\ 5 & 6 & 7 \end{bmatrix}$, press [2nd] <MATRX> (on the TI-82/83, press

[MATRX]) , cursor over to highlight **EDIT** and press [ENTER] to select [1:[A]].

Matrix A has two rows and three columns. At the blinking cursor type [2], press [ENTER], type [3], and press [ENTER]. The values for the first row of the matrix can now be entered. Pressing [ENTER] after each entry will progress you through each row from left to right, or you may use the arrow keys to move to a desired location on the screen. Return to the home screen ([2nd] <QUIT>) and press [2nd] <MATRX> [1:[A]] [ENTER] to display the matrix (remember, on the TI-82/83, just press [MATRX], not [2nd]). Your display should correspond to the one at the right.

```
[A]
      [[2 3 4]
       [5 6 7]]
```

Now enter matrix $B = \begin{bmatrix} 1 & -5 & 3 \\ -6 & 7 & 2 \end{bmatrix}$. In order to add or subtract

```
[A]+[B]
       [[3  -2 7]
        [-1 13 9]]
```

matrices, the dimensions of the matrices must be the same. At the home screen, press [2nd] <MATRX> [1:[A]] [+] [2nd] <MATRX> [2:[B]] [ENTER]. Your display should look like the one at the right.

Now multiply: [A] · [B]. The screen display should correspond to the one at the right. In order to perform matrix multiplication, the dimensions must correspond as follows: Since [A] is a 2 x 3 matrix, it can only be multiplied by a 3 x n matrix.

```
ERR:DIM MISMATCH
1:Goto
2:Quit
```

Enter the matrix $C = \begin{bmatrix} 4 & 5 \\ -2 & 3 \\ 7 & 4 \end{bmatrix}$ and multiply [A] · [C]. Your

```
[A]*[C]
       [[30 35]
        [57 71]]
```

display should correspond to the one at the right.

Directions: Enter the following matrices in the calculator:

$$A = \begin{bmatrix} 2 & 3 & 4 \\ 5 & 6 & 7 \end{bmatrix} \qquad B = \begin{bmatrix} 1 & -5 & 3 \\ -6 & 7 & 2 \end{bmatrix} \qquad C = \begin{bmatrix} 4 & 5 \\ -2 & 3 \\ 7 & 4 \end{bmatrix} \quad D = \begin{bmatrix} 3 & 1 \\ -1 & 3 \end{bmatrix}$$

1. Using matrix A and matrix C, determine if matrix multiplication is a commutative operation.

2. What requirements are necessary for matrix multiplication to be defined?

3. Compute ([A] + [B])[C] and record the resulting matrix:

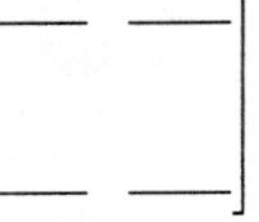

Compute [A][C] + [B][C] and record the resulting matrix:

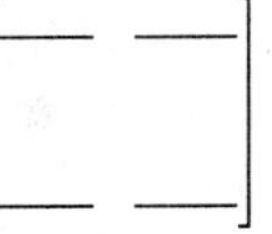

Is matrix multiplication distributive from the right? ________

NOTE: If the name of a matrix is followed by an open parenthesis it <u>does</u> <u>not</u> indicate implied multiplication! A multiplication symbol must be used when parentheses <u>follow</u> a matrix name.

4. Compute [C]([A] + [B]) and record the resulting matrix:

This must be entered as [C] * ([A] + [B]).

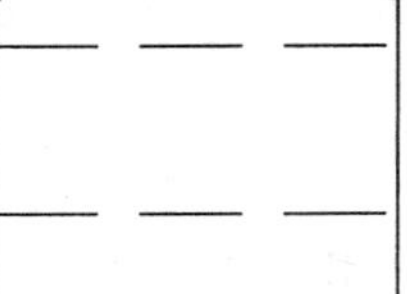

Compute [C][A] + [C][B] and record the resulting matrix:

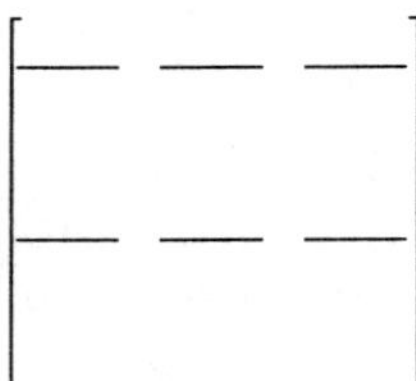

Is matrix multiplication distributive from the left? ________

5. Compute each of the following and record the resulting matrices:

[D]([A] + [C])

[D][A] + [D][C]

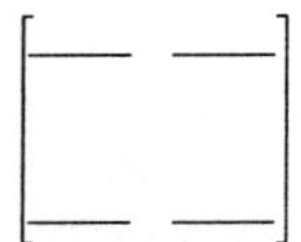

In general, is matrix multiplication distributive?______

What requirement is necessary to ensure distributivity?

6. **Application**: The Golden Oldies: Records, Tapes and CDs chain has a warehouse in Kentucky and one in Tennessee. Use matrix addition and multiplication to determine the total value of the inventory for each "oldies" group that is listed. Each LP (long playing record) is valued at $7, each cassette tape at $9 and each CD (compact disc) is valued at $17.

Warehouse	GROUP	LPs	TAPES	CDs
KY Warehouse	"The Beagles"	125	250	275
	"Herman's Hideaways"	80	300	115
	"Peter, Piper, and Pepper"	75	185	200
TN Warehouse	"The Beagles"	200	180	200
	"Herman's Hideaways"	125	150	165
	Peter, Piper, and Pepper"	50	90	125

Record your matrix problem below, perform the indicated operation with your calculator and record your response to the problem.

The inverse of matrix A has the characteristic that $[A] \cdot [A]^{-1}$ is equal to the identity matrix. The multiplicative identity matrix is a square matrix that has ones along the diagonal (from upper left to lower right) and zeroes everywhere else. In the multiplication of real numbers, a number x has an inverse $1/x$ ($x \neq 0$) and the property that

$x \cdot 1/x = 1$ where 1 is the multiplicative identity. Multiplicative inverses exist <u>only</u> for <u>some</u> square matrices.

To compute the inverse of matrix A, first enter $A = \begin{bmatrix} 2 & 5 & 4 \\ 1 & 4 & 3 \\ 1 & -3 & -2 \end{bmatrix}$. At the home screen,

enter A^{-1} by keystroking **[2nd] <MATRX> [1: [A]] [x^{-1}]**
[ENTER]. A^{-1} is displayed at the far right.

$$A^{-1} = \begin{bmatrix} -1 & 2 & 1 \\ -5 & 8 & 2 \\ 7 & -11 & -3 \end{bmatrix}$$

<table>
<tr><td>TI-85/86</td><td>THE "x^{-1}" IS LOCATED ABOVE THE "EE" KEY ON THE FACE OF THE CALCULATOR.</td></tr>
</table>

NOTE: If this matrix needs to be stored for future reference, it can be stored in one of the other available matrix locations. To store A^{-1} in the matrix [E] location, type **[2nd] <MATRX> [1: [A]] [STO▶] [2nd] < MATRX> [5: [E]] [ENTER]**.

Multiply $A \cdot A^{-1}$ to display the 3 x 3 identity matrix. This matrix should be numerically equivalent to the identity matrix displayed by pressing **[2nd] <MATRX> [▶] [5:identity]** **[3]**(for a 3 x 3 identity matrix) **[ENTER]**.

<table>
<tr><td>TI-85/86</td><td>THE **IDENTITY** FUNCTION IS FOUND UNDER THE **OPS** SUBMENU OF **MATRX**, **[2ND] <MATRX>** **[F4](OPS) [F3](IDENT)**.</td></tr>
</table>

Because the inverse of A exists, A is called a nonsingular matrix. If the inverse of A did not exist then A would be called a singular matrix.

If the inverse of a matrix exists, the multiplication of the matrix times its inverse is a commutative operation. (Test this idea using A and A^{-1}.)
 Matrix multiplication can be used to rewrite linear systems in matrix form and then solved.

Example 1: Solve the linear system $\begin{array}{l} x + 4y = 2 \\ 3x - 2y = 6 \end{array}$ using matrices.

Solution: The linear system $\begin{array}{l} x + 4y = 2 \\ 3x - 2y = 6 \end{array}$ would be rewritten in matrix form as

$\begin{bmatrix} 1 & 4 \\ 3 & -2 \end{bmatrix} \cdot \begin{bmatrix} x \\ y \end{bmatrix} = \begin{bmatrix} 2 \\ 6 \end{bmatrix}$. Then proceeding with the matrix algebra:

$$\begin{bmatrix} 1 & 4 \\ 3 & -2 \end{bmatrix}^{-1} \cdot \begin{bmatrix} 1 & 4 \\ 3 & -2 \end{bmatrix} \cdot \begin{bmatrix} x \\ y \end{bmatrix} = \begin{bmatrix} 1 & 4 \\ 3 & -2 \end{bmatrix}^{-1} \cdot \begin{bmatrix} 2 \\ 6 \end{bmatrix}$$

$$I_2 \cdot \begin{bmatrix} x \\ y \end{bmatrix} = \begin{bmatrix} 1 & 4 \\ 3 & -2 \end{bmatrix}^{-1} \cdot \begin{bmatrix} 2 \\ 6 \end{bmatrix}$$

$$\begin{bmatrix} x \\ y \end{bmatrix} = \begin{bmatrix} 1 & 4 \\ 3 & -2 \end{bmatrix}^{-1} \cdot \begin{bmatrix} 2 \\ 6 \end{bmatrix}$$

Therefore, to solve a linear system with matrices, the matrix formed by the constants should be multiplied by the inverse of the coefficient matrix.

Enter $A = \begin{bmatrix} 1 & 4 \\ 3 & -2 \end{bmatrix}$, $B = \begin{bmatrix} 2 \\ 6 \end{bmatrix}$ and multiply $A^{-1} \cdot B$.

The solution matrix $A^{-1} \cdot B$ is displayed at the right. $\begin{bmatrix} 2 \\ 0 \end{bmatrix}$

The solution to the linear system $\begin{array}{c} x + 4y = 2 \\ 3x - 2y = 6 \end{array}$ is the ordered pair (2, 0).

$\blacklozenge$

Example 2:   Solve the system $\begin{array}{c} x + y + z = 7 \\ x + 2y + z = -1 \\ 2x + y + z = 2 \end{array}$ using matrices.

Solution: First, record in matrix form:

$$\begin{bmatrix} 1 & 1 & 1 \\ 1 & 2 & 1 \\ 2 & 1 & 1 \end{bmatrix} \cdot \begin{bmatrix} X \\ Y \\ Z \end{bmatrix} = \begin{bmatrix} 7 \\ -1 \\ 2 \end{bmatrix}$$

Using the MATRX EDIT screen, enter matrix A as $\begin{bmatrix} 1 & 1 & 1 \\ 1 & 2 & 1 \\ 2 & 1 & 1 \end{bmatrix}$ and matrix B as $\begin{bmatrix} 7 \\ -1 \\ 2 \end{bmatrix}$

Use the calculator to compute $[A]^{-1} * [B]$.

The solution matrix is displayed as $\begin{bmatrix} -5 \\ -8 \\ 20 \end{bmatrix}$, which translates to the ordered triple(-5,-8,20).

$\blacklozenge$

What happens when the entries in the inverse of the coefficient matrix are decimal approximations? The determinant of the matrix illustrates the response.

The determinant is the numerical value assigned to a square matrix. More specifically, the determinant of a given matrix A is the denominator value for all the entries of the inverse of matrix A.

Suppose A is the 2 x 2 matrix $\begin{bmatrix} 4 & 6 \\ 8 & -9 \end{bmatrix}$.

Use the calculator to compute $[A]^{-1}$. Your display should agree with the one below. You can use the cursor arrows to cursor through each decimal entry.

$$\begin{bmatrix} .1071428571 & .0714285714 \\ .0952380952 & -.0476190476 \end{bmatrix}$$

Now compute the determinant of A. Press [2nd] <MATRX>, (cursor right to highlight
MATH), [1:det] [2nd] <MATRX> [1: [A]] [ENTER]. The determinant of A is displayed as -
84. The - 84 is the denominator of the decimal approximations displayed in [A]$^{-1}$.

TI-85/86	TO COMPUTE THE DETERMINANT OF A, PRESS [2ND] <MATRX> [F3](MATH) [F1](det) [2ND] <M1>(NAMES) [F1](A) [ENTER].

Multiplying each entry in the inverse matrix by the determinant value
will yield the corresponding numerator values. Press [2nd] <MATRX>
(highlight MATH) [1:det] [2nd] <MATRX> [1: [A]] [*] [2nd]
<MATRX> [1: [A]] [x^{-1}] [ENTER]. Your screen should correspond to
the one at the right.

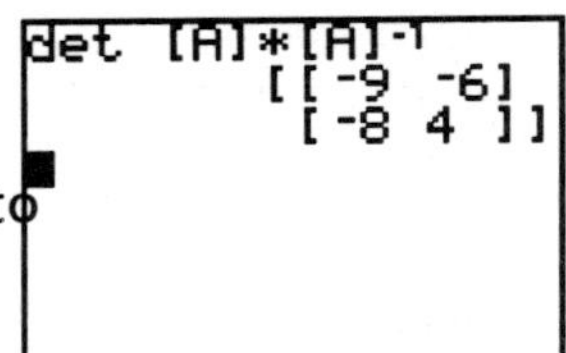

Since this matrix is the result of -84 · [A]$^{-1}$, we can conclude that the
decimal approximations in [A]$^{-1}$ are represented by the following fractions:

$$\begin{bmatrix} \dfrac{-9}{-84} & \dfrac{-6}{-84} \\[2ex] \dfrac{-8}{-84} & \dfrac{4}{-84} \end{bmatrix} = \begin{bmatrix} \dfrac{3}{28} & \dfrac{1}{14} \\[2ex] \dfrac{2}{21} & -\dfrac{1}{21} \end{bmatrix}$$

Confirm this with the calculator by computing [A]$^{-1}$ again, but this time
use the ▶FRAC command. Your display should correspond to the one
at the right.

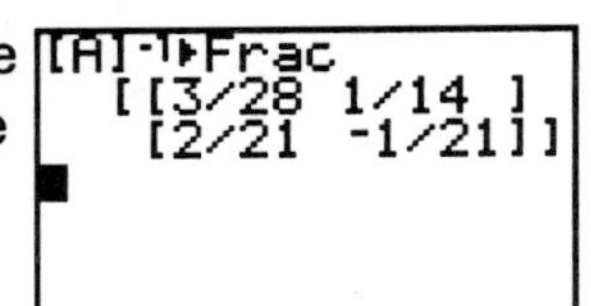

EXERCISE SET CONTINUED

Directions: Rewrite each system in matrix multiplication form and solve by using the inverse
of the coefficient matrix. Express answers in fractional form.

7.
$$\begin{aligned} 3x - 2y &= 7 \\ 5x + y &= 3 \end{aligned}$$

$$\begin{bmatrix} & \\ & \end{bmatrix} \cdot \begin{bmatrix} \\ \end{bmatrix} = \begin{bmatrix} \\ \end{bmatrix}$$

Solution matrix: $\begin{bmatrix} \\ \end{bmatrix}$

Ordered pair (_____ , _____)

8.
$$\begin{aligned} 8x + 2y &= 7 \\ 3x + 12y &= 5 \end{aligned}$$

$$\begin{bmatrix} & \\ & \end{bmatrix} \cdot \begin{bmatrix} \\ \end{bmatrix} = \begin{bmatrix} \\ \end{bmatrix}$$

Solution matrix: $\begin{bmatrix} \\ \end{bmatrix}$

Ordered pair (_____ , _____)

9.
$$4x + 2y + 6z = 2$$
$$-7x - 3y + 3z = 4$$
$$3x + 6y + 9z = 6$$

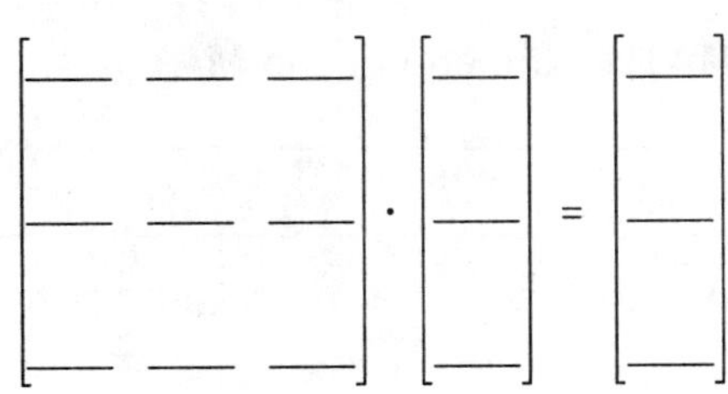

Solution matrix:

Ordered triple: (_____ , _____ , _____)

10.
$$3x + 2y - z = 1$$
$$2x - y + 3z = 5$$
$$x + 3y + 2z = 2$$

Solution matrix:

Ordered triple: (_____ , _____ , _____)

11.
$$x + y + 4z = 2$$
$$2x - y + z = 1$$
$$3x - 2y + 3z = 5$$

Solution matrix:

Ordered triple: (_____ , _____ , _____)

✍12. What happens when the coefficient matrix is singular (i.e. has no inverse)?

$$\begin{bmatrix} 1 & -4 & 1 \\ 2 & -7 & -2 \\ 3 & -11 & -1 \end{bmatrix}$$ is a singular matrix. Compute the determinant. (determinant = _____)

Explain <u>why</u> a matrix with no inverse would have a zero determinant (or conversely why a zero determinant would mean the matrix is singular).

When the coefficient matrix is singular, Gaussian elimination must be used to solve the system since the coefficient matrix would have no inverse.

Example 3: Solve the system $\begin{aligned} x - 4y + z &= 1 \\ 2x - 7y - 2z &= -1 \\ 3x - 11y - z &= 2 \end{aligned}$ using matrices.

Solution: The matrix formed from the coefficients of the system

$\begin{aligned} x - 4y + z &= 1 \\ 2x - 7y - 2z &= -1 \\ 3x - 11y - z &= 2 \end{aligned}$ has no inverse. Thus, Gaussian elimination will be used to

solve the system. The matrix formed by both the coefficients and the constants is an augmented matrix. Enter the augmented matrix in the calculator as matrix B now:

$$\begin{bmatrix} 1 & -4 & 1 & 1 \\ 2 & -7 & -2 & -1 \\ 3 & -11 & -1 & 2 \end{bmatrix}$$

NOTE: Because the rows represent individual equations: a) any row may be multiplied by a non-zero number, b) any two rows can be interchanged, and c) any two rows can be added together. Row operations are found under the **MATH** submenu of the **MATRX** menu

Perform row operations on the above matrix in order to place the matrix in reduced row eschelon form :

$$\begin{bmatrix} 1 & - & - & - \\ 0 & 1 & - & - \\ 0 & 0 & 1 & - \end{bmatrix}$$

a. Objective: Row 1, column 1 should have an entry of **1** with the remainder of the column being zeroes.

<table>
<tr><td>mathematical operation:</td><td>-2 * Row 1 + Row 2 →
(i.e.replaces) Row 2</td><td rowspan="2">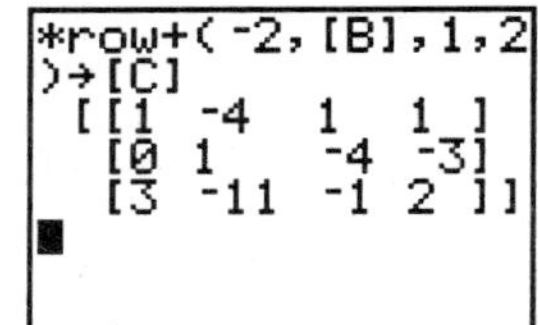</td></tr>
<tr><td>calculator operation: ***row** + (-2, [B],1,2) **STO▸ [C]**</td><td></td></tr>
</table>

NOTE: Row operations do not change the matrix stored in memory! The new matrix must be stored each time row operations are performed. The new matrix was stored in [C] to preserve the original matrix.

<table>
<tr><td>mathematical operation: -3 * Row 1 + Row 3 → Row 3</td><td rowspan="2">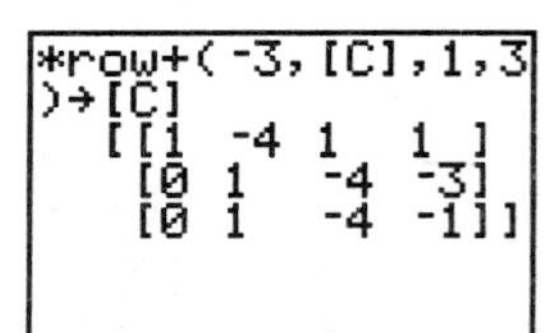</td></tr>
<tr><td>calculator operation: *** row** + (-3, [C],1,3) **STO▸ [C]**</td></tr>
</table>

b. Objective: Row 2, column 2 should have an entry of **1** with the remainder of the
column being zeroes.

mathematical operation: -1 * Row 2 + Row 3 → Row 3

calculator operation: ***row +** (-1, [C],2,3)

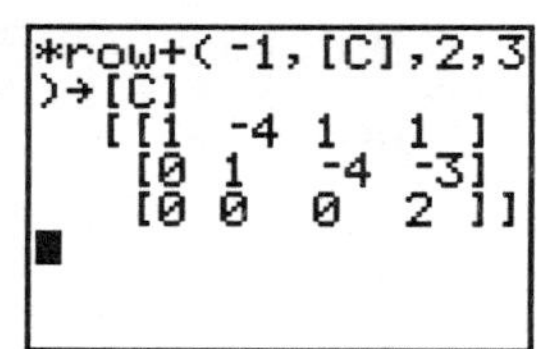

c. The matrix yields the remaining equations:

$$x - 4y + z = 1$$
$$y - 4z = -3$$
$$0x + 0y + 0z = -2$$

The last equation indicates that there is no solution to this system. (Had there
been a solution you would need to "back" substitute at this point to find the
values of x and y.)

d. The solution is the empty set.

Remember: There is a unique solution to an $n \times n$ system if and only if the coefficient
matrix is nonsingular.

♦

NOTE: Not all of the available row operations were used. The following row
operations indicate the order in which information is to be entered.

➤➤   **rowSwap** (matrix, row 1, row 2)   swaps row 1 and row 2
| TI-85/86 | **rSwap** (MATRIX, ROW 1, ROW 2) |

➤➤   **row +** (matrix, row 1, row 2)   adds row 1 and row 2 and stores result in row 2
| TI-85/86 | **rAdd**(matrix, row 1, row 2) |

➤➤   ***row** (value, matrix, row)   multiplies a row by the indicated value
| TI-85/86 | **multR**(VALUE, MATRIX, ROW) |

➤➤   ***row +** (value, matrix, row 1, row 2)   multiplies the matrix row 1 by the indicated
value, adds this product to row 2 and
stores the result in row 2
| TI-85/86 | **mRAdd** (VALUE, MATRIX, ROW 1, ROW 2) |

Directions: Use matrix row operation to perform Gaussian elimination to solve the following systems. Record the row operations as was done in Example 2.

13.
$$2x + y + 2z = 1$$
$$x - 2y + 3z = 4$$
$$2x - 3y + z = 0$$

Solution: _________________

14.
$$4x - 4y - 3z = 2$$
$$4x + 3y - 3z = 0$$
$$4x + 6y - 3z = 1$$

Solution: _________________

15.
$$x + y + 9z = 8$$
$$x + 3y - z = 0$$
$$x + 6y - 7z = 0$$

Solution: _________________

16.
$$2x + 3y - z = -2$$
$$x + 2y + 2z = 8$$
$$5x + 9y + 5z = 2$$

Solution: _________________

17.   **Application:** The atomic number lead is four more than three times the atomic number of iron. If the atomic number of lead is decreased by twice the atomic number of iron the result is the atomic number of zinc which is 30. Find the atomic numbers of lead and iron.

APPLICATION TO COMPLEX NUMBERS

The set of all matrices of the form $\begin{bmatrix} a & b \\ -b & a \end{bmatrix}$ with the usual operations of addition and multiplication of matrices and the set of all complex numbers a + bi with their operations of addition and multiplication are algebraically equivalent (i.e. isomorphic). If the complex number a + bi is identified with the matrix $\begin{bmatrix} a & b \\ -b & a \end{bmatrix}$ and the complex number c + di with the matrix $\begin{bmatrix} c & d \\ -d & c \end{bmatrix}$ then the matrix sum $\begin{bmatrix} a & b \\ -b & a \end{bmatrix} + \begin{bmatrix} c & d \\ -d & c \end{bmatrix} = \begin{bmatrix} a+c & b+d \\ -(b+d) & a+c \end{bmatrix}$ is identified with the complex sum (a + bi) + (c + di). This complex sum is equal to (a + c) + (b + d)i. Thus the complex product (3 - 2i)(4 + 5i) could be performed as the matrix multiplication of

$\begin{bmatrix} 3 & -2 \\ 2 & 3 \end{bmatrix} \cdot \begin{bmatrix} 4 & 5 \\ -5 & 4 \end{bmatrix}$. This multiplication yields $\begin{bmatrix} 2 & 7 \\ -7 & 2 \end{bmatrix}$. Thus the product of the complex numbers is 2 + 7i.

EXERCISE SET CONTINUED

Directions: Use matrix operatons to perform the following complex number computations. Express each complex number as a matrix and display the matrix (with fractional entries where applicable) and the corresponding complex number that result from the matrix computation.

18.   (4 - 3i) + (7 - 2i)

19.   (5 + 6i) - (-7 - 3i)

20.   (5 - 7i) (3 - 4i)

21.   Find the reciprocal (multiplicative inverse) of 1 + 2i.

22. $\dfrac{4 + 7i}{3 - 2i}$ (HINT: Use multiplicative inverses.)

Solutions: 1. no   **2.** Although the product may be defined, number of columns in matrix A equals number of rows in matrix B, AB does not always equal to BA.

3. Both matrices should be $\begin{bmatrix} 65 & 37 \\ 33 & 70 \end{bmatrix}$. Matrix multiplication is distributive from the right.

4. Both matrices should be $\begin{bmatrix} 7 & 57 & 73 \\ -9 & 43 & 13 \\ 17 & 38 & 85 \end{bmatrix}$. Matrix multiplication is distributive from the left.

5. Both matrices should yield a dimension error . A matrix is distributive through multiplication provided the dimension requirements are satisfied. However, A(B + C) does not always equal to (B + C)A.

6. "Beagles": $14220, "Herman's Hideaways": $10,245, "Peter, Piper & Pepper": $8875

7. (1,-2)   **8.** (37/45, 19/90)   **9.** (-7/13, 6/13, 7/13)   **10.** (6/7, -2/7, 1)

11. (-3/2, -5/2, 3/2) **12.** determinant = 0 **13.** (-1, -1/7, 11/7) **14.** ∅; Answers may vary.

15. (-20/3, 8/3, 4/3) **16.** ∅ **17.** Lead = 82, Iron = 26

18. 11 - 5i   **19.** 12 + 9i **20.** -13 - 41i **21.** $(1 + 2i)^{-1}$ = 1/5 - (2/5)i

22. -2/13 + (29/13)i NOTE: $\dfrac{A}{B} = A \div B = A \cdot \dfrac{1}{B} = A \cdot B^{-1}$

UNIT 36
COMBINATORICS AND PROBABILITY

*Unit 18 is a prerequisite for this unit. Answers appear at the end of the unit.

COMBINATORICS

This section will examine the use of the calculator to compute permutations and combinations.

A permutation is an ordered arrangement of objects. Symbolically, the number of permutations of n things taken r at a time is expressed as P(n, r) or $_nP_r$, and the algebraic formula for computing permutations is $\dfrac{n!}{(n-r)!}$. **ORDER** is important when computing permutations. Consider the following example.

Example 1: Suppose the manager of a little league baseball team has 15 players. How many different batting orders consisting of 9 players can he arrange?

Solution: We want the permutation of 15 people taken 9 at a time. To use the calculator to do the computation access the probability sub-menu under the MATH menu. Press **[MATH]** and cursor over to **PRB**.

TI-85/86 PRESS [2ND] <MATH> [F2](PROB) TO DISPLAY BOTH THE PERMUTATION FORMULA AND THE COMBINATION FORMULA USED IN THE SECOND PROBLEM.

Because the value for n (which is 15) must be entered before the formula, return to the HOME screen and enter 15. Now press **[MATH]**, cursor over to highlight **PRB** and press **[2:nPr]**. We then enter 9 because we want the permutation of 15 people taken 9 at a time. Pressing **[ENTER]** gives the solution of 1,816,214,400. Your screen should match the one at the right.

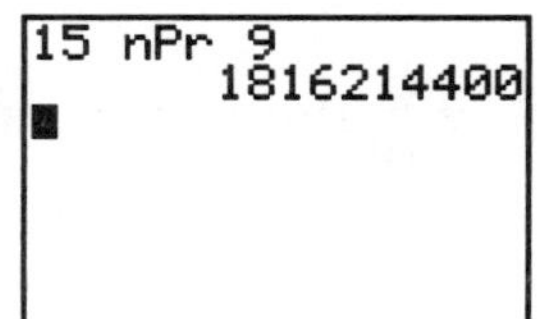

◆

A combination is an **UNORDERED** arrangement of objects. Symbolically, the combinations of n things taken r at a time is written as $\dbinom{n}{r}$, C(n,r) or nCr and is represented by the formula $\dfrac{n!}{r!(n-r)!}$. This formula is also known as the binomial coefficient and is used to determine the coefficient of the term whose variable factors are $a^r b^{n-r}$ in the expansion of $(a+b)^n$.

Pascal's triangle is commonly used to display the coefficients of the expansions of $(a+b)^n$. The portion of the triangle displayed represents the expansions of $(a + b)^n$ where $0 \le n \le 6$:

$$
\begin{array}{ccccccccccccc}
 & & & & & & 1 & & & & & & \\
 & & & & & 1 & & 1 & & & & & \\
 & & & & 1 & & 2 & & 1 & & & & \\
 & & & 1 & & 3 & & 3 & & 1 & & & \\
 & & 1 & & 4 & & 6 & & 4 & & 1 & & \\
 & 1 & & 5 & & 10 & & 10 & & 5 & & 1 & \\
1 & & 6 & & 15 & & 20 & & 15 & & 6 & & 1
\end{array}
\qquad
\begin{array}{l}
(a + b)^0 \\
(a + b)^1 \\
(a + b)^2 \\
(a + b)^3 \\
(a + b)^4 \\
(a + b)^5 \\
(a + b)^6
\end{array}
$$

By hand, complete the seventh row of Pascal's triangle for the coefficients of the expansion of $(a + b)^7$:

_____ _____ _____ _____ _____ _____ _____ _____

and then perform the computations of nCr where $n = 7$ and $0 \le r \le 7$ to verify the coefficients of the expansion. The value for *n* must be entered before the formula. At the home screen, enter the number **7** and then press **[MATH]**, cursoring over to highlight **PRB**, press **[3:nCr]** and then enter the number **0** for the value of r.

Your screen should correspond to the one at the right. This value of 1 is the coefficient of the term whose variable factors are $a^r b^{n-r}$. The coefficient has been placed in the appropriate blank with the corresponding nCr formula beneath it. Continue this process, always letting $n = 7$, successively letting $r = 7, 6, 5, \ldots, 1$ and recording the appropriate nCr beneath the term to complete the coefficients of the expansions of $(a + b)^7$.

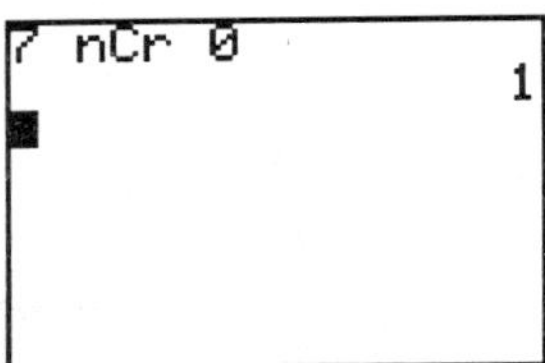

$$\underline{}a^7 b^{7\text{-}7} + \underline{}a^6 b^{7\text{-}6} + \underline{}a^5 b^{7\text{-}5} + \underline{}a^4 b^{7\text{-}4} + \underline{}a^3 b^{7\text{-}3} + \underline{}a^2 b^{7\text{-}2} + \underline{}a^1 b^{7\text{-}1} + \underline{\mathbf{1}}a^0 b^{7\text{-}0}$$
$$_7C_0$$

Since the definition of a combination is an unordered arrangement of objects, the formula nCr can also be used for the following types of problems.

Example 2: Students in an English class must choose four books to read from a list of seven. How many choices are possible?

Solution: We want the combination of seven books considered four at a time. To compute with the calculator, your screen should correspond to the one at the right. This shows that thirty five choices are possible.

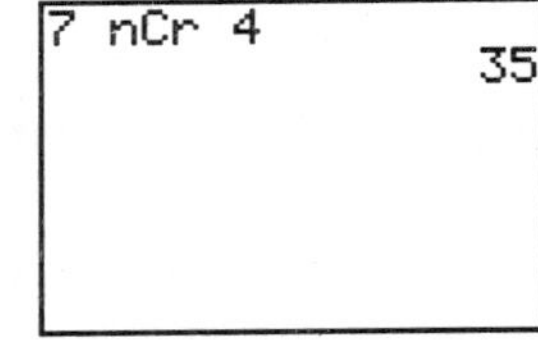

◆

EXERCISE SET

Directions: Use the calculator to compute the permutations and combinations.

1.　Nashville, Tennessee has 149 telephone prefixes. Compute the number of possible phone numbers that have these 149 prefixes followed by four digits selected from the numbers zero through nine.

2. The Technology Committee is to be staffed from the seventeen members of the Mathematics and Computer Science Department. Letting $Y = {}_{17}C_X$, generate a table that represents the total possible committee sizes (from a committee of 1 to a committee of 17 members).

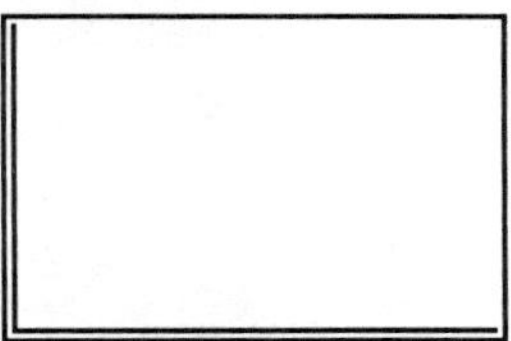

TI-85 USERS WILL NEED TO **TRACE** ON THE **ZINT** SCREEN.

X	Y
1	
2	
3	
4	
5	
6	
7	
8	
9	
10	
11	
12	
13	
14	
15	
16	
17	

Display the graph in the WINDOW [0,18.8] by [0,25000] and **TRACE** along the graph to observe the values displayed. Sketch the display.

a. When tracing, not all of the Y-values are defined. Why not?

b. Explain why this particular viewing window was selected.

3. Determine the expansion of $(a + b)^{17}$ using the calculator for coefficients and record the expansion below.

4.   How many bridge hands containing 13 cards can be dealt containing 4 hearts, 4
      spades, 3 diamonds, and 2 clubs?

The probability of an event **E** occurring in a given sample space **S** is given by $P(E) = \dfrac{N(E)}{N(S)}$

where **N** is the number of distinct and equally likely outcomes. The following example
demonstrates the use of combinatorics on the calculator as they apply to probability.

Example 3: A hand of 13 cards is chosen randomly from a standard deck of 52 cards (no
jokers included). What is the probability the hand contains at least one king?

Solution: The sample space **S** consists of all possible groups of 13-card hands from this 52-
card deck. Thus, $N(S) = C(52,13)$. If E represents the hand that contains at least one king,
then $\overline{E}$ (some texts refer to the complement of E as E') represents the hand that contains
no kings. The number $N(\overline{E})$ is the number of remaining 13-card hands from a 48 card deck
(what remains after the four kings have been removed). Thus, $N(\overline{E}) = C(48,13)$.
Therefore,

$$P(E) = 1 - P(\overline{E})$$
$$= 1 - \dfrac{C(48,13)}{C(52,13)}$$

which corresponds to the given calculator display:

```
1-(48 nCr 13)/(5
2 nCr 13)
          .696182473
```


Directions: Use the calculator for the computation of probability. Record the screen display
to justify your work.

An intermural athletics committee of seven is to be selected by "drawing straws" in a
fraternity and its sister sorority. The fraternity has 24 members and the sorority has
35 members. Find the probability that the committee will contain:

5.   7 sorority members

6.   4 fraternity and 3 sorority members

7.   4 sorority and 3 fraternity members

8.   at least one fraternity member.

A binomial experiment consists of repeated trials whose only outcomes are success or failure. Each trial is independent of the others. If the probability of success in one trial is P then the complement of this event, $1 - P$, would be the probability of failure. The probability of r successes in n trials is defined by the expression $nCr \cdot P^r(1-P)^{n-r}$. This expression can be compared to the general terms of a binomial expansion, $nCr \cdot a^r b^{n-r}$, making the terms in a binomial expansion equivalent to the probabilities of r successes in n trials, where $0 \le r \le n$.

Example 4:   A student takes a five question true/false quiz. If he guesses at all the answers, what is the probability that
     a. all five answers will be wrong?
     b. exactly four are wrong?
     c. exactly three are wrong?
     d. exactly two are wrong?
     e. exactly one is wrong?
     f. all five answers are correct?

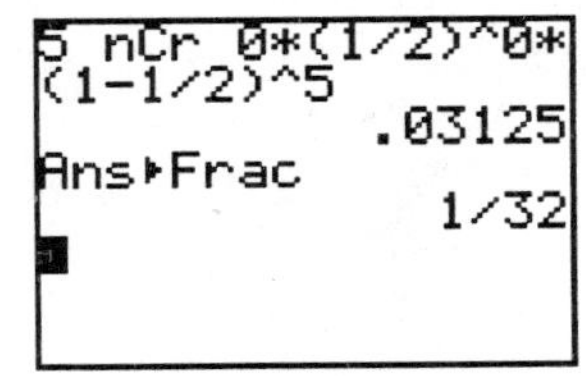

Solution: a. If all five answers are wrong then the number of successes r is zero in $n = 5$ trials. The probability of success in one trial is 1 in 2. Thus $nCr \cdot P^r(1-P)^{n-r} = {_5C_0}(1/2)^0(1 - 1/2)^{5-0} = 1/32$ or $.03125$. Your calculator screen should correspond to the one at the right.

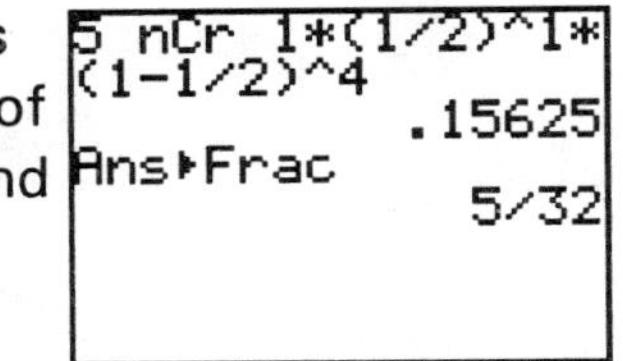

b. Keep in mind that the problem is expressed in terms of failures, thus four wrong answers means a success of **one** right answer, $r = 1$. Your screen should correspond to the one at the right.

c. Using **a.** and **b.** as your model, find the results of **c - f.** Record your screen display in the space below.
     ◆

9. In Exercise 2, the equation $Y = {}_{17}C_X$ was used to generate a table representing possible committee arrangements for each possible committee size **X** where $1 \le X \le 17$. Write the equation necessary to generate a table for the preceding binomial probability example that displays the probabilities of successes r where $0 \le r \le 5$. Remember, $r = 0$ would be the probability that all five answers are wrong and $r = 0$ would be the probability that all 5 answers are correct. Verify your table values with the values determined in the example.

equation:_______________________________

X	Y1
0	
1	
2	
3	
4	
5	

10. If the final exam for History 2010 is a 100-question true/false exam, what is the probability that a student who guesses on every question will make a 70%?

SOLUTIONS:

1. $149({}_{10}P_4)$ **2.** a. Because combinations represent subsets, there will be no non-integer values. b. The horizontal space width formula was used to determine the Xmin and Xmax, the maximum y value in the table was used to establish Ymax and the graph was restricted to the first quadrant because only positive integers are valid to combinations.

3. $a^{17} + 17a^{16}b + 136a^{15}b^2 + 680a^{14}b^3 + 2380a^{13}b^4 + 6188a^{12}b^5 + 12376a^{11}b^6 + 19448a^{10}b^7 + 24310a^9b^8 + 24310a^8b^9 + 19448a^7b^{10} + 12376a^6b^{11} + 6188a^5b^{12} + 2380a^4b^{13} + 680a^3b^{14} + 136a^2b^{15} + 17ab^{16} + b^{17}$

4. ${}_{13}C_4 \bullet {}_{13}C_5 \bullet {}_{13}C_3 \bullet {}_{13}C_2 \approx 1.14044073 \text{ E } 10$

5. $\dfrac{{}_{35}C_7}{{}_{59}C_7}$ **6.** $\dfrac{{}_{35}C_3 \cdot {}_{24}C_4}{{}_{59}C_7}$ **7.** $\dfrac{{}_{35}C_4 \cdot {}_{24}C_3}{{}_{59}C_7}$ **8.** $1 - \dfrac{{}_{35}C_7}{{}_{59}C_7}$

9. $Y1 = 5 \text{ nCr } X \cdot (1/2)^X \cdot (1-1/2)^{5-X}$

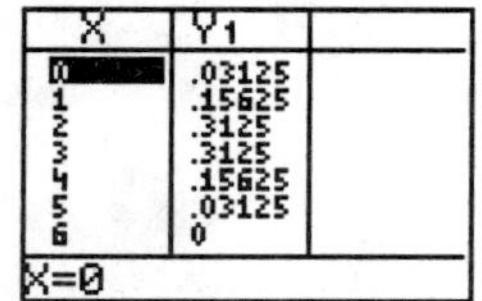

10. $2.317069058 \times 10^{-5}$

*Unit 17 is prerequisites for this unit. Answers appear at the end of the unit.

Descriptive statistics is the collection, presentation and summarization of data. This unit examines paired data information and the type of plots available on the calculator for displaying this data.

1.   The STAT feature can be used as a tool for graphing ordered pairs of numbers. Unlike the graph screen, or TABLE feature, STATPLOT allows the plotting of individual ordered pairs of numbers. The graph screen and TABLE feature are dependent on functions being entered on the **Y=** screen.
     NOTE: Before proceeding, delete all entries on the **Y=** screen.

TI-85/86	ACCESS THE y(x)= SCREEN BY PRESSING [GRAPH] [F1](y(x)=) AND CLEAR ALL ENTRIES.

2.   The ordered pairs (2,3), (-5,7), (4,-2), (0,-4) and (5,8) can be arranged in a table of values:

X	Y
2	3
-5	7
4	-2
0	-4
5	8

     Two lists will be created in the **STAT** menu to represent each list in the table of values.

TI-85	TI-85 USERS GO TO THE APPENDIX (PG.253).

TI-86	TI-86 USERS GO TO THE APPENDIX (PG.255).

3.   Begin by pressing [STAT] and [1:Edit]. Use the ▶ arrow key to cursor over to the sixth list (L6) to observe the content of each of the lists. Data must be cleared from the lists before beginning any problem.

4.   There are several approaches to clearing lists. For demonstration purposes, suppose that only the first and second lists (L1 and L2) are to be cleared. To clear these lists, press **[2nd] <QUIT>** to return to the home screen, press **[STAT] [4:ClrList] [2nd] <L1> [,] [2nd] <L2> [ENTER]**.

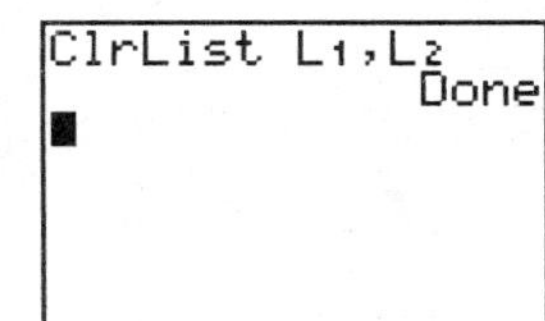

Another approach is to press **[STAT] [1:Edit]**, use the ▲ key to highlight L1 and press **[CLEAR]**. Cursor down below L1 to see the list clear.

These approaches are useful for clearing specific lists while leaving other lists intact. If the intent is to clear **all** lists then the quickest approach is to press **[2nd] [MEM]** **<4:ClrAllLists>** **[ENTER]**.

Press **[2nd] <MEM>** **[2:Delete] [3:List]** and placing the marker in front of each list press **[ENTER]**.

5.   The data (ordered pairs in the table of values) can be entered in the lists on the **Edit** screen in two ways. Examine the first approach by using the data listed in #2.

   a. Press **[STAT] [1:Edit]**. Enter each X value, in order, in the L1 list.
   (Press **[2] [ENTER] [(-)] [5] [ENTER] [4] [ENTER] [0] [ENTER] [5] [ENTER]**.)

   b. Use the ► arrow key to cursor to the L2 list and enter the Y values. Double check the paired data to be sure they are the same as the original table of values.

   c. Press **[2nd] <STATPLOT>** **[1:Plot 1]**. Press **[ENTER]** to turn on the first plot. Use the ▼ arrow key to highlight the first icon after **Type** and press **[ENTER]**.

Use the ▼ arrow key to highlight the first entry (L1) after **Xlist** (press **[ENTER]**), the ▼ and ► arrow keys to highlight the second entry (L2) after **Ylist** (press **[ENTER]**), and finally, the ▼ arrow key to highlight the first entry after **Mark** (press **[ENTER]**). At this point, continue reading at letter **d**.

Use the ▼ arrow key followed by **[2nd]<L1>** to enter **L1** after **Xlist** if L1 is not already displayed. Then use **[▼]** to **Ylist** and press **[2nd] <L2>** to display **L2** after **Ylist**. Finally, use the ▼to highlight the first entry after **Mark**. Plotted data points can be represented by □, +, or a •. We selected □ for easy visibility.

   d. Set the standard viewing window to view the ordered pairs.

   e. Press **[TRACE]**; the screen at the right is displayed. P1:L1,L2 is displayed in the upper left corner of the screen, indicating that Plot 1 is turned on and the ordered pairs are graphed from lists L1 and L2. The ◄ and ► arrow keys allow movement from point to point, displaying the ordered pair values at the bottom of the screen. The display is called a Scatter Plot.

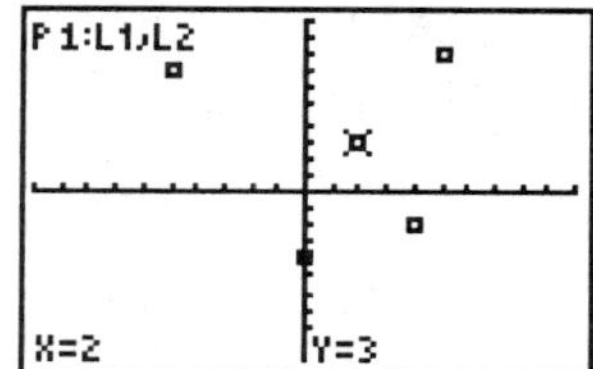

Displayed in the upper right corner of the screen will be **P1**, indicating that Plot 1 is turned on.

   f. Ordered pairs can be plotted using either the **ZOOM** menu, or by entering values on the **WINDOW** screen. If entering values on the WINDOW screen, be sure that the X values in the L1 list are between the Xmin and Xmax of the WINDOW and that the

Y values in L2 list are between the Ymin and Ymax of the WINDOW.

g. Press **[ZOOM] [9:ZoomStat]** and the points are replotted with the calculator automatically resetting the viewing WINDOW to include all the data points.

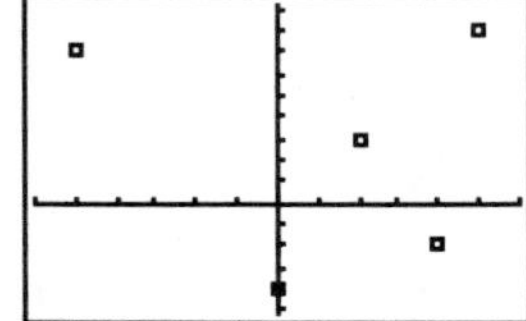

6. Clear lists **L1** and **L2** using one of the methods in #4.

7. A new table of values will now be entered using a different approach.

a. The table of values is ___X___Y___.   To enter the entire set of values at once, return to the home screen. Use the **{ }** to list all the X values and **STO**re them in L1.

X	Y
3	6
-2	1
4	-2

To accomplish this, press **[2nd] < { > [3] [,] [(-)] [2] [,] [4] [2nd] < } > [STO▸] [2nd] <L1> [ENTER]**.

<table><tr><td>TI-85/86</td><td>TI-85: THE BRACES, **{ }**, ARE FOUND UNDER THE **LIST** MENU BY PRESSING **[2ND] <LIST>**. TO LOCATE **xSTAT** AND **ySTAT**, PRESS **[2ND] <VARS> [MORE] [MORE] [F3](STAT)** AND USE THE ▼ KEY TO PLACE THE MARKER AT **xSTAT**. PRESS **[ENTER]** TO DISPLAY THE SCREEN AT THE RIGHT.

TI-86: THE BRACES, **{ }**, ARE FOUND UNDER THE **LIST** MENU BY PRESSING **[2ND] <LIST>**. **xSTAT** AND **ySTAT** CAN BE ACCESSED FROM THE **LIST** MENU BY PRESSING **[F3](NAMES)**.</td><td></td></tr></table>

b. Store the list of Y values in L2. Your screen should now look like the one at the right.

c. Check **Plot 1** under the **STATPLOT** menu to be sure all items are selected as in #5c.

d. Press **[ZOOM] [9:ZoomStat]**. Sketch the graph displayed.

<table><tr><td>TI-85/86</td><td>TI-85: THERE IS NO FEATURE COMPARABLE TO ZoomStat. **RANGE** VALUES MUST BE SET BY HAND.

TI-86: PRESS **[GRAPH] [F3](ZOOM) [MORE] [F5](ZDATA)**.</td></tr></table>

NOTE: Remember to clear the lists in **L1** and **L2** before beginning a new problem. (See #4 for directions.)

8. Enter the table of values in **L1** and **L2** under the **STAT/EDIT** menu and display the data points using the **ZoomStat** viewing WINDOW. Sketch the graph displayed.

X	Y
1	3
2	-5
0	0
-1	-2
-3	2
5	1

9.   **Application:** On the first five days of January, the daily highs were 22°, 19°, 12°, 17° and 13°. Let days 1 through 5 be the X values and the temperatures be the Y values.

a. Display the data points using the **ZoomStat** Window. Sketch the display.

b. Now display this same data, connecting the data points with a straight line. This is called the **XYline.** To do this, return to the **STATPLOT** menu by pressing **[2nd] <STATPLOT>** and **[1:Plot1]**. Plot1 is already turned ON, cursor down to **TYPE** and right once, to the second icon. Press **[ENTER]** to select the icon (which represents the **XYline**). Press **[GRAPH]** and sketch your screen display. This **XYline** graph clearly displays the temperature pattern over the 5 day period.

TI-86	TO DISPLAY THE XYLINE PRESS [2ND] <STAT> [F3](PLOT) [F1](PLOT1), CURSOR DOWN TO TYPE AND PRESS [F2](XYLINE). PRESS [GRAPH] TO DISPLAY THE GRAPH.

10.   Using {-3, -2, -1, 0, 1, 2, 3} for the list of X values, create a table of values for the equation Y = 3X - 2. Use the **STOre** feature to accomplish this. Using braces, enter the list of X values on the home screen and store in L1. Because the X values are now in L1, evaluate 3L1 - 2 (which will create the values for the list L2) and store those evaluations in L2. Remember, colons are used to separate each command. Before pressing "enter", your screen should look like the one at the above right. Press **[ENTER]** to activate the command then press **[STAT] [1:Edit]** to view the lists L1 and L2.

Calculator screen (top right):
```
{-3, -2, -1,0,1,2,
3}→L1:3L1-2→L2
```

TI-85/86	THE TI-85/86 SCREEN WILL LOOK LIKE THE ONE AT THE RIGHT BEFORE PRESSING "ENTER" TO ACTIVATE THE COMMAND. REFER TO THE TI-85/86 NOTE IN #7A FOR DIRECTIONS ON LOCATING THE STAT VARIABLES.

Calculator screen (right):
```
{-3, -2, -1,0,1,2,3}→xS
tat:3 xStat-2→yStat█
```

a. Display the data points as a **Scatter Plot** graphed in the **ZoomStat** viewing WINDOW and sketch the screen display.

b. Now display this same data with an **XYline**. Sketch the screen display.

✍11. Summarize what you have learned in this unit. Include the following:
    a. how to clear lists in **STAT**
    b. how to enter data in the lists
    c. how to select the Plot display
    d. explanation of the difference between Scatter Plots and XYlines.

 7d. **8.** **9a.** **9b.** **10a.** **10b.**

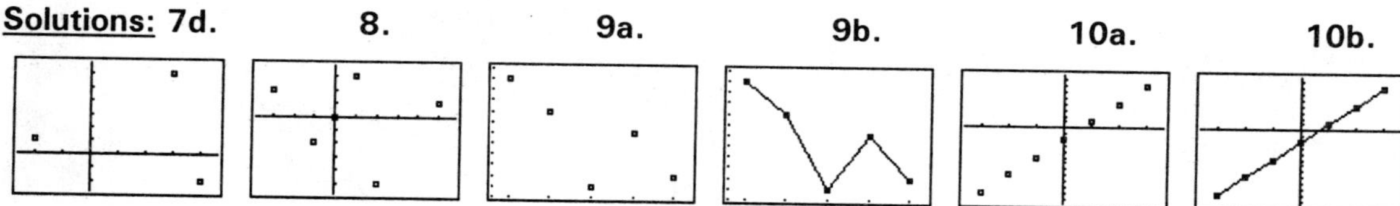

11. Answers may vary.

3.   Begin by pressing **[STAT]** and **[F2](EDIT)**. Each list in the TI-85 must be named (notice the cursor is a blinking "A" indicating alpha mode). The default names for the first two lists are **xStat** and **yStat**. Press **[ENTER]** twice to accept these names for the first two lists of data. Each list must have a different name.

4.   The list can be cleared by pressing **[STAT] [F2](EDIT) [ENTER] [ENTER]** to display the lists xStat and yStat followed by **[F5](CLRxy)**.

5.   a. Data must be entered pairwise (data from #2 on pg. 247), pressing **[ENTER]** after each entry except the last. Displayed on the screen at the right are the first two ordered pairs of values.

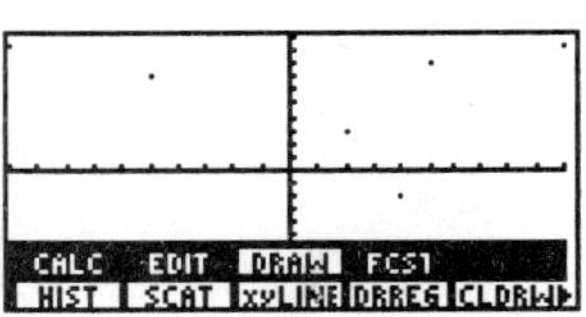

Data can be sorted in ascending order based on either the x value or y value. Press **[F3](SORTX)** to order the pairs based on the x values.

b. Pressing **[2nd] <M3>(DRAW)** displays the submenu that allows the drawing of scatter plots (SCAT), xyLINEs, and histograms (HIST). Press **[F2](SCAT)** to display the scatterplot at the right. It may be necessary to press **[EXIT]** ONCE to remove the top menu line or **[CLEAR]** to remove both menu lines so that the entire scatterplot is visible. These points are displayed in the standard viewing window.

Because these points are activated from the **DRAW** menu, **TRACE** cannot be used. The TI-85 is not graphing these points. The viewing window (**GRAPH/RANGE**) must be determined by hand, being sure that these values fall between xMin and xMax and that the y values fall between yMin and yMax. There is no ZoomStat feature on the TI-85 nor are there any options such as the □ or + for the scatterplot display.

6.   Press **[F5](CLDRW)** to clear the drawn pixel points before proceeding to the next problem.

☞   RETURN TO THE CORE UNIT STEP #7 (PG.249) AND COMPLETE THE UNIT .

3.   Begin by pressing **[2nd]** **<STAT>** and **[F2] (EDIT)**. The three lists displayed must be cleared of data before beginning any problem.

4.   To clear the lists, use the up arrow key to highlight **xStat** (or **yStat** or **fStat**) and press **[CLEAR]**. Cursor down into the list to see it actually clear.

5a,b.   Data is entered item by item as in the TI-82 (items #5a and b from the main unit) with the only difference being that the lists are named **xStat** and **yStat** rather than **L1** and **L2**. Enter the data from page 247 (#2) now by entering each value in the x column, pressing **[ENTER]** after each entry. Cursor over to the yStat column and enter the y-values, pressing **[ENTER]** after each entry.

    Note: Data can be sorted in ascending order based on either the **xStat** values or **yStat**. Return to the home screen. Press **[2nd] <LIST>** **[F5] (OPS)** **[MORE] [MORE]** **[F5] (Sortx)**. The prompt displayed is **Sortx**. Enter a parenthesis and press **[2nd] <M3>(NAMES)**. The list to be sorted is entered first and the list that corresponds to it is entered second. Press **[F2] (xStat)** **[,]** **[F3] (yStat)** **[ )] [ENTER]**. The word **DONE** should appear to indicate the process is complete. Return to **STAT EDIT** screen to view list.

5c.   Press **[2nd] <STAT>** **[F3] (PLOT)** **[F1] (PLOT1)** **[ENTER]** to highlight **ON** on the **PLOT1** screen. Your **STAT PLOT** is now on. Cursor down to **Type** and press **[F1] (SCAT)**; cursor down to **Xlist Name** and press **[F1] (xStat)**; cursor down to **Ylist Name** and press **[F2] (yStat)**; cursor down to **Mark** and press **[F1] (□)**. Press **[GRAPH] [F3]** **(ZOOM) [MORE]** and **[F5] (ZDATA)** to display the scatter plot of the data. Because the data was displayed from the **GRAPH** menu, you can use the **TRACE** feature.

    Note: The data can be displayed from the **DRAW** menu in the following manner: Return to the Home Screen. Press **[2nd] <STAT>** **[F4] (DRAW)**. Your display can be changed from this menu only by pressing **CLDRW** and returning to the defaults you set in **PLOT**.

6.   Clear the xStat and yStat lists as directed in #4.

☞   RETURN TO THE CORE UNIT STEP #7 (PG.249) AND COMPLETE THE UNIT .

UNIT 38
FREQUENCY DISTRIBUTIONS

*Unit 37 is a prerequisite for this unit. Answers appear at the end of the unit.

This unit will continue to look at methods for presenting and summarizing data. The last unit examined scatter plots and XYlines as means of presenting two variable statistics. One variable data can be presented as a histogram or a box-and-whisker plot. A histogram is a vertical bar graph with no spaces between the bars. The horizontal axis (X-axis) displays the values (L1) and the vertical axis (Y-axis) displays the frequencies (L2) at which the L1 values occur. A box-and-whisker plot summarizes data using the extreme data values, the median and the upper and lower quartiles.

CONSTRUCTING HISTOGRAMS

The following steps should be followed when displaying one variable statistics as a histogram. Read through these steps: **DO NOT** implement the steps until you have been given a problem with data.

a. Construct a frequency table of the data:

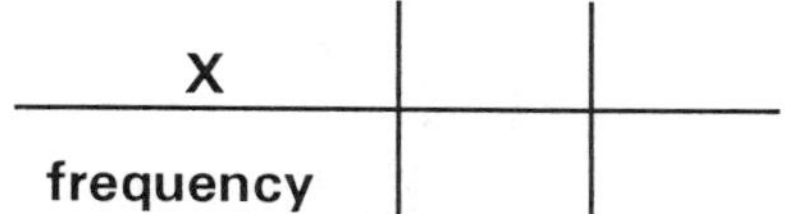

X		
frequency		

b. Delete entries on the **Y** = screen and clear all lists L1- L6 by pressing [2nd] <MEM> [4:ClrAllLists].

Press [2nd] <MEM> [2:Delete] [3:List], place the marker beside each list, press [ENTER] to delete.

AFTER DELETING ALL y(x) = ENTRIES, RETURN TO THE HOME SCREEN BY [2ND] <QUIT>. CLEAR THE **XLIST** AND **YLIST** BY PRESSING [STAT] [F2](EDIT) [ENTER] [ENTER], FOLLOWED BY [F5](CLRxy) UNTIL ONLY X_1 AND Y_1 ARE DISPLAYED.

ON THE TI-85, THE **YLIST** WILL BE DESIGNATED AS THE FREQUENCY LIST. SKIP STEPS 3 AND 4 (THE INFORMATION REQUIRED IN STEP 4 WILL BE APPLIED WHEN YOU ARE READY TO DISPLAY THE 1-VARIABLE STATISTICS) AND PROCEED AT STEP 5.

CLEAR THE LISTS AS DIRECTED IN #4 IN THE TI-86 GUIDELINES ON PAGE 263.
PRESS [2ND] <STAT> [F3] (PLOT) [F1] (PLOT1) [ENTER] TO HIGHLIGHT **ON**. CURSOR DOWN TO **TYPE** AND PRESS [F4] (HIST). CURSOR DOWN TO **X**LIST **N**AME = AND PRESS [F1] (xSTAT) TO ENTER xSTAT AS YOUR FIRST DATA LIST. CURSOR DOWN TO **F**REQ = AND PRESS [F2] (ySTAT) TO ENTER ySTAT AS YOUR FREQUENCY LIST. PROCEED TO STEP #5.

c. Turn on **STATPLOT** and set up for a histogram. Press [2nd] <STATPLOT> [1:Plot 1] (**ON** is highlighted), cursor to **Type** and highlight the third icon (fourth icon on the TI-82), press [ENTER], cursor down and ensure L1 is specified for **Xlist** and L2 for **Frequency**. Your screen display should correspond to the one at the right. (TI-82 note follows.)

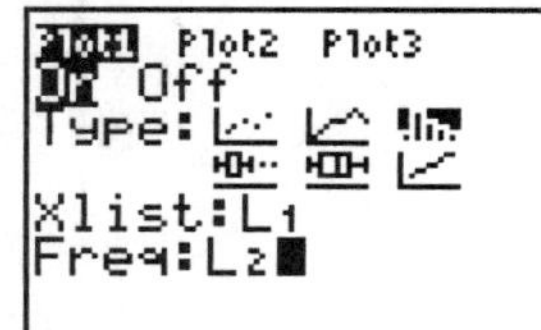

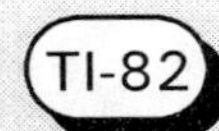

Set the calculator for one variable statistics by pressing **[STAT] [▶]** (for CALC menu) **[3:SetUp]**. L1 should be highlighted for the **Xlist** and L2 for the **Frequency**. Use the cursor to highlight the desired entry and press **[ENTER]**.

d. Enter the data from your frequency table into the calculator's STAT list display by pressing **[STAT] [1:Edit]** and entering the values in the L1 column and their respective frequencies in the L2 column.

e. Set the WINDOW: Be sure that the X values in the frequency table fall between the Xmin and Xmax, that the Ymin = 0 and Ymax is greater than the largest frequency value. The Xscl will determine how many values are grouped in each vertical bar. The calculator **requires** that $\frac{Xmax - Xmin}{Xscl} \leq 47$. Set the Xscl = 1 unless otherwise noted. Yscl will equal 1 since it represents frequencies. Press **[GRAPH]**. You may TRACE on the histogram; note that the number **n** will indicate how many values are in the bar that the trace cursor is located on.

TI-85/86   THESE CALCULATORS REQUIRE THAT $\frac{xMax - xMin}{xScl} \leq 63$.

---

f. Translate the calculator's histogram to a hand drawn and <u>labeled</u> histogram.

Example 1: A 10 point quiz in an intermediate algebra class yielded the following scores:
10,9,6,7,8,8,9,8,7,7,6,7,8,8,8,7,9

Construct a histogram to represent the frequency distribution. **(The lettered steps correspond to each of the lettered steps listed previously.)**

Solution:

a. Begin by constructing a frequency table:

X	10	9	8	7	6
frequency	1	3	6	5	2

b. Clear all lists L1 through L6.

TI-85/86   TI-85 USERS SKIP STEPS #3 AND #4.
TI-86 USERS SKIP ONLY STEP #4.

---

c. Turn on **STATPLOT 1** and set up for a histogram display.

TI-82 users set the calculator for one variable statistics by pressing **[STAT] [▶]** (for CALC menu), **[3:SetUp]**. L1 should be highlighted for the **Xlist** and L2 for **Frequency**. Use the cursor to highlight the desired entry and press **[ENTER]**.

d. Enter the data from your frequency table into the calculator's STAT list display by pressing **[STAT] [1:Edit]** and entering the grades in the L1 column and their respective frequencies in the L2 column.

e. Set the WINDOW values as follows: Xmin = 6, Xmax = 11, Xscl = 1, Ymin = 0, Ymax = 7, Yscl = 1. Press **[GRAPH]** to display the histogram at the right.

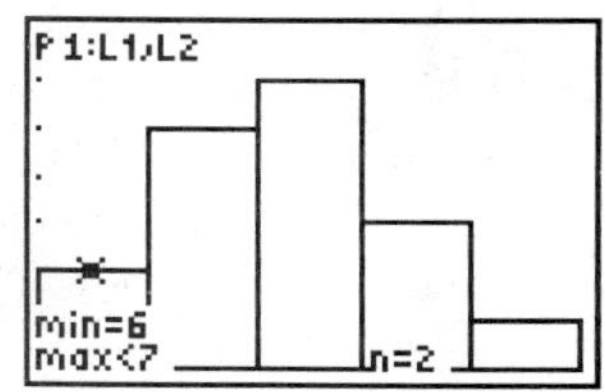

PRESS **[GRAPH] [F2](RANGE)** TO ENTER THE WINDOW VALUES. RETURN TO THE **STAT/DRAW** MENU AND SELECT **[F1](HIST)**. PRESS **[EXIT]** TO REMOVE THE TOP LINE OF MENU DISPLAY.

TI-86
PRESS **[GRAPH] [F2] (WIND)**, ENTER THE DESIGNATED VALUES AND PRESS **[F5](GRAPH)** TO DISPLAY THE HISTOGRAM.

◆

Drawn by hand, the histogram would look like the one at the right. Press **[TRACE]** and trace on the calculator graphed histogram to verify that the height *n* of each bar corresponds to the height of the bars in the hand drawn histogram.

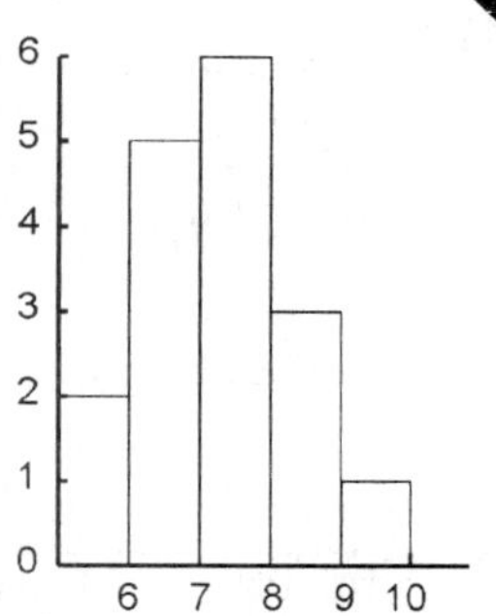

To display the statistical features of these grades, press **[STAT] [▶]** (to highlight **CALC**) and select **[1:1-Var Stats]**. Displayed on the home screen is **1-Var Stats**. The calculator is now set to display one variable statistics. Enter **(L1, L2)** after the **1-Var Stat** prompt. See the screen at the right. Pressing **[ENTER]** displays the statistical data.

TO DISPLAY THE ONE VARIABLE STATISTICS, PRESS **[F1](CALC) [ENTER] [ENTER] [F1](1-VAR)**. THE TI-85 WILL NOT COMPUTE THE MEDIAN.

TO DISPLAY **ONE**VARIABLE STATS, PRESS **[2ND] <STAT> [F1] (CALC) [F1] (ONEVA)** FOLLOWED BY **[( ]** AND **[2ND] <LIST> [F3] (NAMES) [F2] (xSTAT) [,] [F3] (ySTAT)** FOLLOWED BY **[ )]**. YOUR DISPLAY WILL READ: **ONEVAR (xSTAT, ySTAT)**. PRESS **[ENTER]**.

Notice the arrow pointing down at the bottom of the screen. Continuing to cursor down displays more statistical information. Some of the statistical features about these grades that would be of interest are specified below:

$\bar{x} \approx 7.8$ ($\bar{x}$ is the arithmetic average or mean)

median = 8   ("Med" is the number that divides ranked data into two equal groups. Cursor down to display it.)

$\sigma x \approx 1.06$ (σx = standard deviation: measures the dispersion of the data about the mean)

A frequency distribution is not appropriate when there is a wide variance among the data. The example below illustrates this using course scores. In this case, a histogram representing the number of As, Bs, etc. would be more informative. The data will only be entered in L1.

NOTE: Data can be sorted in ascending order by pressing [STAT] [2:Sort A(] [2nd] <L1> [ENTER]. When you have frequencies listed in L2, you would need to enter Sort A(L1,L2) so that the frequencies are sorted with their respective values.

If you do not use a frequency distribution then the frequency value must be set to **1** under **STATPLOT 1** and **STAT/CALC SetUp.** (Do this now.)

Example 2: The scores below are the final course scores for a college mathematics class:
83, 78, 82, 91, 84, 75, 84, 96, 80, 68, 79, 78, 85, 89, 95, 82, 88, 74, 76, 73, 74, 64, 66, 93, 69, 87
Set up a histogram to represent the grade distribution.

Solution: The local college uses a 10 point grading scale (i.e. A: 90-100, B: 80-89, etc.) and the public schools use an 8 point grading scale (i.e. A: 93-100, B: 85-92, C: 77-84, D: 69-76, F: below 69). We will construct two histograms: the distribution of letter grades for a 10 point scale and the distribution of letter grades for an 8 point scale.

10-point scale: The Xscl will determine which values (grades) are grouped in each bar. For the first histogram, set the WINDOW values as follows: Xmin=60, Xmax =101, Xscl =10, Ymin = 0, Ymax =10, Yscl = 1. Ymax may need to be increased if any one grade group has more than 10 scores in it. Press **[GRAPH]**. Your displayed histogram should correspond to the pictured screen. TRACE to determine the number of As is 4, Bs is 10, Cs is 8 and Ds is 4.

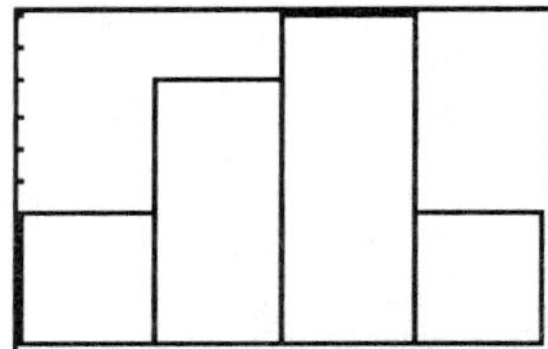

The hand drawn histogram would look like the one pictured:

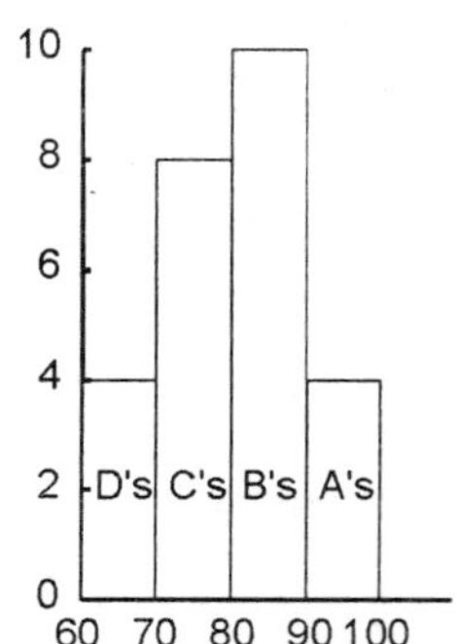

8-point scale: For your second histogram, set the WINDOW values as follows: Xmin: 60, Xmax: 101, Xscl: **8**, Ymin: 0, Ymax: 10, Yscl: 1. Ymax may need to be increased if any one grade group has more than 10 scores in it. Both your calculator generated and hand drawn sketch should correspond to those displayed.

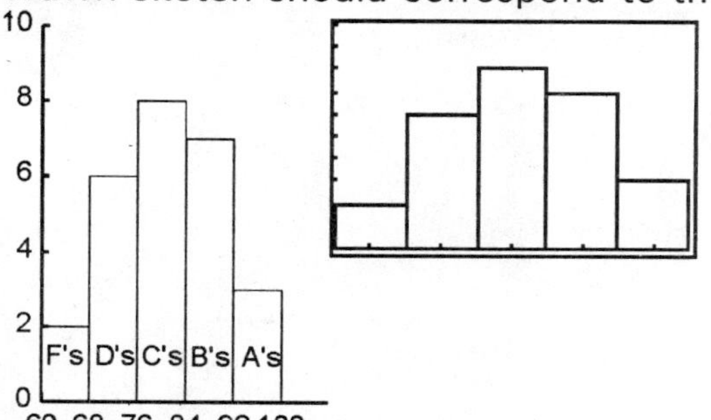

The grade distribution is as follows:
As = 3, Bs = 7, Cs = 8, Ds = 6, and Fs = 2

EXERCISE SET

Directions: For each problem complete the following:
    a. organize the data in a frequency table,
    b. enter the data in the calculator using the STAT feature,
    c. sort the data in ascending order
    d. copy the display of the histogram (Set your own WINDOW values; record the values used. Be sure to set both the Xscl and Yscl at 1.)
    e. construct a hand drawn and labeled histogram, and
    f. record the indicated information.

1. A freshmen girl's P.E. class is surveyed. The following shoe sizes were recorded:
8,7,8,8,7,7,7,6,5,6,6,6,7,9,5,7,10,9,9,8,7,7,7,6,6,8,6,7,6

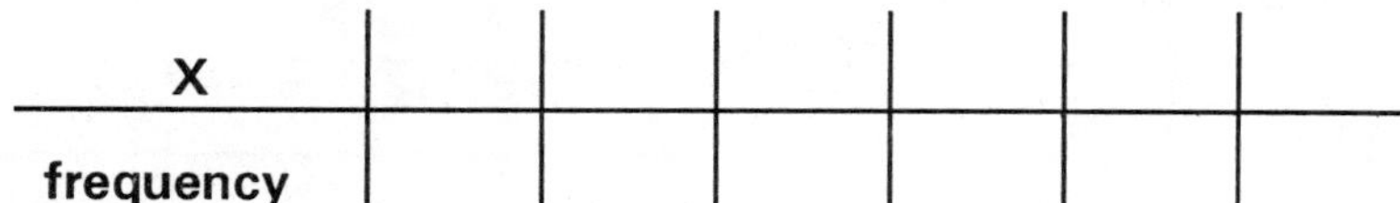

Calculator display:

WINDOW :

```
WINDOW
Xmin=
Xmax=
Xscl=
Ymin=
Ymax=
Yscl=
```

Hand Drawn (and labeled) Sketch:

$\bar{x}$ ≈ __________ (nearest whole size)

σx ≈ __________ (nearest hundredth)

median = __________

2. The daily highs for the month of February were recorded as:

68, 68, 73, 72, 73, 73, 68, 68, 73, 69, 69, 73, 73, 72, 73, 73, 68, 68, 69, 69, 69, 70, 69, 73, 70, 70, 70, 69, 71

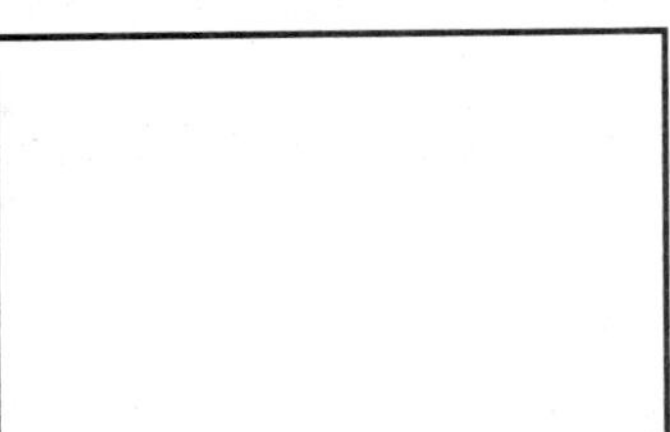

Calculator display:

WINDOW:

```
WINDOW
Xmin=
Xmax=
Xscl=
Ymin=
Ymax=
Yscl=
```

Hand Drawn (and labeled) Sketch:

What is the average temperature? _______
(to the nearest tenth of a degree)

What is the median temperature? _______

CONSTRUCTING BOX-AND-WHISKER PLOTS

TI-85 THE TI-85 DOES NOT HAVE A BOX-AND-WHISKER PLOT FEATURE.

The following steps should be followed when displaying one variable statistics as box-and-whisker plots. Read through these steps: **DO NOT** implement the steps until you have been given a problem with data.

a.   Construct a frequency table of the given data.

b.   Delete all entries on the **Y=** screen and clear all lists L1 through L6.

c.   Turn on **STATPLOT** and set up for box-and-whisker plots. Press **[2nd]** <STATPLOT> **[1:Plot1]** (**ON** should be highlighted), cursor down to **TYPE** and over to highlight the fifth icon, press **[ENTER]**, cursor down enter L1 for **Xlist** and L2 for **Frequency**. Your screen display should correspond to the one at the right. (TI-86 note follows.)

TI-82   Under **Type**, the third icon should be highlighted for boxplot. Press **[ENTER]** to highlight L1 for **Xlist** and L2 for **Frequency**.

d.   Enter the data from your frequency table into the L1 and L2 columns of the **STAT-Edit** screen.

e.   Set the WINDOW: Be sure that your Xlist values (L1) fall between the Xmin and Xmax and that the Ymin and Ymax guarantee your plot will be displayed in the first quadrant of the graph.

f.   Press **TRACE** and trace on the box-and-whisker plot to display the data summary.

NOTE: If data was previously graphed as a histogram, then only the **STATPLOT** screen needs to be changed to alter the way data is displayed.

Example 3: Redisplay the data in Example 1 as a box-and-whisker plot instead of a histogram.

Solution: Steps a and b will be the same as for a histogram. At step c, turn on **STATPLOT 1** and set up for a box-and-whisker plot. Steps d through e will be the same as for a histogram except that Xmin has been decreased by one so that the box plot is distinct from the X-axis. The Ymin and Ymax are not relevant to the boxplot, however, Ymin must still be less than Ymax. Press **GRAPH** to display the pictured screen.

Note: If three plots are highlighted on the **Y =** screen, then the boxplot of Plot 1 will be displayed in the upper section of the screen, Plot 2 will be displayed in the middle section and Plot 3 will be displayed in the lower section of the screen.

Press [TRACE]. The median is the first piece of information displayed. Tracing to the left yields the following two pieces of information:

$Q1 = 7$:   The lower quartile (median of the lower half of the data) is represented by Q1.

minX = 6:   This is the minimum entry in the **Xlist**.

Tracing to the right, past the median, yields the following information:

$Q3 = 8.5$:   The upper quartile (median of the upper half of the data) is represented by Q3.

maxX = 10   This is the maximum entry in the **Xlist**.

The interquartile range can now be computed: $Q3 - Q1 = 1.5$.

Directions: Display the given data as a box-and-whisker plot. Record the indicated information.

3.   Using the data information from EXERCISE 1 on the freshmen girl's P.E. class, sketch the box-and-whisker plot displayed. You may use the same WINDOW values as you did for the histogram.

        Median:__________

        Lower quartile:__________

        Upper quartile:__________

        Interquartile range:__________

4.   Using the data information from EXERCISE 2 on the daily highs for the month of February, sketch the box-and-whisker plot displayed. You may use the same WINDOW values as you did for the histogram.

        Median:__________

        Lower quartile:__________

        Upper quartile:__________

        Interquartile range:__________

5.   Using data information from Example 2 on the test scores, sketch the box-and-whisker plot displayed. You may use the same WINDOW values as you did for the histogram.

        Median:__________

        Lower quartile:__________

        Upper quartile:__________

        Interquartile range:__________

<u>**Solutions:**</u>

Exercise 1.

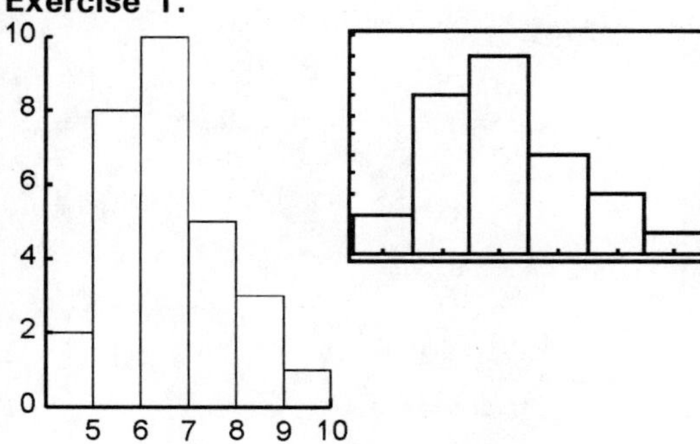

[5,11,] by [0,11]

$\overline{x} = 7$ $\sigma x \approx 1.16$ median = 7

Exercise 2.

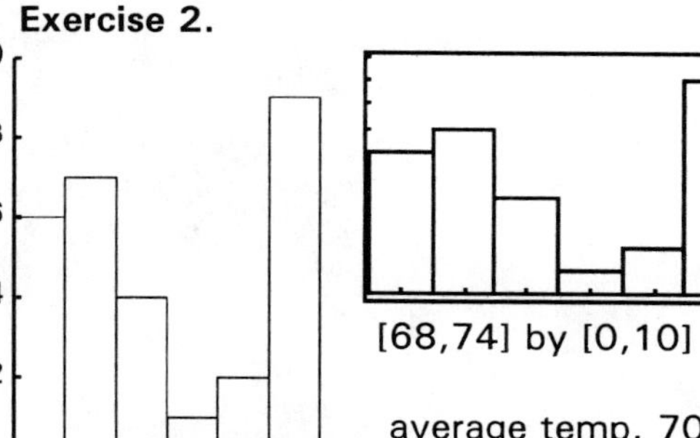

[68,74] by [0,10]

average temp. 70.4
median temp. 70

Exercise 3. median = 7, lower quartile = 6, upper quartile = 8, interquartile range = 2

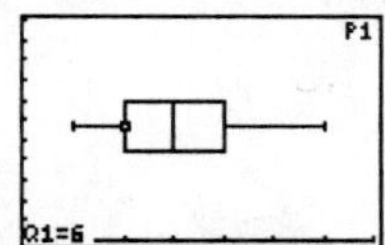

Exercise 4. median = 70,
lower quartile = 69, upper quartile = 73,
interquartile range = 4

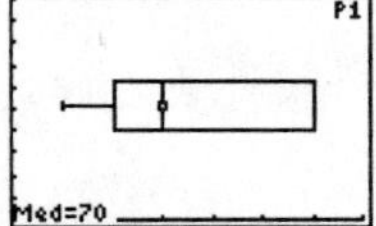

Exercise 5. median = 81,
lower quartile = 74, upper quartile = 87,
interquartile range = 13

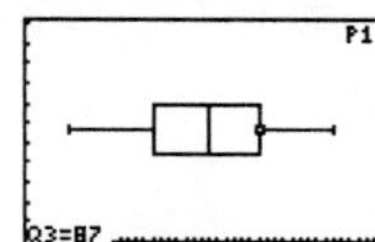

UNIT 39
LINE OF BEST FIT

*Unit 38 is a prerequisite for this unit. Answers appear at the end of the unit.

This unit determines the relationship between two variable quantities (data points) in different types of problems. If the data points lie in a straight line then the equation of the line passing through these points can be used to predict unknown data points. The line of best fit, defined by a linear regression equation, is a line that comes "close" to passing through all the data points. This unit explores only LINEAR REGRESSIONS. (In situations where the relationship is <u>not</u> linear, students should examine one of the following types of regression equations: quadratic, cubic, quartic, exponential or power.)

STEP-BY-STEP PROCEDURE FOR REGRESSION EQUATIONS

TI-85 USERS GO TO THE APPENDIX (PG.273).

a.   Clear all data from lists L1 - L6 and delete all entries from the **Y=** screen.

b.   Enter paired data in lists L1 and L2.

c.   Press **[STAT] [▶]** (to highlight **CALC**) **[2:2-Var Stats] [ENTER]**. On the home screen "2-Var-Stats" will be displayed. Press **[ENTER]** to display statistics.

(TI-82)   Press **[STAT] [▶]** (to highlight CALC) **[3:Setup]**. Under 2-Var Stats highlight L1 for the Xlist and L2 for the Ylist.

TI-86   PRESS **[2ND] <STAT> [F1] (CALC) [F2] (TwoVa) [( ]** FOLLOWED BY **[2ND] <LIST> [F3] (NAMES) [F2] (xStat) [,] [F3] (yStat)** AND **[ )]**. PRESS **[ENTER]** TO ACTIVATE.

d.   Turn **ON STATPLOT 1**, highlight the first icon under **TYPE** (scatter plot), L1 for Xlist and L2 for Ylist. There are three choices for the type of Mark; select " □ " for visual clarity.

e.   Press **[ZOOM] [9:ZoomStat]** to view the scatter plot. If you have a mechanical pencil, use a piece of the lead - or a similarly short piece of spaghetti, to approximate the "line of best fit". The next steps will instruct the calculator to compute the equation of this "line of best fit" and graph it.

TI-86   PRESS **[GRAPH] [F3] (ZOOM) [MORE] [F5] (ZDATA)**.

f.   Press **[STAT] [▶] [4:LinReg(ax+b)] [ENTER]**.

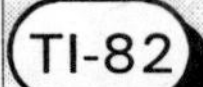

(TI-82)   Press **[STAT] [▶] [5:LinReg(ax+b)] [ENTER]**.

On the home screen the equation y = ax + b, along with the corresponding variable values, will be displayed. Directions for copying this equation to the **Y=** screen are in the next step.

PRESS [2ND] <STAT> [F1] (CALC) [F3](LINR) [ENTER].

g.   Press [VARS] [5:Statistics] [▶] [▶] (to highlight EQ for equation) **[1: RegEQ]**.

Press **[Y=]** (You must be at the Y= screen to copy the equation. Clear all Y= entries now.) **[VARS] [5:Statistics] [▶] [▶]** (to highlight EQ for equation) **[7:RegEQ]**.

PRESS **[GRAPH] [F1]** (Y(X)). TO COPY THE LINEAR REGRESSION EQUATION TO THE Y(X)= SCREEN, PRESS **[2ND] <CATLG-VARS> [MORE] [MORE] [F4] (STAT)**, CURSOR DOWN TO RegEq AND PRESS **[ENTER]**.

h.   Press **[GRAPH]** to display the scatter plot and the "line of best fit". The viewing WINDOW was automatically set to ZoomStat in #5. As long as the **STATPLOT** is turned **ON** you will be able to TRACE on the scattered data points and on the line using the ▲ or ▼ keys to move from data point to line. You will not be able to TRACE beyond the ZoomStat viewing window unless you edit the WINDOW values. For this reason, we will examine information relating to the line by using the TABLE feature.

PRESS **[2ND] <GRAPH> [M5] (GRAPH)** TO VIEW THE SCATTERPLOT AND THE LINE OF BEST FIT. NOTE: USE THE TABLE FEATRUE TO PREDICT VALUES AS INDICATED FOR THE TI-82/83.

Example: The following table gives the winning time (in seconds) for the Men's 400-Meter Freestyle swimming event in the Olympics from 1904 to 2000. The Olympics were not held during the years 1916, 1940, and 1944 due to the two World Wars. Find a linear regression equation to predict the winning times for each of those years.

YEAR	1904	1908	1912	1920	1924	1928	1932	1936	1948	1952
TIME	376.2	336.8	324.4	326.8	304.2	301.6	288.4	284.5	281	270.7

1956	1960	1964	1968	1972	1976	1980	1984	1988	1992	1996	2000
267.3	258.3	252.2	249	240.3	231.9	231.3	231.2	227	225	228	213.6

(Source: *The New York Times 1998 Almanac*)

Solution: Lettered steps correspond to the numbers listed on the previous page.

a.   Clear data from lists.

b.   Enter years in L1 list and winning time in the L2 list.

c.   Go to **STAT/CALC**, press **[2:2-Var Stats]** and enter (L1,L2) after 2-Var Stat on the home screen and press **[ENTER]**.

Go to **STAT/CALC Setup** and under 2-Var Stats highlight L1 for Xlist and L2 for Ylist.

PRESS **[2ND] <STAT> [F1] (CALC) [F2] (TwoVa) [( ]** FOLLOWED BY **[2ND] <LIST> [F3] (NAMES) [F2] (xStat) [,] [F3] (yStat)** AND **[ )]**. PRESS **[ENTER]** TO ACTIVATE.

d. Turn on **STATPlot 1** (be sure all others are OFF). Your
STATPLOT screen should look like the one displayed.

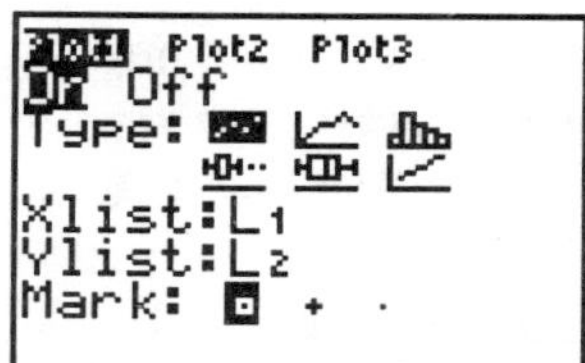

 The TI-82 display will have L1 hightlighted after **Xlist**
and L2 highlighted after **Ylist**.

e. Press **[ZOOM] [9:ZoomStat]** to view the scatter plot. Your
display should correspond to the one at the right. Place your
pencil lead on the screen to approximate the line of best fit.
Draw in this line on the screen at the right.

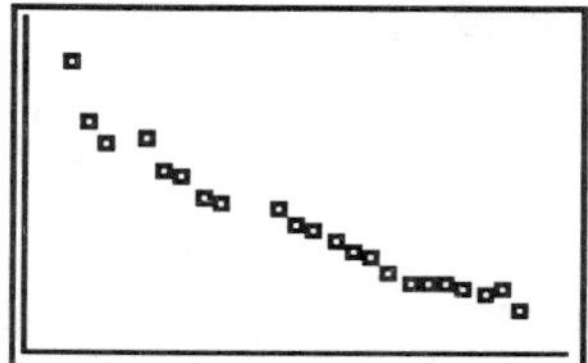

TI-86   PRESS **[GRAPH] [F3] (ZOOM) [MORE] [F5] (ZDATA)**.

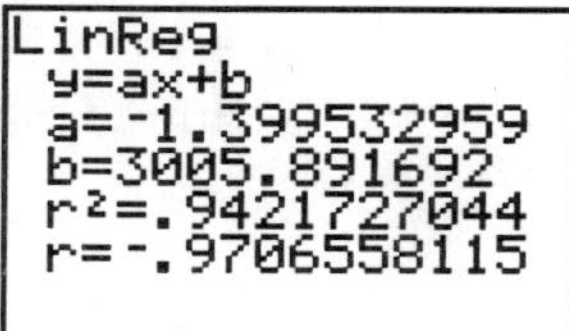

f. Determine the linear regression equation (the equation for your
line of best fit) by pressing **[STAT] [▸] [4:LinReg(ax + b)]**
[ENTER] at the home screen. The following information should
be displayed on your home screen.

The correlation coefficient on the displayed screen is r≈ -.97. This is the measure of
tendency for our variables X and Y to be related in a linear way. The fact that **r** is negative
indicates our data points lie near a "falling" line. The value of r is always between -1 and 1,
with values close to zero indicating that there is no tendency to a linear relationship. (Some
calculator versions may not display the r^2 value.)

Note: The correlation coefficient is not automatically displayed on the regression equation screen
because the calculator default mode is set to **Diagnostic Off**. One way to determine the **r**
value is to press **[VARS] [5:Statistics] [▸] [▸]** to highlight **EQ**, and **[7:r] [ENTER]**. Or, if the
r-value is always needed then the calculator should be reset to **Diagnostic On**: press **[2nd]**
<CATALOG> [▾] until the pointer marks **Diagnostic On**, **[ENTER]** to copy the command to
the home screen and **[ENTER]** again to activate the command.

TI-86   PRESS **[2ND] <STAT> [F1] (CALC) [F3](LiNR) [ENTER]**.

g.   Go to the **Y=** screen. To copy this equation to the **Y=** screen, press **[VARS]**
[5:Statistics] [▸] [▸] [1:RegEq].

TI-86   PRESS **[GRAPH] [F1] (Y(X))**. TO COPY THE LINEAR REGRESSION EQUATION TO THE **Y(X)=** SCREEN, PRESS
[2ND] <CATLG-VARS> [MORE] [MORE] [F4] (STAT), CURSOR DOWN TO **RegEq** AND PRESS
[ENTER].

h. Press **[GRAPH]** to view the scatter plot and the line of best fit.
See screen display at the right.

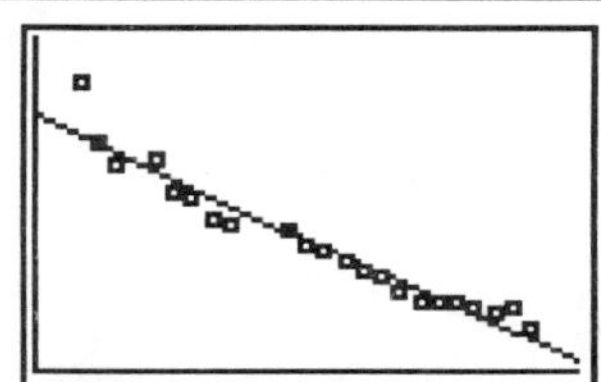

TI-86   PRESS **[2ND] <GRAPH> [M5] (GRAPH)** TO VIEW THE SCATTERPLOT AND THE LINE OF BEST FIT. NOTE:
USE THE TABLE FEATRUE TO PREDICT VALUES AS INDICATED FOR THE TI-82/83.

To predict the winning times in the missing years, set the table to start at 1916 (increment by 4 - Olympics occur every 4 years). Press **[2nd] <TABLE>**, scroll to view the predicted winning times for the missing Olympic years. The winning time in 1916 would have been 324.39, in 1940: 290.80 and in 1944: 285.20. This regression line comes "close" to passing through the set of points. Therefore, your predictions are approximations at best.

TI-85	TI-85 users are reminded to use **evalF** - press **[2nd] <CALC>** - to make predictions.

EXERCISE SET

Directions: For each problem, follow the steps on the first page of this unit. Sketch the graph screen that displays the scatter plot and the calculator's line of best fit. Use the TABLE to answer the question(s) posed.

1.   The chart below gives the year and winning time (in seconds) for the Men's 1000-Meter Speed Skating event. Predict the winning time for the 2006 Winter Olympics.

YEAR	1976	1980	1984	1988	1992	1994	1998
TIME	79.32	75.18	75.8	73.03	74.85	72.43	70.64

(Source: *The New York Times 1998 Almanac)*

Graph display:

Winning time prediction for the 2006 Winter Olympics:_______

What is the value (to the nearest hundredth) of the correlation coefficient?________

2. A survey was taken of notably tall buildings on the east coast, mid-west, and west coast. The survey compares height, in feet from sidewalk to roof, to number of stories (beginning at street level). If the Empire State building has 102 stories, use the given information to predict the height of the building.
(Source: The World Almanac and Book of Facts 1995, actual height is 1250 ft.)

BUILDING	STORIES	HEIGHT (FT.)
Baltimore U.S. Fidelity and Guaranty Co.	40	529
Maryland National Bank (Baltimore, MD)	34	509
Sears Tower (Chicago, IL)	110	1454
John Hancock Building (Chicago, IL)	100	1127
Transamerica Pyramid (San Francisco, CA)	48	853
Bank of America (San Francisco, CA)	52	778

Graph display:

Predicted height of the Empire State Building:______

What is the value (to the nearest hundredth) of the correlation coefficient?________

3. The statistics in the table below represent dollars per 100 pounds for cattle. Predict the price per 100 lb. for the years 2004 - 2006.

YEAR	1940	1950	1960	1970	1975	1979	1980	1984
PRICE	7.56	23.30	20.40	27.10	32.20	66.10	62.40	57.30

1985	1986	1987	1988	1989	1990	1991	1992	1993
53.70	52.60	61.10	66.60	69.50	74.60	72.70	71.30	72.60

1994	1995	1996
66.70	61.80	58.70

Source: The World Almanac and Book of Facts 1998)

Graph display:

Predicted price per 100 lb.:
2004:________
2005:________
2006:________

What is the value (to the nearest hundredth) of the correlation coefficient?________

<u>**Solutions:**</u>

1. Winning time: 68.588, r≈-.89

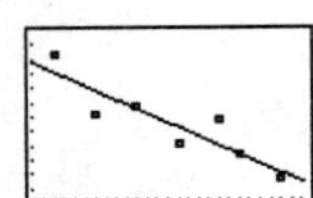

2. height: 1281.1, r≈.95

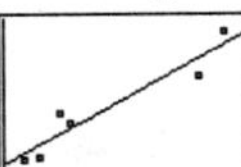

3. r≈.91

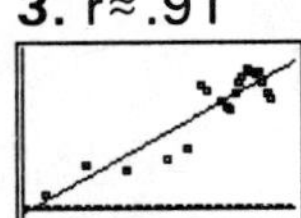

2004: 81.11

2005: 82.30

2006: 83.50

TI-85 GUIDELINES UNIT 39

PROCEDURE FOR REGRESSION EQUATION

a.   Press **[GRAPH]** **[F2](RANGE)** and set range values to correspond to the data to be entered into xlist and ylist.

b.   Clear xlist and ylist on the **STAT/EDIT** screen.

c.   Enter data into xlist and ylist.

d.   Press **[2nd] <M3>(DRAW)** **[F2](SCAT)** to display the scatter plot.

e.   Press **[STAT]** **[F1](CALC)** **[ENTER]** **[ENTER]** **[F2](LINR)** to display the linear regression equation information and the correlation coefficient. To copy this information to the y1 = prompt, press **[GRAPH]** **[F1](y(x)=)** **[2nd] <VARS>** **[MORE]** **[MORE]** **[F3](STAT)**. Place the pointer at **RegEq** and press **[ENTER]**. Press **[2nd] <M5>(GRAPH)** to graph the regression equation. (Note: Press **[CLEAR]** to remove all menu lines.)
Press **[STAT]** **[F3](DRAW)** **[F2](SCAT)** to redisplay the scatter plot.

f.   Because the regression equation is <u>GRAPHED</u> at y1 = , it is possible to **TRACE** on the line.

g.   To make predictions based on the regression equation, you will need to use the **FCST** (forecast) option found under **[STAT]** **[F4]**. It returns a forecasted value for x or y based on the current regression equation. You press **[F5]** to solve.

☞   RETURN TO THE FIRST EXAMPLE IN THE CORE UNIT (PG.268).

273

TROUBLE SHOOTING

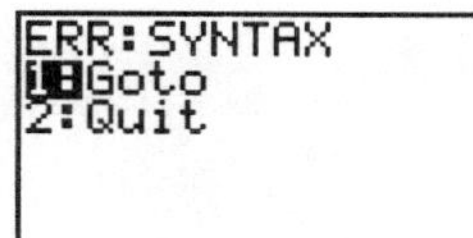

Instructions entered into the calculator were incorrect. Look for misplaced parentheses, use of subtraction sign for the negative sign, and so on.

* * * * *

An expression whose denominator is zero has been entered. Division by zero is undefined.

* * * * *

The values entered in the WINDOW screen are inappropriate.
Be sure that
a. Xmin < Xmax and Ymin < Ymax and
b. you have used the negative key for negative values and not the subtract key.

* * * * *

The value that has been selected for X is not acceptable (i.e. not in the domain). Example: $\sqrt{-4}$

* * * * *

a. The LOWER/LEFT bound is not less than (i.e. to the left of) the UPPER/RIGHT bound on the graph screen. Remember, lower bound refers to an X value that is less than x-coordinate of the point you are determining.
b. When establishing upper and lower bounds, your "guess" was not selected <u>between</u> the established bounds.

* * * * *

* * * * *

```
ERR:SIGN CHNG
1:Quit
```

a. No real root - graph does not intersect the X-axis
between the lower/left and upper/right bounds established.
b. The two graphs do not intersect.
c. The two graphs do intersect but the intersection is not
visible on the display screen.

* * * * *

```
ERR:BAD GUESS
1:Quit
```

When using the CALC menu, the "guess" entered was not
acceptable. Try again, this time entering a guess as close as
possible to the desired point.

* * * * *

```
ERR:DIM MISMATCH
1:Goto
2:Quit
```

a. Check dimension rules for addition, subtraction and
multiplication of matrices.
b. If using the STAT menu, be sure that the lists have the same
number of entries.

* * * * *

```
ERR:STAT PLOT
1:Quit
```

a. Data entered incorrectly
b. Turn STATPLOT OFF if graphing functions.
c. STAT SetUp (**[STAT]** [▶] (CALC) **[3:SetUp]**) not
compatible with data lists and/or STATPLOTS.

* * * * *

```
ERR:STAT
1:Quit
```

$$\frac{Xmax - Xmin}{Xscl} \text{ must be less than or equal to } 47$$

* * * * *